(4.15b) $\tan \frac{1}{2}\theta = \dfrac{1 - \cos \theta}{\sin \theta}$

(4.16) $\tan (A - B) = \dfrac{\tan A - \tan B}{1 + \tan A \tan B}$

(4.17) $2 \sin A \cos B = \sin (A + B) + \sin (A + B)$

(4.18) $2 \cos A \sin B = \sin (A + B) - \sin (A - B)$

(4.19) $2 \cos A \cos B = \cos (A + B) + \cos (A - B)$

(4.20) $2 \sin A \sin B = -\cos (A + B) + \cos (A - B)$

(4.21) $\sin C + \sin D = 2 \sin \frac{1}{2}(C + D) \sin \frac{1}{2}(C - D)$

(4.22) $\sin C - \sin D = 2 \cos \frac{1}{2}(C + D) \sin \frac{1}{2}(C - D)$

(4.23) $\cos C + \cos D = 2 \cos \frac{1}{2}(C + D) \cos \frac{1}{2}(C - D)$

(4.24) $\cos C - \cos D = -2 \sin \frac{1}{2}(C + D) \sin \frac{1}{2}(C - D)$

(6.1) $\log_b N = 1$ if and only if $b^L = N$

(6.2) $\log_b MN = \log_b M + \log_b N$

(6.3) $\log_b M/N = \log_b M - \log_b N$

(6.4) $\log_b N^p = p \log_b N$

(7.1) $a/\sin A = b/\sin B = c/\sin C$

(7.2) $K = \frac{1}{2}ab \sin C$

(7.3) $K = \dfrac{\frac{1}{2}b^2 \sin A \sin C}{\sin B}$

(7.4) $c^2 = a^2 + b^2 - 2ab \cos C$

(7.5) $\cos C = \dfrac{a^2 + b^2 - c^2}{2ab}$

(7.6) $\dfrac{a - b}{a + b} = \dfrac{\tan \frac{1}{2}(A - B)}{\tan \frac{1}{2}(A + B)}$

(7.7a) $\tan \frac{1}{2}A = r/(s - a)$

(7.7b) $\tan \frac{1}{2}B = r/(s - b)$

(7.7c) $\tan \frac{1}{2}C = r/(s - c)$

(7.8) $K = rs$

(7.9) $r = \sqrt{\dfrac{(s - a)(s - b)(s - c)}{s}}$

(10.1) $r = \sqrt{a^2 + b^2}$

(10.2) $\theta = \arctan \dfrac{b}{a}$

(10.3) $(r_1 \, cis \, \theta_1)(r_2 \, cis \, \theta_2) = r_1 r_2 \, cis \, (\theta_1 + \theta_2)$

(10.4) $\dfrac{r_1 \, cis \, \theta_1}{r_2 \, cis \, \theta_2} = \dfrac{r_1}{r_2} \, cis \, (\theta_1 - \theta_2)$

(10.5) $(r \, cis \, \theta)^n = r^n \, cis \, n\theta$

(11.1) $x = r \cos \theta, \, y = r \sin \theta$

Fred W. Sparks
Texas Tech University

Paul K. Rees
Louisiana State University

PLANE
TRIGONOMETRY

SIXTH EDITION

Prentice-Hall, Inc., Englewood Cliffs, New Jersey

PRENTICE-HALL INTERNATIONAL, INC., *London*
PRENTICE-HALL OF AUSTRALIA, PTY. LTD., *Sydney*
PRENTICE-HALL OF CANADA, LTD., *Toronto*
PRENTICE-HALL OF INDIA PRIVATE LIMITED, *New Delhi*
PRENTICE-HALL OF JAPAN, INC., *Tokyo*

13-679167-0

Library of Congress Catalog Card Number: 76-153649

Current Printing (last digit):

10 9 8 7 6 5 4 3 2 1

PRINTED IN THE UNITED STATES OF AMERICA

PREFACE

In preparing the sixth edition of *Plane Trigonometry*, we have kept in mind our goal of presenting a modern version of trigonometry in a manner that can be read and understood by the student of average ability and industry. In order to present a modern book, we have changed the presentation in a number of places.

We begin by discussing positive and negative angles, directed line segments, and the distance formula. This leads into the relation between degrees and radians. We then define and discuss the trigonometric ratios which are used in giving a modern definition of trigonometric functions. We consistently differentiate between a trigonometric function and a function value. The second chapter is on right triangles while the third and fourth deal with identities. The identity for the cosine of the difference of any two angles is the first one in the chapter on function values of a composite angle, and this is followed by function values of $\pi/2$ minus an angle. The other fundamental identities follow readily.

A considerable portion of the chapter on graphs of the trigonometric functions has been rewritten in an attempt to clarify and modernize the concepts. The discussions of periodic functions, range, variation, and graphs of functions are all new.

The work on logarithms and oblique triangles is placed between graphs and trigonometric equations to give the student a rest from analytic work since we feel that he may need a chance to "catch his breath," but one may omit or defer these chapters without loss of continuity.

The chapters on the inverse functions and complex numbers have been modernized. The former begins with a discussion of functions and inverse relations, and the latter presents and discusses complex numbers in the form (a, b) as well as $a + bi$ and the polar form.

We still define and illustrate each new concept before using it. The illustrative examples still clarify the immediately preceding text material and serve as models for the problems that follow. Reference position is still used in discussing the characteristic of the logarithm of a number.

We continue giving problem lists a normal lesson apart and problems in groups of four similar ones. These two features make it easy for even the inexperienced teacher to give a good assignment each day. This edition contains about 1900 problems in 39 exercises. Answers are given in the book to three fourths of the problems and the other answers are available in a supplementary booklet.

There is enough material in the book to afford the instructor a considerable amount of choice in the selection of material to fit the needs of the students he has and the type of course he decides to give.

We are grateful to those persons mentioned in the preface to the first (1937) edition and to those who have used the book and made suggestions over the years. This sixth edition is a better book than it would have been without the help of these people.

FRED W. SPARKS

PAUL K. REES

CONTENTS

THE GREEK ALPHABET

LETTERS		NAMES	LETTERS		NAMES	LETTERS		NAMES	LETTERS		NAMES
A	α	Alpha	H	η	Eta	N	ν	Nu	T	τ	Tau
B	β	Beta	Θ	θ	Theta	Ξ	ξ	Xi	Υ	υ	Upsilon
Γ	γ	Gamma	I	ι	Iota	O	o	Omicron	Φ	ϕ	Phi
Δ	δ	Delta	K	κ	Kappa	Π	π	Pi	X	χ	Chi
E	ϵ	Epsilon	Λ	λ	Lambda	P	ρ	Rho	Ψ	ψ	Psi
Z	ζ	Zeta	M	μ	Mu	Σ	σ s	Sigma	Ω	ω	Omega

DEFINITIONS AND THEOREMS FROM PLANE GEOMETRY

1. In a right triangle, the two sides which inclose the right angle are called the *sides* or the *legs* of the triangle. The side opposite the right angle is the *hypotenuse*.

2. *Pythagorean Theorem.* The sum of the squares of the sides of a right triangle is equal to the square of the hypotenuse.

3. The sum of the two acute angles of a right triangle is 90°.

4. The sum of the three angles of a triangle is 180°.

5. If the acute angles of a right triangle are 30° and 60°, the hypotenuse is twice the shorter side.

6. If two sides of a triangle are equal, it is called an isosceles triangle. If all three sides are equal, it is an equilateral triangle.

7. In an isosceles triangle, the angles opposite the equal sides are equal. In an equilateral triangle, each angle is equal to 60°.

8. If two angles of a triangle are equal, the sides opposite them are equal and the triangle is isosceles.

9. If the three angles of a triangle are respectively equal to the three angles of another, the triangles are similar.

10. In two similar triangles, the ratios of the pairs of sides opposite the equal angles are equal.

11. The radius of a circle is perpendicular to the tangent at the point of tangency.

12. Tangents from an external point to a circle are equal and make equal angles with the straight line joining the point to the center of the circle.

1

ANGULAR MEASURE
AND THE
TRIGONOMETRIC FUNCTIONS

1.1. The Word Trigonometry

If we look in a dictionary, we will find that trigonometry is that branch of mathematics that treats the relations among the sides and angles of a triangle. This was the purpose for which trigonometry was invented. It deals with six ratios and has tremendous theoretical as well as practical importance. Both of these aspects of trigonometry will be presented in this book.

1.2. Positive and Negative Angles

A straight line that originates at a fixed point and extends indefinitely far in one direction is called a *ray*. If two rays are drawn from the same point, an angle is formed. In trigonometry, we generally regard an angle as hav-

1

ing been formed or generated by one ray revolving about a fixed point on a stationary ray. The stationary ray is called the *initial side* of the angle, the revolving ray is called the *terminal* side, and the fixed point, the *vertex*. If the direction of rotation is counterclockwise, the angle is said to be a *positive* angle; but if the direction is clockwise, the angle is *negative*.

The magnitude of an angle is the amount of rotation necessary to move a ray through the vertex from the initial side to the terminal side of the angle. We shall often employ a Greek letter to designate an angle. (The Greek alphabet is given preceding page 1.)

In Figure 1.1, the side BA is the initial side of the angle θ, the side BC

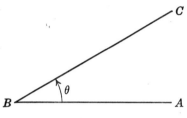

Figure 1.1

is the terminal side, and the point B is the vertex. The direction indicated is counterclockwise; hence, the angle is positive.

1.3. Directed Line Segments

A straight line segment is the portion of a straight line between two points on it. We shall consider two concepts that are associated with a straight line segment. They are the length and the direction. The measure of the length of a line segment is usually expressed in terms of a standard unit. For example, in the British system, a length is usually expressed in inches, feet, yards, or miles; and, in the metric system, lengths are usually expressed in millimeters, centimeters, or meters. In this book, we shall often express a length in terms of some convenient segment that is chosen as a unit. For example, in Figure 1.2, if the segment u is chosen as a unit, the length of AB is 4 units, the length of CD is $4\frac{3}{4}$ units, and the length of EF is $\sqrt{2^2 + 3^2} = \sqrt{13} = 3.606\cdots$ units.

It is not sufficient to say merely that the point B in the figure is four units from A. It is really four units to the right of A, and A is four units to the left of B. In mathematics, we use the terms *positive* and *negative* to designate opposite directions. Hence, if we choose to regard the distance from A to B as positive, the distance from B to A is negative.

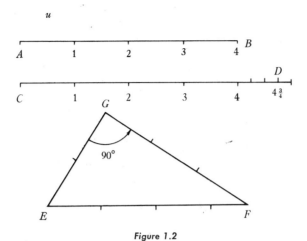

Figure 1.2

In this book, we shall usually consider horizontal and vertical lines as directed. On a horizontal line, the direction to the right is positive and that to the left is negative, while on a vertical line the upward direction is positive and the downward is negative.

1.4.   The Rectangular Cartesian Coordinate System

It is essential for the work that is to follow that we have some method for locating a point in a plane. The method that we shall use is known as the *rectangular Cartesian coordinate system*. To set up this system, we ordinarily draw a horizontal directed line and a vertical directed line in the plane. We call the horizontal line the *X-axis*, the vertical line the *Y-axis*, and their point of intersection the *origin*. We may locate any point in the plane with reference to these axes by stating its distance to the right or left of the *Y*-axis and its distance above or below the *X*-axis. The perpendicular directed distance from the *Y*-axis to the point is called the *abscissa* of the point, the perpendicular directed distance from the *X*-axis to the point is called the *ordinate* of the point, and the length of the line from the origin to the point is called the *radius vector* of the point. The abscissa is positive or negative according as the point is to the right or to the left of the *Y*-axis. The ordinate is positive or negative according as the point is above or below the *X*-axis. The radius vector is always considered positive. We shall use the symbols, x, y, and r to designate the abscissa, the ordinate, and the radius vector, respectively, of a point.

The abscissa and ordinate of a point are known as the *coordinates* of the point and are written as a pair of numbers enclosed in parentheses and separated by a comma, the first designating the abscissa and the second the ordinate.

The coordinate axes divide the plane into four parts called *quadrants*, which are numbered from I to IV in a counterclockwise direction, beginning with the upper right-hand portion. These quadrants will frequently be designated by the symbols Q_1, Q_2, Q_3, and Q_4.

In Figure 1.3, $X'X$ is the X-axis, $Y'Y$ is the Y-axis, and the point O

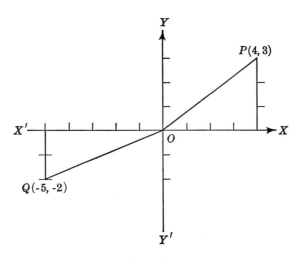

Figure 1.3

is the origin. The point P is 4 units to the right of the Y-axis and 3 units upward from the X-axis. Hence, its abscissa is 4, its ordinate is 3, and we write its coordinates in the form: (4, 3). Likewise, the abscissa of the point Q is -5, its ordinate is -2, and its coordinates are $(-5, -2)$. The radius vector of the point P is the length of the line OP and the length of OQ is the radius vector of Q.

The process of locating a point by means of its coordinates is called *plotting the point*. Since, in the notation (x, y) for the coordinates of a point, the first number always represents the abscissa and the second number represents the ordinate, (x, y) is called an *ordered pair of numbers*.

1.5. The Distance Formula

In order to obtain a formula for the distance between any two points $P_1(x_1, y_1)$ and $P_2(x_2, y_2)$, we shall make use of the Pythagorean theorem and of the distance between two points that are equidistant from a coordinate axis. If the points are the same distance from the X-axis as shown in Figure 1.4, then the length of the segment P_1P_2 is $x_2 - x_1$, since x_2 is the

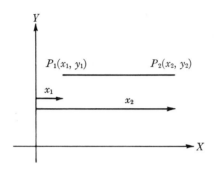

Figure 1.4

distance and direction from the Y-axis to P_2 and x_1 is the distance and direction from the Y-axis to P_1. Similarly, the distance between two points that are equidistant from the Y-axis is the ordinate of the upper one minus the ordinate of the lower one.

EXAMPLES

1. The distance between $(2, 3)$ and $(2, 8)$ is $8 - 3 = 5$.
2. The distance between $(-1, 4)$ and $(5, 4)$ is $5 - (-1) = 6$.

We shall now consider any two points as shown in Figure 1.5. In addition to $P_1(x_1, y_1)$ and $P_2(x_2, y_2)$, the figure shows, as P_3, the point determined

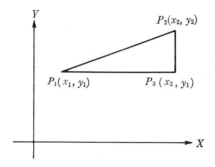

Figure 1.5

by the intersection of a parallel to the X-axis from P_1 and a parallel to the Y-axis from P_2; hence, its coordinates are as shown. Furthermore, P_1P_3 is

$x_2 - x_1$, and P_3P_2 is $y_2 - y_1$. Now, applying the Pythagorean theorem, we see that

$$(P_1P_2)^2 = (P_1P_3)^2 + (P_3P_2)^2$$
$$= (x_2 - x_1)^2 + (y_2 - y_1)^2;$$

consequently, *the distance **d** between* P_1 *and* P_2 *is*

$$d = \sqrt{(x_2 - x_1)^2 + (y_2 - y_1)^2}. \qquad (1.1)$$

EXAMPLE

If we make use of the formula, we see that the distance between $P_1(7, 1)$ and $P_2(3, -2)$ is

$$d = \sqrt{(3 - 7)^2 + (-2 - 1)^2} = \sqrt{16 + 9} = 5.$$

EXERCISE 1.1

Construct each angle given in problems 1 to 8, label each initial and terminal side, and indicate the direction of rotation.

1. $30°, -120°, 315°$
2. $60°, 210°, -135°$
3. $-150°, 180°, 225°$
4. $45°, -330°, 240°$
5. $-60°, 270°, 420°$
6. $135°, -90°, 510°$
7. $495°, -240°, 570°$
8. $660°, 510°, -225°$

Plot each point whose coordinates are given in problems 9 to 12.

9. $(2, 3), (3, -1), (5, 0)$
10. $(6, -3), (-4, -5), (0, -2)$
11. $(-3, -2), (-5, 4), (0, -6)$
12. $(-5, -3), (-2, -4), (-6, 0)$

Using $(3, -5)$ as the starting point, draw the directed line segments described in problems 13 to 20, and find the coordinates of each terminal point.

13. 4 units to the right
14. 2 units upward
15. 3 units downward
16. 5 units to the left
17. 2 units upward, 3 units to the right, 4 units downward
18. 3 units to the left, 2 units downward, 5 units to the right
19. 4 units to the right, 5 units upward, 3 units to the left
20. 5 units downward, 4 units to the left, 2 units upward

Draw the segment between the pair of points given in each of problems 21 to 24 and find the length of each segment.

21. $(3, 2), (7, 5)$
22. $(2, -8), (7, 4)$
23. $(-2, -9), (6, 6)$
24. $(-5, 3), (-2, -1)$

If the points A, B, and C, as given in each of problems 25 to 28 are the coordinates of the vertices of a triangle, find the lengths of the sides.

25. $A(2, 4)$, $B(5, 0)$, $C(1, -3)$ **26.** $A(7, -3)$, $B(2, 9)$, $C(-2, 6)$

27. $A(-6, 1)$, $B(9, -7)$, $C(-3, -2)$ **28.** $A(-4, -1)$, $B(8, 4)$, $C(5, 0)$

1.6. Angular Units

As the measure or length of a line segment is the ratio of the line segment to some standard length chosen as a unit, so is the measure or magnitude of an angle equal to the ratio of the given angle to some other angle that has been chosen as a unit. The unit angles most often used are the degree and the radian. These and certain other units less frequently used will be defined and discussed in this chapter.

The right angle is undoubtedly the oldest and simplest unit of angular measure. Its use is involved in such familiar notions as the change in direction from vertical to horizontal or from a north-south line to an east-west line. It is much too large to use as an ideal scientific tool, although it has been retained in some traditional geometrical theorems whose wording may be traced to antiquity. For purposes of navigation by compass, the right angle was divided into eight "points." A point was an entirely adequate unit for steering a sailing vessel. Astronomers since the time of the ancient Babylonians needed a much finer graduation of the circle. To them we owe the best known of the scientific systems of angular measurement, the *sexagesimal system,* in which the unit of measurement is the *degree.* *The **degree** is an angle that, if placed with its vertex at the center of a circle, intercepts an arc that is equal in length to $\frac{1}{360}$ of the circumference.* A degree is subdivided into 60 min, and the minute into 60 sec. This system of measurement is used in most practical work and in surveying and navigation.

The *circular system* of angular measurement is employed almost exclusively in advanced mathematics and in many branches of scientific work. The unit in the circular system is the *radian,* it is defined below and is illustrated in Figure 1.6.

DEFINITION

*A **radian** is an angle that, if placed with its vertex at the center of a circle, intercepts an arc equal in length to the radius of the circle.*

Other units of angular measure that are frequently used are the *mil* and the *revolution.*

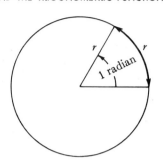

Figure 1.6

The mil is the angle of measure employed by the United States Army in artillery firing. It is the angle subtended by an arc that is $\frac{1}{6400}$ of the circumference, and it is very convenient because a central angle of one mil intercepts an arc that is approximately one yard in length on a circle of radius one thousand yards.

The angular velocity of a revolving wheel is frequently given in revolutions per unit of time, usually in revolutions per minute (abbreviated as rpm).

1.7.   The Relation between Degrees and Radians

Since the circumference of a circle is 2π times the radius, it subtends a central angle of 2π radians. Furthermore, the circumference subtends a central angle of 360°. Therefore,

$$2\pi \text{ radians} = 360°,$$

and, dividing by 2, we see that

$$\pi \text{ radians} = 180°. \tag{1.2}$$

This is a fundamental relation between degrees and radians, and by means of it we can change from degrees to radians and from radians to degrees. For example, dividing each side of (1) by 180, we obtain

$$1° = \frac{\pi}{180} \text{ radians.}$$

Similarly,

$$1 \text{ radian} = \frac{180}{\pi} \text{ degrees.}$$

The angles 30°, 45°, and their multiples are in constant use, and such an angle can be rapidly converted into radians by expressing it as a fractional part of 180° and then replacing 180° by π radians. Other angles can be converted into radians in a similar manner.

EXAMPLES

1. $30° = \dfrac{180°}{6} = \dfrac{\pi}{6}$ radians.

2. $45° = \dfrac{180°}{4} = \dfrac{\pi}{4}$ radians.

3. $120° = \dfrac{2(180°)}{3} = \dfrac{2\pi}{3}$ radians.

4. $25° = \dfrac{25(180°)}{180} = \dfrac{5\pi}{36}$ radians.

If an angle is expressed in degrees, minutes, and seconds, we convert the minutes and seconds to a fractional part of a degree and proceed as in the examples above.

EXAMPLE

5. $25°36'48'' = 25°2208''$

$\qquad = 25\frac{2208}{3600}$ degrees $\qquad$ since $3600'' = 1°$

$\qquad = 25.613°$

$\qquad = \dfrac{25.613}{180}\,\pi$ radians

$\qquad = .(14229)(3.1416)$ radians

$\qquad = .44702$ radians.

Angles expressed in radians are often written as fractional parts of π, such as $\pi/6$, $5\pi/3$, and $\pi/12$. Angles expressed in this way are converted to degrees by replacing π radians by 180° and then simplifying the result.

EXAMPLES

6. $\dfrac{\pi}{6}$ radians $= \dfrac{180°}{6} = 30°$.

7. $\dfrac{5\pi}{3}$ radians $= \dfrac{5(180°)}{3} = 300°$.

8. $\dfrac{\pi}{12}$ radians $= \dfrac{180°}{12} = 15°$.

In the work that follows, we shall use both the radian and the degree as units of angular measure. If at any time an angle is given in terms of radians and an approximation to the nearest second is acceptable, we may use 1 radian = 57°17′45″, since

$$1 \text{ radian} = \frac{180°}{\pi}$$

$$= \frac{180°}{3.14159} \quad \text{approximately}$$

$$= 57.2958° \quad \text{approximately}$$

$$= 57°17′45″ \quad \text{approximately.}$$

It is customary to omit the word "radian" when one expresses an angle in terms of radians. For example, we write $\pi/4$ and read, "pi over 4 radians."

EXERCISE 1.2

Express the angle in each of problems 1 to 36 as a constant times π radians.

1. 18°	**2.** 60°	**3.** 6°	**4.** 15°
5. 225°	**6.** 36°	**7.** 45°	**8.** 72°
9. 12°	**10.** 5°	**11.** 9°	**12.** 30°
13. 3°	**14.** 24°	**15.** 20°	**16.** 126°
17. 18°45′	**18.** 11°15′	**19.** 43°20′	**20.** 61°15′
21. 6°40′	**22.** 37°30′	**23.** 25°12′	**24.** 16°12′
25. 1°18′45″	**26.** 32°16′48″	**27.** 2°11′24″	**28.** 5°3′45″
29. 3 right angles		**30.** 5 right angles	
31. 2.3 right angles		**32.** 4.1 right angles	
33. 5 revolutions		**34.** 3 revolutions	
35. 3.7 revolutions		**36.** 4.9 revolutions	

Express the angle in each of problems 37 to 60 in terms of degrees, minutes, and seconds.

37. $\pi/9$	**38.** $\pi/6$	**39.** $\pi/15$	**40.** $\pi/36$
41. $3\pi/5$	**42.** $5\pi/6$	**43.** $7\pi/12$	**44.** $11\pi/15$
45. $\pi/48$	**46.** $7\pi/16$	**47.** $11\pi/24$	**48.** $13\pi/72$
49. $3\pi/32$	**50.** $5\pi/64$	**51.** $2\pi/27$	**52.** $5\pi/108$
53. 3.1 radians	**54.** 4.3 radians	**55.** 2.7 radians	**56.** 5.6 radians
57. 2.73 radians	**58.** 1.84 radians	**59.** 3.87 radians	**60.** 4.61 radians

1.8.   The Central Angle and Arc Length

One of the advantages of the circular system of angular measurement is the fact that we can easily find the length of the intercepted arc when the central angle is measured in radians. For example, if on a circle of radius r a central angle of θ radians intercepts an arc a units in length, then, by the definition of a radian,

$$\frac{a}{r} = \theta, \tag{1}$$

$$a = r\theta. \tag{1.3}$$

Hence, in order to obtain the length of an arc, we multiply the number of radians in the central angle by the radius of the circle.

EXAMPLE

Find the length of the arc on a circle of radius 12.6 in. intercepted by a central angle of 36°.

Solution: We must first express the central angle in terms of radians. We know that

$$1° = \frac{\pi}{180},$$

hence

$$36° = \frac{36}{180}(180°) = \frac{\pi}{5} \text{ radians,}$$

since

$$36/180 = \tfrac{1}{5} \quad \text{and} \quad 180° = \pi \text{ radians.}$$

Then, by (1.3),

$$a = 12.6\left(\frac{\pi}{5}\right) = 7.92 \text{ in.}$$

1.9.   Angular Speed and Linear Speed

This article will be devoted to finding a relation between angular speed and linear speed, but, before doing that, we shall define and illustrate each. If a ray revolves in a plane at a uniform rate around its origin, the *angular speed* of the ray is the angle through which it revolves in a unit of time. Furthermore, the *linear speed* of any point on the ray is the distance that point travels in a unit of time. The angular speed of the minute hand of a clock is 1 revolution, 360°, or 2π radians per hr. If the hand is 4 in. long, its tip moves the distance 8π in. around a circle of radius 4 in. in 1 hr.

If the ray revolves through θ radians in t units of time and if we designate the angular speed in radians per unit of time by ω, then

$$\omega = \theta/t. \tag{1}$$

If P is a point on a ray that is revolving about its origin O and $OP = r$, we can use (1) and the relation $a = r\theta$ to express the linear speed v in terms of r and ω. If we divide each member of $a = r\theta$ by t, we have

$$\frac{a}{t} = \frac{r\theta}{t}. \tag{2}$$

Now a/t is the linear speed of the point P since it is distance divided by time; furthermore, $\theta/t = \omega$ from (1). Therefore, (2) becomes

$$v = r\omega, \tag{1.4}$$

and we see that *the linear speed of a point on a revolving ray is the angular velocity of the ray times the distance of the point from the center of rotation.*

EXAMPLE 1

The tires of an automobile are 36 in. in diameter and are revolving at the rate of 374 rpm. Find the linear speed of the car in mph.

Solution:

$$374 \text{ rpm.} = 374(2\pi) \text{ radians per min} = 748\pi \text{ radians per min.}$$

Then, by (1.4),

$$v = (748\pi) \ 18 \text{ in. per min.}$$

In order to reduce this quantity to miles per hour, we multiply by 60 (the number of minutes in an hour) and then divide by 5280×12 (the number of inches in a mile), thus obtaining

$$\frac{748\pi(18) \times 60}{5280 \times 12} = 40.1 \text{ mph.}$$

EXAMPLE 2

Two wheels, 12 in. and 4 ft in diameter, respectively, are connected by a belt. The larger wheel is revolving at the rate of 60 rpm. Find the linear speed of the belt and the angular velocity in radians per minute of the smaller wheel.

Solution: Since the linear speed of the belt is equal to the linear speed of the larger wheel, we can obtain it by using (1.4). The radius of the wheel is 2 ft, and the angular speed ω is $60 \times 2\pi$. Therefore, in feet per minute, $v = 2 \times 60 \times 2\pi = 240\pi = 754$.

Since the linear speed of a point on the circumference of the smaller wheel is equal to the speed of the belt and since the radius of the smaller wheel is 6 in. $= \frac{1}{2}$ ft, we obtain, by use of (1.4),

$$754 = \omega(\tfrac{1}{2}).$$

Hence

$$\omega = 2 \times 754$$

$$= 1508$$

Note. This result can be checked as follows: The radius of the smaller wheel is $\frac{1}{4}$ that of the larger; hence, its angular speed is 4 times that of the larger. Consequently, $\omega = 4(120\pi) = 480\pi = 1508$.

EXERCISE 1.3

Find the intercepted arc to three digits in problems 1 to 8, and to four digits in problems 9 to 12, if the central angle and radius are as given.

1. .387 radians, 2.36 ft

2. .809 radians, 5.37 ft

3. 1.68 radians, 9.77 cm

4. 4.92 radians, 11.3 cm

5. 2.39 radians, 1.82 meters

6. 3.08 radians, 3.08 meters

7. .87 radians, 11.4 cm

8. .919 radians, 7.94 cm

9. 1.891 radians, 19.02 in.

10. 1.924 radians, 9.037 in.

11. 1.066 radians, 14.92 meters

12. 1.776 radians, 1.968 meters

Find the central angle in radians to three digits in problems 13 to 16, and to four digits in problems 17 to 20. The arc length is given first and is followed by the radius.

13. 2.37 cm, 1.58 cm

14. 56.4 cm, 85.6 cm

15. 9.81 in., 12.3 in

16. .239 ft, .185 ft

17. 25.92 cm, 39.71 cm

18. 39.71 cm, 25.92 cm

19. 100.4 in., 472.3 cm

20. 32.74 cm, 40.01 cm

21. How many radians are in the angle between the hands of a clock at 7 o'clock?

22. Through how many radians does the minute hand of a watch move in 35 min?

23. How far does the tip of a 4-in. hour hand of a clock move in 3 hr and 30 min?

24. The minute hand of a town clock is 4 ft long. How fast does its tip move?

25. If the rear wheels of a wagon are 3.6 ft in diameter, how far will the wagon move if a spoke turns through 36°?

26. How many revolutions will the wheel of problem 25 make if the wagon travels a mile?

27. Find the angular speed in radians per second of the wheel of a car provided the car is traveling 56 mph and the wheel is 28 in. in diameter.

28. A wheel that is 4 ft in diameter is connected by a belt to another wheel that is 6 in. in radius. If the larger wheel revolves at 60 rpm, find the rate of revolution of the smaller wheel and the linear speed of the belt.

29. The pendulum of a grandfather clock is 4 ft long and swings through an arc of 22° per second. How far does its tip move in 1 minute?

30. The minute hand of a clock is 3.5 in. long. How far does its tip move in an hour?

31. A curve on a highway subtends an angle of 31° on a circle of radius 1800 ft. How long will it take a car at 40 mph to round the curve?

32. A curve on an interstate highway is laid out as an arc of a circle whose radius is 1500 ft. What central angle is subtended if it takes 30 sec to round the curve at 60 mph?

33. A weight is being raised by means of a rope that passes over an overhead pulley and then to the drum of a windlass that is 2 ft in diameter. If the drum is revolving at 42 rpm, find the rate at which the weight is rising. Also, find the angular velocity of the pulley if it is 6 in. in diameter.

34. What is the speed in mph of a point on the equator as the earth rotates? Assume the earth's radius to be 4000 mi.

35. An engine has drive wheels 5 ft in diameter and is traveling at 50 mph. Find the angular velocity of the drive wheels in radians per second.

36. What angle is subtended at the sun by the earth? Assume the earth's radius is 4000 mi and that it is 93,000,000 mi from the sun.

37. Find the linear speed in mph of the tip of an 8-ft airplane propeller that is turning 2000 rpm.

38. An isosceles triangle whose equal sides meet at an angle of 24° is inscribed in a circle. Find the radius of the circle if the base of the triangle intercepts an arc of 3.2 in. on the circle.

39. What are the latitude and longitude of a point that is 2700 mi north of the equator and 2100 mi west of the meridian through Greenwich, England? The longitude of Greenwich is 0°, and the earth has a radius of 4000 mi.

40. Tangents are drawn from each end of an arc 26 in. long to a circle of radius 14 in., and then produced until they meet. Find the angle between them to the nearest degree.

1.10. Standard Position of an Angle

In order to be able to define the six ratios that are associated with an angle, and which are the foundation of trigonometry, it is desirable to have the angle in a specified position. We say that an angle is in *standard position* relative to the coordinate axes if the vertex is at the origin and the initial side coincides with the positive X-axis. The terminal side may be in either quadrant or along a positive or negative portion of a coordinate axis. We say that the angle is in the quadrant in which the terminal side lies. Therefore, the angle θ in Figure 1.7 is a positive angle in standard position and is in the first quadrant.

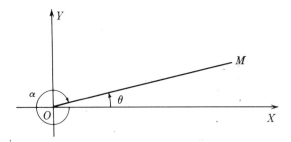

Figure 1.7

Two angles are called *coterminal* if they are in standard position and have the same terminal side. Thus, in Figure 1.7, α and θ are coterminal angles.

The distance from the origin to a point on the terminal side of an angle is called the *radius vector* of the point and is always positive.

1.11. The Trigonometric Ratios

The trigonometric ratios are defined in terms of the coordinates and radius vector of a point on the terminal side of the angle. Before defining these ratios, we shall show that the value of the ratio of any two of the abscissa, ordinate, and radius vector of a point on the terminal side of an angle does not depend on the point that is chosen. In order to do this, we shall consider an angle θ, as in Figure 1.8, and let P and Q be any two points on the terminal side. We construct perpendiculars PR and QS to the X-axis and thereby obtain the similar triangles ORP and OSQ; hence, the ratios of the corresponding sides of these triangles are equal and we have

$$\frac{OR}{OP} = \frac{OS}{OQ} \qquad \frac{RP}{OP} = \frac{SQ}{OQ} \qquad \frac{RP}{OR} = \frac{SQ}{OS}$$

$$\frac{OP}{OR} = \frac{OQ}{OS} \qquad \frac{OP}{RP} = \frac{OQ}{SQ} \qquad \frac{OR}{RP} = \frac{OS}{SQ}.$$

We now call attention to the facts that OR, RP, and OP are the abscissa,

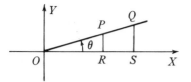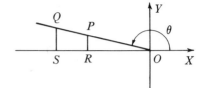

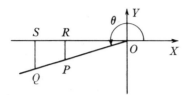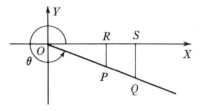

Figure 1.8

ordinate, and radius vector of P and that OS, SQ, and OQ are the coordinates and radius vector of Q. Therefore, the values of the six ratios that can be formed from the coordinates and radius vector of a point on the terminal side of an angle in standard position do not depend on the particular point chosen. Consequently, they depend only on the angle and are called the *trigonometric ratios* of, or associated with, the angle. They are given names and abbreviated in accordance with the following convention.

If θ is an angle in standard position as in Figure 1.9 and if x, y, and r

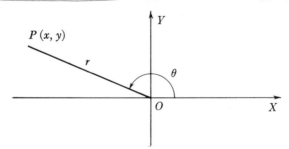

Figure 1.9

are the abscissa, ordinate, and radius vector, respectively, of any point on the terminal side of the angle, then

$$\text{sine } \theta = \frac{y}{r} = \sin \theta, \qquad\qquad \text{cotangent } \theta = \frac{x}{y} = \cot \theta, \, y \neq 0,$$

$$\text{cosine } \theta = \frac{x}{r} = \cos \theta, \qquad\qquad \text{secant } \theta = \frac{r}{x} = \sec \theta, \, x \neq 0,$$

$$\text{tangent } \theta = \frac{y}{x} = \tan \theta, \, x \neq 0, \qquad \text{cosecant } \theta = \frac{r}{y} = \csc \theta, \, y \neq 0.$$

The usual convention of signs is used in determining the sign of a trigonometric ratio of an angle. Thus, the cosine of a second-quadrant angle is a negative number since it is a negative number divided by a positive one, whereas the tangent of a third-quadrant angle is positive since it is the ratio of two negative numbers.

The angle θ may be positive or negative since the rotation can be in a counterclockwise or clockwise direction. It need not be between -2π and 2π since the ray whose position determines the terminal side of the angle may make any number of revolutions before coming to rest. Therefore, θ can represent any number of radians or degrees.

1.12.   The Relation between the Abscissa, Ordinate, and Radius Vector

If $P(x, y)$ is any point in the cartesian plane as in Figure 1.10, if r is its radius vector, and if a perpendicular PQ is drawn to the X-axis, then

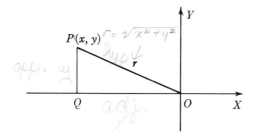

Figure 1.10

$OQ = x$ and $QP = y$ are the sides of a right triangle of which $OP = r$ is the hypotenuse, or one or both of them are the negatives of the sides of such a triangle. Now, making use of the Pythagorean theorem and the fact that the square of a number and of its negative are equal, we find that

$$x^2 + y^2 = r^2 \qquad (1.5)$$

for any point $P(x, y)$. This fact can be stated in words as follows: *The sum of the squares of the coordinates of a point is equal to the square of the radius vector.* This equation can be used to find either a coordinate or the radius vector, if the other two of x, y, and r are given. We will get two values for the unknown for which we are solving but can use only the positive value if r is the unknown; furthermore, only one value of a coordinate can be used if we know the quadrant in which $P(x, y)$ lies.

EXAMPLE 1

Find the radius vector of $P(5, -12)$.

Solution: If we replace x and y in (1.5) by the given values, we have

$$r^2 = 5^2 + (-12)^2 = 25 + 144 = 169.$$

Hence, $r = \sqrt{169} = 13$. We used 13 rather than ± 13 since the radius vector is defined as a positive number.

EXAMPLE 2

Find the abscissa of a point in the second quadrant if its ordinate and radius vector are 15 and 17, respectively.

Solution: If we substitute the given values in (1.5), we obtain

$$x^2 + 15^2 = 17^2.$$

Therefore, $x^2 = 17^2 - 15^2 = 289 - 225 = 64$, and $x = \pm 8$. We must, however, use only -8 since P is in the second quadrant.

EXAMPLE 3

If $A(8, -6)$ and $B(x, 3)$ are on the same line through the origin, find the abscissa and radius vector of B.

Solution: We begin by locating $A(8, -6)$ and $B(x, 3)$ as shown in Figure 1.11, and then draw perpendiculars AC and BD to the X-axis. We thereby have two

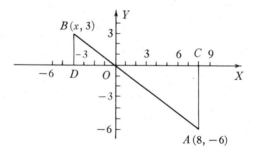

Figure 1.11

similar triangles. In order to find the length of OD, we make use of the fact that corresponding sides are proportional:

$$\frac{DO}{DB} = \frac{OC}{AC}.$$

Putting in the given values, we obtain

$$\frac{DO}{3} = \frac{8}{6}.$$

Thus, $DO = 4$. Accordingly, $x = OD = -4$, and by use of (1.5) we find that $r = OB = \sqrt{(-4)^2 + 3^2} = 5$.

EXERCISE 1.4

Find the one of x, y and r that is missing in each of problems 1 to 20.

1. $x = 3, y = 4$ 2. $x = 5, y = -12$

3. $x = -8, y = 15$ 4. $x = -24, y = -7$

5. $x = -7, r = 25$ 6. $x = 8, r = 17$

7. $x = 15, r = 17$ 8. $x = -24, r = 25$

9. $y = 8, r = 17$ 10. $y = -7, r = 25$

11. $y = -24, r = 25$ 12. $y = 3, r = 5$

13. $x = 4, r = 5$, $P(4, y)$ in the first quadrant

14. $x = -15, r = 17$, $P(-15, y)$ in the third quadrant

15. $x = -5, r = 13, y < 0$ 16. $x = 24, r = 25, y > 0$

17. $y = 6, r = 10, x < 0$ 18. $y = -14, r = 34, x > 0$

19. $y = 8, r = 17$, $P(x, 8)$ in the second quadrant

20. $y = -24, r = 25$, $P(x, -24)$ in the fourth quadrant

In each of problems 21 to 28, the points A and B are on the same line through the origin. Find the unknown coordinate and radius vector of B in each problem.

21. $A(3, 4), B(6, y)$ 22. $A(5, 12), B(x, -24)$

23. $A(-8, 15), B(x, 30)$ 24. $A(-7, 24), B(x, -12)$

25. $A(-5, -12), B(x, 6)$ 26. $A(-15, -36), B(-10, y)$

27. $A(7, -24), B(-14, y)$ 28. $A(12, -5), B(6, y)$

Draw an angle in standard position that satisfies the conditions in each of problems 29 to 40 and find the values of the trigonometric ratios of the angle in each case.

29. $P(5, 12)$ is on the terminal side

30. $P(3, -4)$ is on the terminal side

31. $P(-8, 15)$ is on the terminal side

32. $P(-24, -7)$ is on the terminal side

33. $P(-3, y), y > 0$, is 5 units out on the terminal side

34. $P(8, y), y < 0$, is 17 units out on the terminal side

35. $P(24, y), y > 0$, is 25 units out on the terminal side

36. $P(-5, y), y < 0$, is 13 units out on the terminal side

37. $P(x, -8), x > 0$, is 17 units out on the terminal side

38. $P(x, -24)$, $x < 0$, is 25 units out on the terminal side

39. $P(x, 5)$, $x > 0$, is 13 units out on the terminal side

40. $P(x, 3)$, $x < 0$, is 5 units out on the terminal side

1.13. Sets

We shall use the terminology and operations of sets from time to time in this book; hence, a brief introduction to sets.

A *set* is a collection of well-defined objects or concepts that are called *elements*. By "well-defined" we mean that there is a criterion for determining whether an element belongs to the set or does not belong to it. We indicate that a is an element of the set A by writing $a \in A$, and that a is not an element of A by $a \notin A$.

The sets S and T are *equal* if, and only if, each element of each set is an element of the other. If each element of S is also an element of T, then S is a *subset* of T; furthermore, if there is an element of T that is not an element of S, then S is a *proper subset* of T. We indicate that S is a subset of T symbolically by $S \subseteq T$, and that S is a proper subset of T by $S \subset T$. The set that contains no elements is called the *empty* or *null set* and is symbolized by $\varnothing$.

A set may be designated by listing its elements or by giving a rule that determines the elements. Thus the set S, which consists of the odd positive integers less than 10, may be indicated by

$$S - \{1, 3, 5, 7, 9\} \tag{1}$$

or by

$$S = \{x \mid x \text{ is an odd positive integer less than } 10\}.$$

The latter is read "S is the set of all x such that x is an odd positive integer less than 10."

If S is given by (1) and $T = \{1, 3, 7\}$, then T is a proper subset of S and we write $T \subset S$.

The set of elements that is common to the sets S and T is called their *intersection* and is designated by $S \cap T$. If $S \cap T = \varnothing$, the sets S and T are said to be *disjoint*. The set of elements that is in S or in T is called the *union* of S and T and is designated by $S \cup T$. If $S = \{1, 3, 5, 7, 9\}$ and $R = \{2, 3, 5, 6\}$ then $S \cup R = \{1, 2, 3, 5, 6, 7, 9\}$ and $S \cap R = \{3, 5\}$.

The *complement* of the set S with respect to T consists of those elements of T that are not elements of S and is indicated by $T - S$. Symbolically,

$T - S = \{x \mid x \in T \text{ and } x \notin S\}$. Thus the complement of $S = \{2, 3, 5\}$ with respect to $T = \{1, 2, 3, 4, 5\}$ is $T - S = \{1, 4\}$.

1.14. Functions

If there is a rule or agreement such that for each x in a set there corresponds one or more values of y, and if D is the set of elements x and R is the set of elements y, then the set of ordered pairs $\{(x, y)\}$ is said to form a *relation* with domain D and range R. Thus, $\{(x, y) \mid y^2 = x\}$ is a relation, since for each positive value of x there correspond two real values of y. One of them is positive and the other is negative. If $x = 9$, then $y = 3$ or -3.

In mathematics there is a special type of relation that is of particular interest. It is the function.

*If in a relation no two ordered pairs have the same first element and different second elements, then the relation is called a **function**.*

Consequently, in a function, each element of D is matched with exactly one element of R. A function is ordinarily designated by the statement $f = \{(x, y) \mid y = f(x)\}$. The equation $y = f(x)$ is often called the *defining equation*, and $f(x)$ is a *function value*. The relation $r = \{(x, y) \mid y^2 = x\}$ is not a function, since there are two values in the range for at least one value in the domain. The relation $\{(x, y) \mid y = 2x - 1\}$ is a function, since no two pairs with the same first element have different second elements. The domain and range are both all real numbers. In the function $f = \{(x, y) \mid y = \sqrt{x - 1}\}$, the domain is $\{x \mid x - 1 \geq 0\}$ and the range is $\{y \mid y \geq 0\}$.

EXAMPLE

If $f = \{(x, y) \mid y = 3x^2 - 2x + 1\}$, and $D = \{5, 4, 2\}$, find the ordered pairs that constitute f.

Solution: To find the function value $y = 3x^2 - 2x + 1$ corresponding to a given value in D, we replace x by that value. Hence, if $x = 5$, then $y = 3 \cdot 5^2 - 2 \cdot 5 + 1 = 66$, and the number pair in the function is $(5, 66)$. We can find the other number pairs similarly, and then know that $f = \{(5, 66), (4, 41), (2, 9)\}$.

1.15. The Trigonometric Functions

In this section we shall consider a unit circle with its center at the origin, and shall show that six different ordered pairs of numbers can be associated with a point P on the circumference. For this purpose we construct the

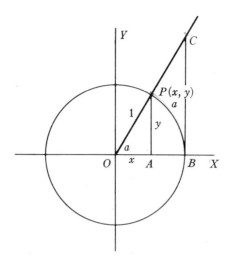

Figure 1.12

unit circle in Figure 1.12, choose a point $P(x, y)$ on the circumference, draw a ray through O and P, and construct the perpendiculars AP and BC.

We designate the length of the arc BP by a, and, since the radius of the circle is 1, the number of radians in the central angle BOP is $a/1 = a$. Thus a is the measure of the length of the arc in linear units and of the central angle in radians. Also, since the coordinates of P are (x, y), it follows that $OA = x$, $AP = y$, $\sin a = y/1$, and $\cos a = x/1$. Hence, $y = \sin a$, and $x = \cos a$. Furthermore, since triangles OAP and OBC are similar, it follows that

$$\frac{BC}{OB} = \frac{y}{x} = \tan a,$$

and

$$\frac{OC}{OB} = \frac{OP}{x} = \frac{1}{x} = \frac{1}{\cos a} = \sec a, \qquad \text{since } OP = 1.$$

Therefore

$$BC = \tan a, \quad OC = \sec a, \qquad \text{since } OB = 1.$$

Now we form the following pairs of numbers, each of which is associated with the point P:

$$(a, y) = (a, \sin a)$$

$$(a, x) = (a, \cos a)$$

$$(a, \overline{BC})^* = (a, y/x) = (a, \tan a)$$

$$(a, \overline{OC}) = (a, 1/x) = (a, \sec a).$$

* By $\overline{BC}$ we mean the length of the line segment BC.

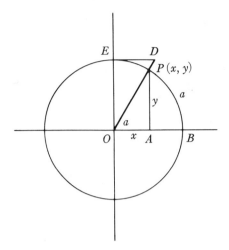

Figure 1.13

By use of Figure 1.13, we obtain two additional pairs, as follows: Since triangles OED and AOP are similar, we have

$$\frac{ED}{OE} = \frac{OA}{AP} = \frac{x}{y} = \cot a,$$

and

$$\frac{OD}{OE} = \frac{OP}{AP} = \frac{1}{y} = \frac{1}{\sin a} = \csc a \qquad \text{since } OP = 1 \text{ and } y = \sin a.$$

Since $OE = 1$, then $ED = \cot a$, and $OD = \csc a$. Thus we have the two following additional pairs that are associated with P:

$$(a, \overline{ED}) = (a, \cot a) \qquad \text{and} \qquad (a, \overline{OD}) = (a, \csc a).$$

In article 1.11 we stated that the value of a trigonometric ratio depends upon the coordinates of a point P on the terminal side of the angle θ, and that θ can represent any real number of radians. Consequently the argument in this article is not affected if P makes any number of revolutions about O before coming to rest at the point indicated in the figure. Therefore the length of the arc traced by P and the number of radians in the angle generated by the ray OP may be any real number. This number is positive or negative according as P moves in a counterclockwise or clockwise direction around the circle.

Now we consider the ordered pair $(a, y) = (a, \sin a)$. As P moves around the circle, a varies over a set of real numbers, and y varies over a set of numbers between -1 and 1 inclusive, since $|y| \leq 1$.* Furthermore, one and only one value of y corresponds to each value of a. Consequently we have a set of ordered pairs $\{(a, \sin a)\}$ that satisfies the definition of a function. This set is called the *sine function*. The domain is the set of all real numbers and the range is the set $\{y| -1 \leq y \leq 1\}$.

By a similar argument we can show that as P moves around the circle, each of the other five pairs of numbers associated with P generates a function. Thus we have the six *trigonometric functions*. We shall tabulate these functions and show the domain and range of each. Below the tabulation we shall justify the statements made about the domain and range. R stands for the set of all real numbers.

Function name	Function	Domain	Range
sine	$(a, \sin a)$	R	$\{y \mid -1 \leq y \leq 1\}$
cosine	$(a, \cos a)$	R	$\{x \mid -1 \leq x \leq 1\}$
tangent	$(a, \tan a)$	$R - \left\{ \dfrac{n\pi}{2} \,\middle|\, n \text{ is an odd integer} \right\}$	R
cotangent	$(a, \cot a)$	$R - \{n\pi \mid n \text{ is an integer}\}$	R
secant	$(a, \sec a)$	$R - \left\{ \dfrac{n\pi}{2} \,\middle|\, n \text{ is an odd integer} \right\}$	$R - \{n \mid -1 \leq n \leq 1\}$
cosecant	$(a, \csc a)$	$R - \{n\pi \mid n \text{ is an integer}\}$	$R - \{n \mid -1 \leq n \leq 1\}$

Now, $\sin a = y$ and $\cos a = x$, and values of x and y exist for all values of a. Therefore, the domain of the sine and cosine functions is R. Furthermore $-1 \leq y \leq 1$ and $-1 < x \leq 1$. Thus the range is as specified in the tabulation.

Since $\tan a = y/x$ and $\sec a = 1/x$, a value of neither function exists if $x = 0$. Therefore the numbers $n\pi/2$, where n is an odd integer, must be deleted from the domain of the tangent and secant functions. Thus the domain is as tabulated above. Furthermore, $\cot a = x/y$ and $\csc a = 1/y$. Thus, y cannot be zero, so the numbers $n\pi$, where n is an integer, must be deleted from the ranges of the cotangent and cosecant functions.

If in $\tan a = y/x$, $y = 0$, then $\tan a = 0$. However, as $|y|$ increases from zero toward 1, $|x|$ decreases from 1 toward zero. Hence $|y|/|x|$ increases continuously from zero and becomes greater than any number

---

* The symbol $|y|$ is called the absolute value of y, and is defined thus: if $y > 0$, $|y| = y$; if $y = 0$, $|y| = 0$; and if $y < 0$, $|y| = -y$.

chosen in advance. Hence if x and y have the same sign, tan a varies continuously over the set of all positive real numbers, and if x and y have different signs, tan a varies over the set of all negative real numbers. Hence the range of the tangent function is the set of all real numbers.

Similarly, it follows that the range of the cotangent function is the set of all real numbers.

Next we consider the range of the secant and cosecant functions. Since sec $a = 1/y$, and $-1 \leq y \leq 1$, it follows that $1/|y|$ is never less than 1. However, as $|y|$ decreases from 1 toward zero, $|1/y|$ increases continuously over the set of all real numbers greater than or equal to 1 if y is positive, and $1/y$ decreases continuously over all real numbers less than -1 if y is negative. Hence the range of the secant function is $R - \{n \mid -1 \leq n \leq 1\}$. By use of a similar argument we can show that the range of the cosecant function is also $R - \{n \mid -1 \leq n \leq 1\}$.

Since the second number in each ordered pair in a function is called the *function value*, we shall refer to the second number in each of the trigonometric functions as a *trigonometric function value*. Thus we may refer to sin a as a trigonometric function value or as a *trigonometric ratio*. Similar statements can be made with respect to cos a, tan a, cot a, sec a, and csc a.

1.16. Given One Trigonometric Ratio: To Find the Others

If we know the value of one of the trigonometric ratios, we can, by means of the definition of the ratios and equation (1.5), calculate two values for each of the other ratios except the one that is the reciprocal of the given one.

✓EXAMPLE 1

Find each of the other trigonometric ratios if sin $\theta = 3/5$.

Solution: If we make use of the definition of the trigonometric ratio sin θ, we have

$$\sin \theta = \frac{y}{r} = \frac{3}{5}.$$

Consequently, there is a point on the terminal side with ordinate 3 and radius vector 5 as in Figure 1.14. Hence by (1.5), we have $x^2 + 3^2 = 5^2$. Therefore, $x^2 = 25 - 9 = 16$ and $x = \pm 4$. Consequently,

$$\cos \theta = \pm 4/5, \qquad \sec \theta = \pm 5/4,$$
$$\tan \theta = \pm 3/4, \qquad \csc \theta = 5/3.$$
$$\cot \theta = \pm 4/3,$$

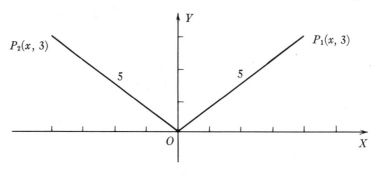

Figure 1.14

If, in addition to one of the trigonometric function values, we know the quadrant in which the given angle or angles that corresponds to the given number lies, there is only value for each of the other ratios.

√EXAMPLE 2

Evaluate each of the other trigonometric ratios if $\cot \theta = 8/15$ and θ is in the third quadrant.

Solution: By use of the definition of $\cot \theta$ and the given value, we know that there is a point on the terminal side with $x = -8$ and $y = -15$. Consequently, by (1.5), we have $(-8)^2 + (-15)^2 = r^2$. Therefore, $r^2 = 289$ and $r = \pm 17$, but we must use $+17$ since r is positive. Consequently,

$$\sin \theta = -15/17, \qquad \sec \theta = -17/8,$$
$$\tan \theta = 15/8, \qquad \csc \theta = -17/15.$$
$$\cos \theta = -8/17,$$

EXERCISE 1.5

1. If $D = \{x \mid x$ is an integer with $2 \leq x \leq 5\}$, find the set of ordered pairs $\{(x, f(x)) \mid f(x) = 2x + 3\}$

2. If $D = \{x \mid x$ is an integer with $-1 \leq x \leq 3\}$, find the set of ordered pairs $\{(x, f(x)) \mid f(x) = 3x - 4\}$.

3. Find the set of ordered pairs $\{(x, f(x)) \mid f(x) = x^2 - 1\}$ if $D = \{x \mid x$ is an integer with $-3 \leq x \leq 2\}$.

4. Find the set of ordered pairs $\{(x, f(x)) \mid f(x) = x^2 + x + 2\}$ if $D = \{x \mid x$ is an integer with $-4 \leq x \leq 1\}$.

In each of problems 5 to 8, find the range of the function f.

5. $f = \{(x, y) \mid y = 3x + 2\}$, $D = \{x \mid -2 \leq x \leq 4\}$

6. $f = \{(x, y) \mid y = 2x - 5\}$, $D = \{x \mid -3 \leq x \leq 5\}$

7. $f = \{(x, y) \mid y = x^2 + x\}$, $D = \{x \mid 2 \leq x \leq 7\}$

8. $f = \{(x, y) \mid y = 1 - x^2\}$, $D = \{x \mid 1 \leq x \leq 6\}$

Determine whether the relation in each of problems 9 to 12 is a function and give the reason for your decision.

9. $\{(x, y) \mid x^2 + y = 18\}$, $D = \{x \mid x \text{ is an integer with } -1 \leq x \leq 4\}$

10. $\{(x, y) \mid x + y^2 = 18\}$, $D = \{x \mid x \text{ is an integer with } -1 \leq x \leq 4\}$

11. $\{(x, y) \mid x - y^2 = 4\}$, $D = \{x \mid 4 \leq x \leq 9\}$

12. $\{(x, y) \mid x^2 - y = 4\}$, $D = \{x \mid 4 \leq x \leq 9\}$

Construct an angle in standard position that satisfies the conditions in each of problems 13 to 20 and find the other function values.

13. $\cos \theta = 5/13$, θ in the fourth quadrant

14. $\tan \theta = 8/15$, θ in the third quadrant

15. $\csc \theta = 25/24$, θ in the second quadrant

16. $\cot \theta = 4/3$, θ in the first quadrant

17. $\sin \theta = -7/25$, θ in the third quadrant

18. $\sec \theta = -17/8$, θ in the second quadrant

19. $\tan \theta = -3/4$, θ in the fourth quadrant

20. $\cos \theta = -15/17$, θ in the third quadrant

In each of the following problems, one function value is given. Find the others.

21. $\csc \theta = 5/4$ 22. $\sec \theta = -17/8$

23. $\cot \theta = -24/7$ 24. $\tan \theta = 8/15$

25. $\cos \theta = 3/5$ 26. $\sin \theta = -24/25$

27. $\tan \theta = -4/3$ 28. $\cos \theta = 7/25$

1.17. The Trigonometric Function Values of 30°, 45°, and Their Multiples

The computation of the trigonometric function values of angles in general is beyond the scope of this book. However, by making use of certain theorems from plane geometry, given in the Reference Material preceding page 1 of this book, we can compute the numerical values of the functions of 30°, 45°, and of integral multiples of these angles.

We shall first consider an angle of 30° and shall construct the angle in standard position as in Figure 1.15. On the terminal side of the angle, we lay off a distance OP equal to two units and then drop a perpendicular PM to the X-axis. Since angle MOP is 30°, then angle MPO is 60° (by State-

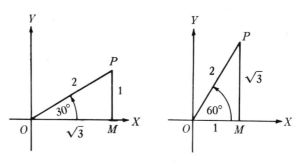

Figure 1.15   Figure 1.16

ment 3 in the Reference Material). Therefore, by Statement 5 in the Reference Material, PM is one half of OP, and hence is equal to one unit. Furthermore, by the Pythagorean theorem (Statement 2):

$$OM = \sqrt{(OP)^2 - (MP)^2}$$
$$= \sqrt{4 - 1}$$
$$= \sqrt{3}.$$

Hence, by the definition of the trigonometric ratios, we have

$$\sin 30° = \frac{1}{2}, \qquad \cot 30° = \frac{\sqrt{3}}{1} = \sqrt{3},$$

$$\cos 30° = \frac{\sqrt{3}}{2}, \qquad \sec 30° = \frac{2}{\sqrt{3}} = \frac{2\sqrt{3}}{3},$$

$$\tan 30° = \frac{1}{\sqrt{3}} = \frac{\sqrt{3}}{3}, \qquad \csc 30° = \frac{2}{1} = 2.$$

If we construct an angle of 60° in standard position as in Figure 1.16 and lay off a distance OP equal to two units on the terminal side and drop a perpendicular PM to the X-axis, we obtain the right triangle OMP. By Statement 3 in the Reference Material, angle $OPM = 30°$. Hence, by Statement 5 in the Reference Material, the shorter side, OM, is one half of OP and is therefore equal to 1. Furthermore, by the Pythagorean theorem, $MP = \sqrt{3}$. Hence, the trigonometric function values of 60° are

$$\begin{cases} \sin 60° = \dfrac{\sqrt{3}}{2}, & \cot 60° = \dfrac{1}{\sqrt{3}} = \dfrac{\sqrt{3}}{3}, \\[2ex] \cos 60° = \dfrac{1}{2}, & \sec 60° = \dfrac{2}{1} = 2, \\[2ex] \tan 60° = \dfrac{\sqrt{3}}{1} = \sqrt{3}, & \csc 60° = \dfrac{2}{\sqrt{3}} = \dfrac{2\sqrt{3}}{3}. \end{cases}$$

In order to compute the trigonometric function values of 45°, we construct an angle of 45° in standard position, as in Figure 1.17, and lay off

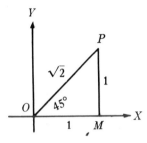

Figure 1.17

the distance OM equal to one unit on the X-axis. At M, we erect a perpendicular to the X-axis and extend it until it meets the terminal side of the angle at P. By Statement 3 in the Reference Material, angle $OPM = 90° - 45° = 45°$; hence, triangle OMP is isosceles, and $OM = MP = 1$. Consequently, $OP = \sqrt{1^2 + 1^2} = \sqrt{2}$. Therefore, the trigonometric function values of 45° are

$$\begin{cases} \sin 45° = \dfrac{1}{\sqrt{2}} = \dfrac{\sqrt{2}}{2}, & \cot 45° = \dfrac{1}{1} = 1, \\[2ex] \cos 45° = \dfrac{1}{\sqrt{2}} = \dfrac{\sqrt{2}}{2}, & \sec 45° = \dfrac{\sqrt{2}}{1} = \sqrt{2}, \\[2ex] \tan 45° = \dfrac{1}{1} = 1, & \csc 45° = \dfrac{\sqrt{2}}{1} = \sqrt{2}. \end{cases}$$

The function values of any multiple of 30° or 45° whose terminal side does not coincide with one of the coordinate axes may be found by a method analogous to that used above. For example, to compute the function values of 240°, we construct the angle in standard position as in Figure 1.18, lay off the distance OP equal to two units on the terminal side, and draw the perpendicular PM to the X-axis. Now, angle $MOP = 240° -$

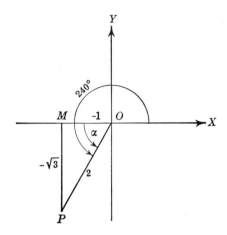

Figure 1.18

$180° = 60°$, and thus the angle $MPO = 30°$. Therefore, the lengths of OM and MP are 1 and $\sqrt{3}$, respectively. Thus, since P is in the third quadrant, its coordinates are $(-1, -\sqrt{3})$. Hence,

$$\sin 240° = -\frac{\sqrt{3}}{2}, \qquad \cot 240° = \frac{-1}{-\sqrt{3}} = \frac{\sqrt{3}}{3},$$

$$\cos 240° = -\frac{1}{2}, \qquad \sec 240° = -\frac{2}{1} = -2,$$

$$\tan 240° = \frac{-\sqrt{3}}{-1} = \sqrt{3}, \qquad \csc 240° = -\frac{2}{\sqrt{3}} = -\frac{2\sqrt{3}}{3}.$$

1.18. Function Values of Quadrantal Angles

Before discussing the function values of quadrantal angles, we wish to call attention to the role of zero in division. If the dividend is zero and the divisor is not zero, the quotient is zero; and furthermore, if the divisor is zero, the quotient does not exist as a unique number.

An angle in standard position whose terminal side coincides with one of the coordinate axes is called a *quadrantal angle*. Thus, 0°, 90°, 180°, 270°, and all angles coterminal with these are quadrantal angles.

We shall illustrate the process of obtaining the trigonometric function values of a quadrantal angle by calculating the functions of 270°. Since the point P on the terminal side of the angle in Figure 1.19 is on the

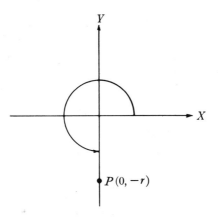

Figure 1.19

lower half of the Y-axis, we have $x = 0$ and $y = -r$. Hence, the values of the trigonometric functions of $270°$ are

$$\sin 270° = \frac{y}{r} = \frac{-r}{r} = -1,$$

$$\cos 270° = \frac{x}{r} = \frac{0}{r} = 0,$$

$$\tan 270° = \frac{y}{x}$$

$$= \frac{-r}{0}, \qquad \text{which does not exist,}$$

$$\cot 270° = \frac{x}{y} = \frac{0}{-r} = 0,$$

$$\sec 270° = \frac{r}{x}$$

$$= \frac{r}{0}, \qquad \text{which does not exist,}$$

$$\csc 270° = \frac{r}{y} = \frac{r}{-r} = -1.$$

EXERCISE 1.6

In each of problems 1 to 12, find the function values of the given number.

1. $3\pi/4$ **2.** $2\pi/3$ **3.** $7\pi/6$ **4.** $\pi/2$

5. $4\pi/3$ **6.** $5\pi/4$ **7.** $5\pi/2$ **8.** $11\pi/6$

9. π **10.** $5\pi/6$ **11.** $-\pi/4$ **12.** $5\pi/3$

Prove that the statement in each of problems 13 to 36 is true by use of the function values of 30°, 45° and their multiples.

13. $\sin^2 240° + \cos^2 120° = 1$ **14.** $1 + \tan^2 135° = \sec^2 315°$

15. $\csc^2 150° = 1 + \cot^2 330°$ **16.** $\sin^2 (-120°) + \cos^2 300° = 1$

17. $\sin 120° = 2 \sin 300° \cos 120°$ **18.** $\cos 240° = \cos^2 300° - \sin^2 120°$

19. $\sin 240° = 2 \sin 60° \sin 330°$ **20.** $\cos 300° = -\cos^2 120° + \sin^2 240°$

21. $\csc 60° - \cot 240° = \cot 60°$ **22.** $1 - \cos 120° = \cot 210° \cos 330°$

23. $\sin 150° = \sin 90° \cos 60° + \cos 270° \sin 45°$

24. $\cos 210° = \cos 90° \cos 240° + \sin 270° \sin 120°$

25. $\cos 330° + \cos 240° \tan 210° = \cot 240°$

26. $4 \sin^3 30° = 3 \sin 150° + \cos 180°$

27. $\sin 180° = 2 \sin 120° + \sin 240° \sec 300°$

28. $\sin 315° = 4 \sin^3 135° + 3 \sin 225°$

29. $\sin 120° = \sqrt{\dfrac{1 - \sin 330°}{2}}$

30. $\cos 330° = \sqrt{\dfrac{1 - \cos 240°}{2}}$

31. $\tan 300° = \dfrac{\tan 60° + \cot 210°}{1 - \tan 300° \cot 330°}$

32. $\dfrac{\cos 225°}{\sin 135°} + \dfrac{\sin 315°}{\cos 135°} = 3 \sin 180°$

33. $\cot 240° = \dfrac{\cos 330°}{1 + \cos 300°}$

34. $\cot 240° + \cot 150° = \dfrac{\cos 300°}{\sin 210° \sin 120°}$

35. $\dfrac{\sin 225° - \sin 135°}{\cos 135° - \cos 315°} = \cot 225°$

36. $\tan 300° = \dfrac{1 - \cos 240°}{\cos 210°}$

Determine which of the following statements are true and which are false. Show that each decision is correct.

37. $\sin 30° + \sin 90° = 2 \sin 60°$ **38.** $2 \cos 45° = \cos 90°$

39. $\sin 60° + \cos 150° = \cos 90°$ **40.** $\sin 90° + \cos 90° = -\cos 180°$

41. $\cos 120° - \cos 90° = \sin 210°$ **42.** $\cos 90° - \cos 120° = -\cos 30°$

43. $\cos 150° \sec 210° = 1$

44. $\cos 120° \tan 300° = \sin 60°$

45. $\sin (90° + 60°) = \sin 60°$

46. $2 \cos (90° + 60°) = \cos 300°$

47. $\tan (180° + 60°) = \tan 60°$

48. $\cos (180° + 30°) = \cos 30°$

49. If $0 < \theta < \pi/2$, then $\sin \theta < \tan \theta$

50. If $0 < \theta < \pi/4$, then $\sin \theta < \cos \theta$

51. If $0 < \theta < \pi/2$, then $\csc \theta < \cot \theta$

52. If $\pi/2 < \theta < \pi$, then $\cos \theta < \sec \theta$

2

SOLUTION OF
RIGHT
TRIANGLES

2.1. Introduction

In this article, we shall discuss the application of trigonometry to the right triangle, and we shall show how to calculate the unknown parts of such a triangle when one side and another part are known. The following notation will be used throughout the chapter. We shall designate the angles of a right triangle by the capital letters A, B, and C, with C placed at the vertex of the right angle as in Figure 2.1, and we shall use the small letters a, b, and c to designate the sides that are opposite the angles A, B, and C, respectively. Note that C always designates the right angle, and thus c always represents the hypotenuse. Since angle C is always equal to 90°, there are five parts of the right triangle, A, B, a, b, and c that can vary. The following example illustrates the method we shall use:

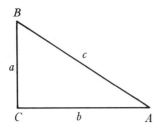

Figure 2.1

EXAMPLE

If, in a right triangle, $A = 36°$, $a = 1200$ ft, find side c. Given sin $36° = .5878$.

Solution: We shall consider b as the initial side of angle A and draw a figure approximately to scale representing the given triangle with angle A in standard position (Figure 2.2). Now we notice that the coordinates of the vertex B are

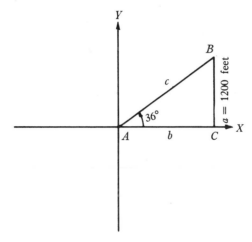

Figure 2.2

$(b, 1200)$ and the radius vector is c. Hence, by the definition of the trigonometric ratios, we have

$$\sin 36° = \frac{1200}{c}, \tag{1}$$

and we can calculate the length of c by the following method:

$c (\sin 36°) = 1200,$　　obtained by multiplying both members of (1) by c,

$c (.5878) = 1200,$　　substituting the given value for sin $36°$,

$$c = \frac{1200}{.5878},$$　　dividing both members by .5878,

$\qquad = 2042$ ft,　　correct to four places.

It should be noted that only the first four digits were given in the value of sin θ, and that we obtained only the first four digits in the value of c. This is an application of the idea of significant figures that is explained in the next article.

2.2. Significant Digits and Computation

In this chapter, we shall be concerned with computation based on numbers that are approximations. We shall be interested in the justifiable degree of accuracy and how to obtain it. The digits known to be correct in a number obtained by measurement or any other approximate number are called *significant digits* and are not affected by the position of the decimal point. The position of the decimal point is determined by the unit that is used. Thus the measurements, 7.38 meters and 738 centimeters, both contain three significant digits.

Numbers obtained by measurement are usually approximate, because their accuracy depends upon the instrument used in measuring and upon the skill and care of the user. Measurements made by a scale, or ruler, graduated in centimeters and millimeters can be made to the nearest tenth of a centimeter (or to the nearest millimeter). Thus, the statement that the radius of a circle has been found, using such a scale, to be 8.7 cm means that the actual length of the radius is between 8.65 cm and 8.75 cm, including 8.65 but not 8.75. Symbolically, 8.65 $\leq$ radius $<$ 8.75.

In any quantity that expresses a measurement or an approximate value, regardless of the position of the decimal point, the figures known to be correct are called **significant.** Except when otherwise specifically indicated, the digits 1, 2, 3, 4, $\cdots$ 9 in a number are always significant, as are the zeros between any two of these digits. Zeros whose only function is to place the decimal point are never significant. Zeros at the right end of the number may or may not be significant, depending upon the knowledge of their accuracy. If the radius of the earth is given as approximately 4000 mi, the zeros in this number are not significant. On the other hand, if the length of a lot is given as 140 ft, the zero would probably be significant, although we cannot be sure without knowing more about its actual length.

The ambiguity connected with the significance of the final zero is removed if the number is written in *scientific notation.* In this system of notation, a number is expressed as the product of a factor greater than or equal to 1 and less than 10 and an integral power of 10. The first factor contains the significant figures in the number. Thus, the number 140 is written $1.40(10)^2$ or $1.4(10)^2$, depending upon whether the final zero is or is not significant. Similarly, the scientific notation form of each of the numbers 125000 and .00627 is $1.25(10)^5$ and $6.27(10)^{-3}$, respectively, if the zeros in 125000 are not significant. _Hereafter, we shall assume that the final zero or zeros are significant, unless otherwise stated._

Frequently, not all of the figures obtained in the result of a computation are significant. In such cases, the digits that are not significant are replaced by zeros or are dropped entirely, and the rightmost digit retained is either increased by one or is left unchanged, depending upon the situation. This procedure is called _rounding off_ a number. We shall use the following rule:

RULE

The process of rounding off a number to **n** significant digits consists of the following steps:

1. _If the nth digit in the number is to the **left** of the decimal point, we replace every digit between it and the decimal point by zero and drop all digits to the right of the decimal point._
2. _If the nth digit is to the **right** of the decimal point, we drop all digits after the nth._
3. _If the left digit in the portion replaced or dropped is 5, 6, 7, 8, or 9, we increase the nth digit by one. If the left digit in the portion dropped is 0, 1, 2, 3, or 4, the nth digit is unchanged._

EXAMPLES

The following examples illustrate the method of rounding off a number to three significant figures:

1. 32.684 rounds off to 32.7 since the first digit dropped is 8.
2. 3.264 rounds off to 3.26 since the first digit dropped is 4.
3. .8723 rounds off to .872 since the digit dropped is 3.
4. 2375 rounds off to 2380 since the digit dropped is 5 and it is to the left of the decimal point. The zero is not significant. In scientific notation, we write: $2.38(10^3)$.
5. 16268.7 rounds off to 16300 since the first digit dropped is 6 and it is to the left of the decimal point. The zeros are not significant. We write $1.63(10^4)$, in scientific notation.

2.3. Calculations Involving Approximate or Measured Quantities

The result of a calculation involving approximate numbers frequently contains more digits than either of the numbers involved, and the matter of deciding how many of these figures are significant is important. The following examples illustrate the principle and practice usually followed.

We shall let $M = 716.8$ and $N = .024$, and shall investigate the number of significant digits in the product MN. According to the first paragraph of article 2.2, we know

$$716.75 \leq M < 716.85,$$

and

$$.0235 \leq N < .0245.$$

Therefore, the product MN is between $(716.75)(.0235)$ and (716.85) $(.0245)$ or $16.843625 \leq MN < 17.562825$. Hence, we know that the first digit in MN is 1, the second is probably nearer 7 than to either 6 or 8, and we know nothing about the third. Consequently, in the product $MN = (716.8)(.024) = 17.2032$, we drop all digits beyond the second and say that $MN = 17$. Now we note that we obtain the same result by rounding 716.8 off to 717 and then multiplying 717 by .024. Thus, $(717)(.024) = 17.208$, and rounding off to two significant digits, we get 17.

Similarly, the quotient M/N is between $716.85/.0235$ and $716.75/.0245$ or between 30504 and 29255. Hence, we know that the first two digits in M/N are either 29 or 30, and we know nothing about the subsequent digits. The quotient $716.8/.024$ is 29867 to 5 digits, which, rounded off to two digits, becomes 30000 with the last three zeros not significant. Hence, we write $3.0(10^4)$.

These examples, and analogous ones involving finding powers or roots, point to the fact that in multiplication and division, in raising to powers or extracting roots, we are seldom justified in carrying a calculation involving approximate data to more significant digits than the least number in any of the given data and, at times, not that far.

We shall state the rules usually followed in finding the product or quotient of two approximate numbers M and N. In the statement of the rules, we shall refer to the *accuracy* of a number. The degree of accuracy of an approximate number is determined by the number of significant digits in it. That is, in any set of numbers, the one that has the least number of significant digits is said to be the least accurate number of the set.

RULE 1

*If the number of significant digits in **M** and **N** are equal or differ by one, we round off the product **MN** and the quotient **M/N** to the number of significant digits in the less accurate of **M** and **N**.*

RULE 2

*If the number of significant digits in **M** and **N** differ by two or more, we round off the more accurate number so that it contains one more significant digit than the other; then we find the product or quotient and round the* *result off to the number of significant digits contained in the less accurate of M and N.*

EXAMPLE 1

Find the product $(56.216)(24.6)$ and the quotient $24.6/56.216$, where each of the numbers is approximate.

Solution: Since 24.6 has only three significant digits, we round 56.216 off to $3 + 1 = 4$ digits and then proceed as below.

$$(56.22)(24.6) = 1383.012 = 1380 = 1.38(10^3) \qquad \text{to three digits,}$$

and

$$24.6/56.22 = .43757 = .438 \qquad \text{to three digits.}$$

In addition, we add the digits that have the same place value. Hence, the accuracy of a sum depends upon the number of significant digits in the *decimal* portion of the addends. We shall, therefore, use the following rule for adding approximate numbers:

RULE 3

*Round each other addend off to **one more** decimal place than appears in the number having the **least** number of decimal places, add the resulting numbers, and round off the last digit in the sum.*

EXAMPLE 2

Find the sum of 3.6273, 423.1, 47.834, and 532.6.

Solution: The second and last numbers above contain only one decimal place. Hence, we round the other addends off to two decimal places and complete the addition as below.

$$
\begin{array}{r}
3.63 \\
423.1 \\
47.83 \\
\underline{532.6} \\
1007.16
\end{array}
$$

Finally, we round off the last digit in the sum and obtain 1007.2.

EXERCISE 2.1

Round off the number in each of problems 1 to 8 to four, three, two, and one significant digits. Express the rounded-off number in scientific notation if the final zero is not significant.

1. .83271   **2.** 3.1416   **3.** .0091764   **4.** .034726

5. 79.836   **6.** 582.36   **7.** 7.2975   **8.** 84.716

Express each of the measurements in problems 9 to 16 in terms of its limits.

9. 8.37   **10.** 78.23   **11.** 6.4   **12.** .073

13. 370   **14.** $4.3(10^2)$ **15.** $5.9(10^{-1})$ **16.** $6.00(10^2)$

Perform the indicated operations in problems 17 to 44 and round each result off to the correct number of significant digits. Use scientific notation if needed to indicate that a final zero is not significant.

17. (3.2)(2.7)   **18.** (5.3)(1.8)

19. (41.2)(1.3)   **20.** (.627)(3.6)

21. (78.3)(2.94)   **22.** (7.05)(68.2)

23. (58.06)(32.5)   **24.** (891.4)(5.29)

25. 38 ÷ 21   **26.** 49 ÷ 76

27. 827 ÷ 96   **28.** 628 ÷ 45

29. 986 ÷ 123   **30.** 68.9 ÷ 321

31. 1.812 ÷ 537   **32.** 10.66 ÷ 48.3

33. 2.879 + 3.61   **34.** 78.593 + 785.93

35. 3.7263 + 372.63   **36.** 4.9096 + 3.27

37. 8.492 − 2.381   **38.** 23.764 − 17.82

39. 48.2987 − 41.30   **40.** 7.8968 − 2.32

41. 71.2059 − 22.397 − 41.72 + 8.425

42. 963.87 − 237.1456 + 103.4405 − 577.7446

43. 83.725 + 76.83 + 91.2637 − 248.347

44. 119.0347 + 42.54 + 37.915 − 46.6025

2.4. Tables of Trigonometric Function Values

With the exception of the angles discussed in articles 1.17 and 1.18 and certain other angles obtained from these by the processes of arithmetic, the method of computing the values of the trigonometric ratios of an angle is beyond the scope of this book. However, tables of values have been constructed for the convenience of those who wish to use them and are included in most trigonometry texts and handbooks of mathematics. Table III in the back of this book contains the values of sines, cosines, tangents, and cotangents of acute angles. These angles are given at intervals of 10 min, and the function values are correct to four significant digits. We need tables that give a greater degree of accuracy than this in certain types of work, and there are many handbooks that contain such tables.

2.5. Given an Angle: To Find Its Function Values

Referring to Table III, we see that angles from 0° to 45° are listed in the column on the left side of the page, and those from 45° to 90° are listed in the column on the right. The function values of an angle are in the same line as the angle. If we wish to find a function value of an angle given by the table, we first locate the angle in the left- or right-hand column. Then, if the angle is between 0° and 45°, we look in the column with the desired function name at its head; but if the angle is between 45° and 90°, we must look in the column having the desired function name at its foot.

For example, in order to find the tangent of 24°30', we look on the third page of Table III in the same line with 24°30' and in the column headed by **Tan,** and we find .4557. Hence, we write tan 24°30' = .4557.

As a second example, we shall find the cosine of 76°20'. This angle is between 45° and 90°; so we look in the right-hand column until we find 76°20', then across the page until we are in the column with **Cos** at its foot, and in line with 76°20' we find .2363. This is correct to only four significant digits; nevertheless, we write cosine 76°20' = .2363.

2.6. Given a Function Value: To Find the Corresponding Angle

If a function value of an angle in the first quadrant is given, the process of finding the angle is the inverse of the procedure in the preceding article. We must first seek the given value in a column that has the name of the specified function at its head or at its foot. If this value appears in the table, the desired angle will be found in the same line as the given value and on the left or right side of the page, according as the name of the given function is at the head or the foot of the column in which the given value was found.

As an illustration, let us suppose that sin θ = .5736, with θ in the first quadrant, and we wish to find θ. We look in the columns headed by **Sin** until we find .5736. The angle on the left side of the page in line with .5736 is 35°. Hence, θ = 35°.

Let us next consider the problem: Given sin θ = .8307, find θ. The number .8307 cannot be found in the columns headed by **Sin,** so we look through each of those with **Sin** at its foot until we find .8307. In line with this number in the column of angles on the right, we find 56°10'. Hence, θ = 56°10'.

2.7. Interpolation

Since the tables list only angles from $0°$ to $90°$ that can be expressed in multiples of 10 min, we cannot obtain the function values of an angle such as $28°13'$ directly from the table. In such cases, we resort to a method of approximation known as *linear interpolation*. Before explaining the method, we call attention to the following fact, which follows from the definition of the trigonometric functions in article 1.11:

As the angle **increases** *from $0°$ to $90°$, the* **sine** *and* **tangent** *of the angle* **increase,** *and the* **cosine** *and* **cotangent** *of the angle* **decrease.**

We shall illustrate the interpolation process by finding sin $28°13'$. The two angles in the table that are nearest $28°13'$ are $28°10'$ and $28°20'$ and their sines are .4720 and .4746, respectively. Since $28°13'$ is between $28°10'$ and $28°20'$, and the sine of an angle less than $90°$ increases with the angle, sin $28°13'$ is between .4720 and .4746. We start with the smaller of the two angles, $28°10'$, and find what must be added to sin $28°10' = .4720$ to get sin $28°13'$. Since

$$28°20' - 28°10' = 10'$$

and

$$28°13' - 28°10' = 3',$$

we see that $28°13'$ is $\frac{3}{10}$ of the way from $28°10'$ to $28°20'$. We assume that sin $28°13'$ is $\frac{3}{10}$ of the way from sin $28°10'$ to sin $28°20'$. Now,

$$\sin 28°20' - \sin 28°10' = .4746 - .4720 = .0026,$$

and $\frac{3}{10}$ of $.0026 = .0008$. Therefore, sin $28°13' = \sin 28°10' + .0008 = .4720 + .0008 = .4728$.

We recommend the following diagrammatic form for the above procedure:

$$c = \tfrac{3}{10}(.0026) = .0008.$$
$$\sin 28°13' = .4720 + .0008 = .4728.$$

In the diagram above, the arrows connect the numbers that are subtracted. The differences are placed at the right and left of the arrows, and c is the correction that must be added to sin 28°10′ to get sin 28°13′.

Note. Our assumption for interpolation that the differences on the left and right are proportional is not quite correct, but for small differences the results are within the limit of required accuracy.

As a second example, we shall find cot 36°42′. The two angles in the table that are nearest 36°42′ are 36°40′ and 36°50′. We shall start with the smaller of these angles, and, since the cotangent of an angle less than 90° decreases with the angle, we shall find the correction c that must be subtracted from cot 36°40′ to obtain cot 36°42′. The steps in the interpolation are shown diagrammatically:

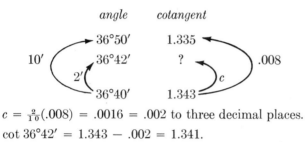

$c = \frac{2}{10}(.008) = .0016 = .002$ to three decimal places.

cot 36°42′ = 1.343 − .002 = 1.341.

We call attention to the fact that the two angles used in interpolation are the two angles in the table that are nearest the given angle, and that the given angle is between these two. Hence, the desired function value will be between the function values of the two angles. When the reader has finished the interpolation process, he should check his result to see if it satisfies this condition.

As a third example, we shall find θ (in the first quadrant) if cos θ = .6453. The decimal fraction .6453 is not listed in the column of cosines, but the two entries nearest it are .6450 and .6472. Therefore, we shall use these values and the corresponding angles and arrange the information in the usual diagram. Note that, in the diagram, we write the cosines in the left column with the smallest cosine, .6450, at the bottom.

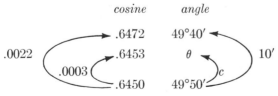

$c = (.0003/.0022)(10′) = (\frac{3}{22})(10′) = 1′$ to the nearest minute.

$\theta = 49°50′ − 1′ = 49°49′.$

As a final example, we shall show the steps without explanation for obtaining θ when $\tan \theta = .4561$.

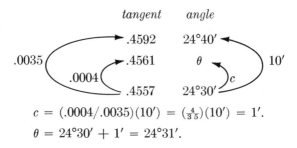

$$c = (.0004/.0035)(10') = (\tfrac{4}{35})(10') = 1'.$$

$$\theta = 24°30' + 1' = 24°31'.$$

When the given trigonometric function value contains fewer than four significant digits, we can obtain only an approximate value of the angle. For example, if $\sin \theta$ is given as .24, then we know that $.235 \leq \sin \theta < .245$; and thus, from the tables, we find that θ is between $13°30'$ and $14°20'$, and this information is sufficient to establish only approximately the number of degrees in θ. Hence, we take θ to be $14°$. Similarly, if the trigonometric function value is given to three significant digits, we can get the value of the angle to the nearest multiple of 10 min. Thus, if $\sin \theta = .236$, we take the value of θ to be $13°40'$.

We shall see in article 2.9 that the trigonometric function values of an angle in a right triangle are usually no more accurate than the lengths of the sides of the triangle. That is, the number of significant digits in the trigonometric function values of an angle in a triangle is usually the same as the number in the length of the side having the least accuracy. Hence, we have the following guide, which is usually reliable for obtaining the value of the angle.

Significant digits in the sides of the triangle	Significant digits in the values of the trigonometric functions of the angles	Value of the angle to the
2	2	nearest degree
3	3	nearest multiple of ten minutes
4	4	nearest minute

For angles near $0°$ or near $90°$, a general rule of this sort might well be modified. For example, to five decimal places, $\cos 10' = \cos 0' = 1.00000$. Indeed, it takes a change of as much as a third of a minute from zero to change the cosine from unity by as much as one digit in the eighth decimal place. Near 0°, knowing the cosine to even seven significant places fails to identify the angle within an error of three minutes. It takes an angle of at least 1°50′ to obtain a change of one unit in the fourth decimal place in the cosine value.

We shall assume that an angle written as a specified number of degrees, given without any minutes or with an integral multiple of 10 minutes, is correct to the nearest degree, 10′, or minute, according as the sides are given to two, three, or four significant digits.

EXERCISE 2.2

Find the function value called for in each of problems 1 to 24.

1. sin 24°	**2.** cos 31°	**3.** tan 44°	**4.** cot 9°
5. cos 37°10′	**6.** tan 18°20′	**7.** cot 41°50′	**8.** sin 27°40′
9. tan 8°30′	**10.** cot 42°10′	**11.** sin 23°40′	**12.** cos 39°20′
13. cot 65°40′	**14.** sin 88°30′	**15.** cos 76°20′	**16.** tan 51°10′
17. sin 52°20′	**18.** cos 49°50′	**19.** tan 69°30′	**20.** cot 73°50′
21. cos 88°50′	**22.** tan 72°40′	**23.** cot 46°10′	**24.** sin 64°30′

Find the angle in each of problems 25 to 48.

25. cos θ = .9976	**26.** tan θ = .2339	**27.** cot θ = 3.412
28. sin θ = .2476	**29.** cos θ = .9537	**30.** tan θ = .2899
31. cot θ = 2.651	**32.** sin θ = .5324	**33.** cos θ = .9272
34. tan θ = .7720	**35.** cot θ = 1.111	**36.** sin θ = .6777
37. cos θ = .2334	**38.** tan θ = 10.08	**39.** cot θ = .1435
40. sin θ = .6841	**41.** cos θ = .7808	**42.** tan θ = 1.117
43. cot θ = 2.539	**44.** sin θ = .5225	**45.** cos θ = .8225
46. tan θ = .6088	**47.** cot θ = .2617	**48.** sin θ = .9992

Find the function value called for in each of problems 49 to 72 by use of interpolation.

49. tan 31°4′	**50.** cot 23°17′	**51.** sin 18°32′
52. cos 9°16′	**53.** tan 39°41′	**54.** cot 28°28′
55. sin 47°27′	**56.** cos 53°33′	**57.** tan 71°17′
58. cot 84°38′	**59.** sin 49°54′	**60.** cos 89°3′
61. tan 55°42′	**62.** cot 77°29′	**63.** sin 62°26′
64. cos 27°34′	**65.** tan 41°56′	**66.** cot 78°35′

67. sin 72°29′	**68.** cos 34°43′	**69.** tan 64°57′
70. cot 41°53′	**71.** sin 52°18′	**72.** cos 37°43′

Find the angle to the nearest minute in each of problems 73 to 96 by use of interpolation.

73. cot θ = 3.724	**74.** sin θ = .3972	**75.** cos θ = .8375
76. tan θ = .7263	**77.** cot θ = 6.032	**78.** sin θ = .6883
79. cos θ = .4239	**80.** tan θ = 2.704	**81.** cot θ = .7986
82. sin θ = .8117	**83.** cos θ = .2116	**84.** tan θ = 1.763
85. cot θ = 7.615	**86.** sin θ = .1091	**87.** cos θ = .3998
88. tan θ = .9236	**89.** cot θ = .7194	**90.** sin θ = .8083
91. cos θ = .0347	**92.** tan θ = 2.233	**93.** cot θ = .5840
94. sin θ = .4892	**95.** cos θ = .8257	**96.** tan θ = .6349

Find the angle to the degree of accuracy justified by the data in each of problems 97 to 104.

97. sin θ = .2345	**98.** cos θ = .5791	**99.** tan θ = 2.36
100. cot θ = .763	**101.** sin θ = .44	**102.** cos θ = .61
103. tan θ = .84	**104.** cot θ = .77	

2.8. Function Values of an Acute Angle of a Right Triangle

In the example of article 2.1, it was necessary to place the angle A in standard position before we could solve the problem. To avoid this inconvenience, we shall define the sine, cosine, tangent, and cotangent of an acute angle of a right triangle in terms of the sides of the triangle, and then we can apply the principles of trigonometry to a right triangle without the use of a coordinate system. In the definition that follows, we shall use the terms *opposite side* and *adjacent side* to refer to the sides opposite and adjacent to the angle.

If we construct a right triangle ABC with the angle A in standard position, as in Figure 2.3, then the coordinates of the vertex B are (b, a), and the radius vector is c. Hence, by the definition of article 1.11, we have

$$\begin{cases} \sin A = \dfrac{a}{c} = \dfrac{\text{opposite side}}{\text{hypotenuse}}, \\ \cos A = \dfrac{b}{c} = \dfrac{\text{adjacent side}}{\text{hypotenuse}}, \end{cases}$$

$$\begin{cases} \tan A = \dfrac{a}{b} = \dfrac{\text{opposite side}}{\text{adjacent side}}, \\[2ex] \cot A = \dfrac{b}{a} = \dfrac{\text{adjacent side}}{\text{opposite side}}. \end{cases}$$

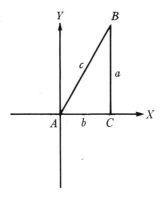

Figure 2.3

These definitions will be used in the problems of this chapter and should be memorized. Obviously,

$$\begin{cases} \sec A = \dfrac{\text{hypotenuse}}{\text{adjacent side}}, \quad \text{and} \quad \csc A = \dfrac{\text{hypotenuse}}{\text{opposite side}}, \end{cases}$$

but, since the tables used in this book do not include them, we shall not employ them in the problems that follow.

Since relations between sides and angles of a geometric figure do not depend on the location of the figure, the ratios between the sides of the triangle will remain the same regardless of the position of the triangle.

Thus, in Figure 2.4, we have

$$\begin{cases} \sin A = \dfrac{a}{c} = \cos B, \qquad \tan A = \dfrac{a}{b} = \cot B, \\[2ex] \cos A = \dfrac{b}{c} = \sin B, \qquad \cot A = \dfrac{b}{a} = \tan B. \end{cases}$$

Comparing the function values of A and B as listed above, we see that any function value of one of these angles is equal to the corresponding *cofunction* value of the other. Furthermore, A and B are complementary angles since they are the acute angles of a right triangle. Therefore, we have the following theorem:

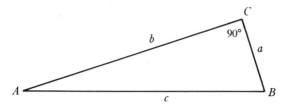

Figure 2.4

THEOREM

Any trigonometric function value of an acute angle is equal to the corresponding cofunction value of the complementary angle.

2.9. The Solution of Right Triangles

The defining equation of each trigonometric function value in the previous article involves two sides and one angle of a right triangle. Hence, if we know two of these three quantities, we can solve for the third. The ability to do this is essential in the solution of right triangles.

Every triangle has three sides and three angles. In a right triangle, one of the angles is fixed, and there are five parts that may vary. If we know the length of a side and one of the other variable parts of a triangle, we may determine the other three parts. The method of procedure is as follows: *Select the trigonometric function value that involves two known parts and the desired unknown. Then the unknown part may be computed by the methods of algebra.* The process of finding the unknown parts of a right triangle is known as *solving the triangle.*

Before solving a triangle, it is always advisable to draw a figure as nearly to scale as practicable.

EXAMPLE 1

In a right triangle, $A = 36°42'$ and $a = 12.63$. Solve the triangle.

Solution: Figure 2.5 is an approximate scale drawing of the triangle determined by the data.

The unknown parts of this triangle are angle B, the side b, and the hypotenuse c. Since angles A and B are complementary, we have

$$B = 90° - A$$
$$= 90° - 36°42'$$
$$= 53°18'.$$

We shall next find the side c. Since the known parts are A and a and $\sin A = a/c$,

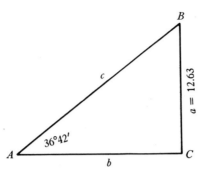

Figure 2.5

we shall use this equation to find c. Since $A = 36°42'$ and $a = 12.63$, we have by definition

$$\sin 36°42' = \frac{12.63}{c}.$$

Hence,

$$c(\sin 36°42') = 12.63,$$

and

$$c = \frac{12.63}{\sin 36°42'}$$

$$= \frac{12.63}{.5974}$$

$$= 21.14 \qquad \text{to four significant figures}$$

In order to find the value of b, we shall use $\tan A$ and obtain

$$\tan 36°42' = \frac{12.63}{b}.$$

Hence,

$$b(\tan 36°42') = 12.63,$$

and

$$b = \frac{12.63}{\tan 36°42'}$$

$$= 16.94.$$

Note. The value of b could have been calculated by use of $\cos A$ and the value of c as computed in the second step of the solution. However, this procedure would have carried along any error that might have been made in the first calculation. For this reason, it is advisable to avoid the use of a computed result in any portion of the solution. In other words, when one step of the solution is completed, return to the given data for the next step, unless this unduly increases the amount of computation necessary.

EXAMPLE 2

Solve the right triangle in which $c = .36$ and $A = 57°$.

Solution: We first draw Figure 2.6 approximately to scale to represent the above

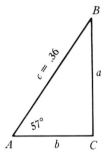

Figure 2.6

data and then compute the unknown parts of the triangle in the order B, a, and b. Since angles A and B are complementary, we have

$$B = 90° - 57° = 33°.$$

In order to find a, we note that $\sin A = a/c$ involves the two known parts of the given triangle and the unknown part a. Hence, we shall use this equation and get

$$\sin 57° = \frac{a}{.36}.$$

In article 2.2, we pointed out that, in any calculation, all numbers are rounded off to one more significant digit than appears in the least accurate number involved. Hence, we shall round $\sin 57° = .8387$ off to three digits and obtain .839. Then we complete the problem as below.

$$a = (.36)(.839)$$
$$= .302$$
$$= .30 \quad \text{to two digits.}$$

We next compute the value of b by use of $\cos A$ rounded off to three digits and obtain

$$\cos 57° = \frac{b}{.36}.$$

Hence,

$$b = (.36) \cos 57°$$
$$= (.36)(.546)$$
$$= .20.$$

EXAMPLE 3

Solve the triangle in which $a = 13.27$ and $b = 26.21$.

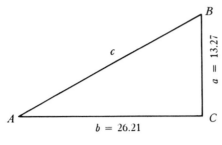

Figure 2.7

Solution: We shall first draw Figure 2.7, and then find the value of angle A. Since $\tan A$ involves the unknown A and the two known parts a and b, we shall use

$$\tan A = \frac{a}{b}$$

$$= \frac{13.27}{26.21}$$

$$= .5063.$$

Hence, using Table III and interpolating, we have

$$A = 26°51'.$$

In order to find c, we shall use $\sin A$* and have

$$\sin 26°51' = \frac{13.27}{c}.$$

Hence,

$$c = \frac{13.27}{\sin 26°51'}$$

$$= \frac{13.27}{.4517}$$

$$= 29.38.$$

Angle $B = 90° - 26°51' = 63°9'$.

---

* It is possible to compute c by use of the Pythagorean theorem. However, this method would involve considerably more computation than the trigonometric method. Hence, we use the computed value of A to complete the solution.

EXAMPLE 4

Given $a = 271$, $c = 428$, solve the triangle.

Solution: In this problem, the sides a and c (Figure 2.8) are given to only three

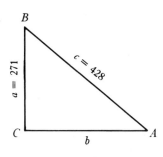

Figure 2.8

significant digits. Therefore, according to the table near the conclusion of article 2.7, we should obtain the angles only to the nearest multiple of 10 min. Since

$$\frac{a}{c} = \sin A,$$

$$\sin A = \tfrac{271}{428}$$

$$= .633.$$

Hence,

$$A = 39°20' \quad \text{(zero not significant)},$$

and

$$B = 90° - 39°20' = 50°40' \quad \text{(zero not significant)}.$$

In order to get the value of b, we start with

$$\cot A = \frac{b}{a},$$

$$\cot 39°20' = \frac{b}{271}.$$

Hence,

$$b = 271(\cot 39°20')$$

$$= 271(1.220)$$

$$= 331.$$

EXERCISE 2.3

Solve the right triangles in problems 1 to 8 without use of tables.

1. $A = 60°$, $c = 16$ 2. $B = 30°$, $c = 32$

3. $A = 45°$, $a = 24$ 4. $B = 45°$, $b = 43$

5. $A = 60°$, $b = 12$ 6. $B = 60°$, $a = 24$

7. $a = 25$, $c = 25\sqrt{2}$ 8. $a = 48$, $b = 48\sqrt{3}$

Solve the right triangles in problems 9 to 36 by finding each missing part to the degree of accuracy justified by the given data.

9. $A = 37°$, $c = 72$ 10. $A = 74°$, $c = 58$

11. $B = 40°$, $a = 25$ 12. $A = 17°$, $a = 66$

13. $B = 63°$, $b = 54$ 14. $A = 46°$, $b = 85$

15. $B = 22°$, $a = 7.3$ 16. $b = 3.9$, $c = 5.3$

17. $b = 723$, $c = 836$ 18. $b = 90.6$, $c = 115$

19. $A = 32°20'$, $c = 3.47$ 20. $A = 63°20'$, $c = 987$

21. $B = 69°40'$, $c = 948$ 22. $B = 22°40'$, $c = 47.7$

23. $A = 56°20'$, $a = 421$ 24. $A = 30°10'$, $a = 140$

25. $B = 41°10'$, $b = 7.63$ 26. $B = 81°30'$, $b = 28.3$

27. $A = 29°30'$, $b = 87.1$ 28. $A = 21°50'$, $b = 9.80$

29. $B = 68°10'$, $a = 9.81$ 30. $B = 28°50'$, $a = 2.39$

31. $a = .428$, $c = .847$ 32. $a = 2.12$, $c = 14.3$

33. $a = .571$, $b = .459$ 34. $a = 4.43$, $b = 6.21$

35. $a = 487$, $b = 519$ 36. $a = .214$, $b = .986$

Solve the right triangles in problems 37 to 56 by use of interpolation.

37. $A = 63°64'$, $c = 16.09$ 38. $A = 70°27'$, $c = .1621$

39. $B = 64°32'$, $c = .7818$ 40. $B = 57°38'$, $c = 4.272$

41. $A = 63°47'$, $a = 9.261$ 42. $A = 32°22'$, $a = .7621$

43. $B = 61°22'$, $b = 5.823$ 44. $A = 41°16'$, $c = 430.6$

45. $a = .9931$, $b = 1.638$ 46. $a = 86.09$, $b = 96.55$

47. $A = 51°42'$, $b = .8163$ 48. $A = 32°38'$, $a = .3179$

49. $B = 32°13'$, $a = .3944$ 50. $B = 78°37'$, $a = 2.175$

51. $a = 45.67$, $c = 62.41$ 52. $a = 235.7$, $c = 371.4$

53. $b = .2383$, $c = .8491$ 54. $b = 42.42$, $c = 46.32$

55. $a = 43.10$, $b = 21.74$ 56. $a = 86.09$, $b = 96.55$

Solve each right triangle in problems 57 to 64, round off the results to the proper number of significant digits, express the measure of the sides in scientific notation, and express each angle to the correct degree of accuracy.

57. $a = 6.2(10^2)$, $c = 1.7(10^3)$ **58.** $b = 7.5(10^2)$, $c = 2.3(10^3)$

59. $a = 3.1(10^2)$, $b = 3.6(10^2)$ **60.** $a = 5.3(10^3)$, $b = 6.4(10^3)$

61. $A = 27°$, $a = 2.7(10^2)$ **62.** $B = 72°$, $b = 7.1(10^2)$

63. $A = 41°10'$, $c = 9.13(10^3)$ **64.** $B = 38°40'$, $c = 6.58(10^3)$

2.10. Angles of Elevation and Depression

If an observer at the point A as in Figure 2.9 is looking at an object at a point B, then the ray AB is called the *line of sight*. Let C be any point on a

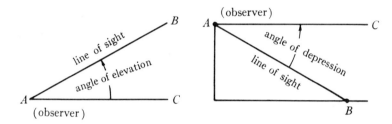

Figure 2.9

horizontal ray through A in the same vertical plane as AB; then the angle CAB between the line of sight and the horizontal is called the *angle of elevation* or the *angle of depression* of B, according as B is above or below A. These definitions are illustrated in Figure 2.9.

In order to illustrate the use of the angle of elevation, we shall solve the following problem:

EXAMPLE

From a point 452 ft in a horizontal line from the base of a building, the angle of elevation of its top is $32°10'$. What is the height of the building?

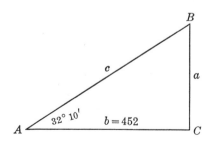

Figure 2.10

Solution: We first construct Figure 2.10 to represent the situation in the problem. In order to solve the problem, we shall use the tangent of the angle A, since $\tan A = a/b$ and this equation involves two known parts and the desired unknown part. By definition,

$$\tan 32°10' = \frac{a}{452};$$

hence,

$$a = 452 \tan 32°10'$$
$$= 452(.6289)$$
$$= 284.2628$$
$$= 284 \quad \text{to three significant digits.}$$

2.11. The Direction of a Line

In surveying and in marine navigation, the direction or *bearing* of a line through a given point is specified by stating the acute angle at which the direction of the line varies to the east or west from the north-south line through the point. For example, in Figure 2.11, the bearing of OB is 50°

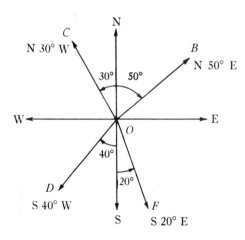

Figure 2.11

east of north, or N 50° E. Similarly, the bearings of the lines OC, OD, and OF are N 30° W, S 40° W, and S 20° E, respectively.

In air navigation, the bearing of a line or course of flight is specified by stating the clockwise angle less than 360° that the line makes with a due northward direction. In Figure 2.12, the directions of OA, OB, OC, and OD are 50°, 120°, 240°, and 300°, respectively.

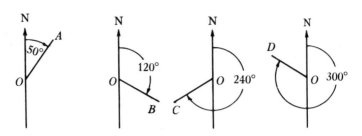

Figure 2.12

EXAMPLE 1

Calculate the area in square feet and find the third side of a plot of land that is bounded as follows: beginning at a certain cypress tree that is marked with two x's, thence 402 ft south, thence 464 ft N 30°10′ E, and thence due west to the starting point. The sketch of this land is shown in Figure 2.13.

Solution: In order to calculate the unknown side a, we shall use the sine of the angle A, since it involves two known parts and the desired unknown part.

$$\sin 30°10′ = \frac{a}{464}.$$

Hence,

$$a = 464 \sin 30°10′$$
$$= 464(.5025)$$
$$= 233.$$

Since this is a right triangle, the area is $\frac{1}{2}$ the base times the altitude; hence, the

$$\text{Area} = \tfrac{1}{2}(402)(233),$$
$$= 4.68(10^4).$$

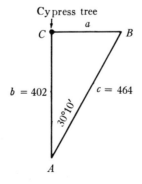

Figure 2.13

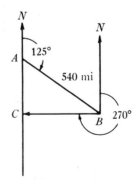

Figure 2.14

EXAMPLE 2

An airplane took off from field A at 125°, flew for 3 hr at 180 mph, and landed at field B. After refueling, it flew on a course 270° and landed at C directly south of A. Find the distance between C and B.

Solution: In Figure 2.14, the lines AB and BC represent the two flights of the airplane. The angle $CAB = 180° - 125° = 55°$. Since the direction 270° is due west, the angle at C is 90°; thus, we can solve the problem by the method of article 2.9. The solution follows:

$$\sin 55° = \frac{CB}{540}.$$

Hence,

$$CB = 540(\sin 55°)$$
$$= 540(.8192)$$
$$= 442 \text{ mi.}$$

don't do **Omit**

2.12. Vectors

The velocity of an airplane in still air is determined by the speed due to the engines and the direction imparted by the rudder. The speed is usually expressed in miles per hour, and the direction is specified by stating the angle that the line of flight makes with a fixed line, or by one of the conventions for describing direction, such as "due east" or "northeast." Thus, we see that the velocity of the airplane has two aspects—a magnitude and a direction—and it is not determined until each of these is specified. Quantities that have both magnitude and direction are known as *vector quantities*. Other examples of vector quantities are *forces, accelerations*, and *displacements*. A vector quantity can be represented graphically by a directed line segment whose length is proportional to the magnitude of the vector quantity and whose direction is that of the given quantity. A directed line segment that represents a vector quantity is called a *vector*. It is proved experimentally in physics that two forces acting simultaneously at the same point are equivalent to a single force acting at that point. The two forces are called the *components*, and the single force is known as the *resultant* of the two components. Furthermore, the two components and their resultant may be represented graphically by three vectors drawn from the same point, the resultant vector being the diagonal of the parallelogram which has the two component vectors as sides. This statement may be verified experimentally for forces and velocities and will be used as the definition of the resultant of two vectors. Thus, in Figure 2.15, the vector AC is the

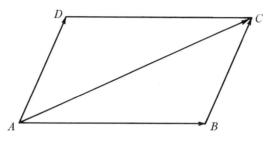

Figure 2.15

resultant of the vectors AB and AD. Furthermore, BC is equal in length and direction to AD, and hence, AC is also the resultant of AB and BC. Thus, we see that we may obtain the resultant of two vectors by constructing the first; then, from its terminus, constructing the second. The resultant of the two is the vector connecting the origin of the first with the terminus of the second.

We shall illustrate the graphical representation of two vectors and their resultant by the following examples.

EXAMPLE 1

If a ball is thrown due east at 15 yd per sec from a car that is traveling due north at the rate of 20 yd per sec, find the velocity of the ball and the direction of its path.

Solution: The vector that represents 15 yd per sec should be $\frac{15}{20} = \frac{3}{4}$ as long as that representing 20 yd per sec. The line OB in Figure 2.16 represents the velocity

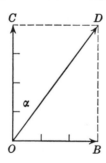

Figure 2.16

the ball would have had if the thrower had not been in motion, the line OC represents the velocity of the car, and the line OD represents the actual velocity of the ball. The length and direction of OD may be determined by the methods of trigonometry, since OCD is a right triangle with two known parts including a side. For example,

$$\tan \alpha = \frac{CD}{OC}$$

$$= \frac{15}{20}$$

$$= .75.$$

Therefore,

$$\alpha = 37°.$$

Furthermore,

$$\cos \alpha = \frac{OC}{OD},$$

or

$$\cos 37° = \frac{20}{OD},$$

and

$$OD = \frac{20}{\cos 37°}$$

$$= \frac{20}{.7986}$$

$$= 25.$$

Therefore, the ball travels N 37° E at 25 yd per sec.

If one of two vectors and their resultant are known, then the other vector can be determined graphically by constructing the given vector and the resultant from the same point and joining their end points. If the two components are perpendicular, the three vectors form a right triangle that can be completely solved by trigonometric methods now available to the reader.

EXAMPLE 2

An airplane is headed due north at a speed of 184 mph. If the wind is blowing from due east with a velocity of 27.3 mph, find the direction and speed of the flight of the airplane.

Solution: We first draw the vector AB in Figure 2.17 to represent 184 mi in a due northward direction and, at its terminus, we construct BC to represent 27.3 mph to the west. Then AC represents the speed and direction of the airplane, and we shall determine these by trigonometric methods.

We have

$$\tan A = \frac{27.3}{184}$$

$$= .148 \qquad \text{to three significant digits.}$$

Figure 2.17

Therefore,

$$A = 8°30' \qquad \text{to the nearest multiple of 10 min.}$$

Consequently, the positive angle that AC makes with the northward direction, or with the direction of AB, is $360° - 8°30' = 351°30'$. Hence, the direction of the flight of the airplane is $351°30'$. Furthermore,

$$\cos 8°30' = \frac{184}{AC},$$

and consequently,

$$AC = \frac{184}{\cos 8°30'}$$

$$= \frac{184}{.989}$$

$$= 186.$$

Hence, the airplane travels 186 mph in the direction $351°30'$.

EXERCISE 2.4

1. Find the length of the altitude of an equilateral triangle with sides 1700 ft long.

2. A tract of land is in the shape of a right triangle. The longest side is 5002 ft and bears N 32°12′ W. How many feet of fence are required if a side runs due north-south?

3. A vertical force of 125 lb and a horizontal force of 104 lb are acting on a body. Find the magnitude and direction of the resultant.

omit for test these word problems 1-40

4. The courthouse is 611 yd south of the cathedral, and this distance subtends an angle of 19°10′ at the eye of an observer who is due west of the cathedral. How far is the observer from the courthouse?

5. A television tower is known to be 475 ft tall. If an observer in the horizontal plane of its base finds the angle of elevation of the top of the tower to be 32°40′, how far is he from the base?

6. What angle does a roof make with the horizontal if the rafters are 21.5 ft long and the roof rises 8.75 ft?

7. In order to find the distance from A to B across a pond from A, a point C is chosen so that BC is 100 yd long and perpendicular to AB. If CA is 125 yd, how far is B from A?

8. The dimensions of a rectangle are 216 and 128 cm. Find the angle between the diagonal and the longer side and also the length of the diagonal.

9. A brace runs from a lower corner of a rectangular wall to the diagonally opposite upper corner. If the wall is 10.75 ft by 16.50 ft, find the length of the brace and the angle it makes with the longer side.

10. The sides of a rhombus are 12.22 cm long and each acute angle is 42°56′. Find the lengths of the diagonals.

11. The elevation at the entrance of a mine is 3202 ft above sea level. The mine shaft is straight and has an angle of depression of 30°10′. Find the elevation of the lower end if the shaft is 200 ft long.

12. Find the length of a chord that subtends a central angle of 26°12′ in a circle of radius 16.45 ft.

13. The boom of a crane is 28.6 ft long. What is the greatest horizontal distance from the crane that can be reached by the boom if the operating manual says that the boom should not be lowered so as to make an angle of less than 45°50′ with the horizontal?

14. The diagonals of a rhombus are 32.16 cm and 19.24 cm in length. Find the sides and the angles.

15. A plot of land is 73.04 ft long and slopes at an angle of 2°12′. The owner is going to build a retaining wall and then fill in so as to have a horizontal surface. How tall must the wall be?

16. A surveyor's field notes describe the boundaries of a tract of land as follows: Beginning at a stone marker, thence due east 1260 yd; thence, N 20°12′ W to a second marker; thence, S 69°48′ W to the starting point. Find the lengths of the unknown sides.

17. On a college campus, the front door of the library building is 243.5 yd southwest of the flagpole, and the front door of the union building is 126.4 yd southeast of the flagpole. Find the direction of the union building from the library.

18. A forester wanted to know how far a forest fire had traveled up a uniform slope. The slope was known to have an angle of elevation of 47°20′. On an aerial photograph the horizontal distance measured 165 ft. How far had the fire traveled?

19. From a point 125 ft from, and in the same horizontal plane with, the base of a building, the angle of elevation of the top and of the foot of a flagpole on its top are 68°41′ and 67°20′. Find the length of the pole.

20. After an explosion in a chemical plant located on Duplantier Street, it was decided to evacuate everyone within an 800-ft radius of the plant. Bratowski Street intersects Duplantier at an angle of 90° and 250 ft from the plant. Would a home on Bratowski 740 ft from the intersection be in the danger zone?

21. In analyzing the stress in a diagonal member of a small bridge truss it was necessary to know the angle the member made with the horizontal roadbed. If the truss was 7.00 ft high, and the diagonal member was 8.08 ft in length, what was the angle?

22. From a corner of a triangular tract, one boundary runs due north 1276 yd and the other bears N 32°14′ E. If the third side runs east and west, find the lengths of the unknown sides.

23. From the top of a hill, the bearings of a church spire in each of two towns on an east-west highway known to be 12 mi apart are N 34° W and N 56° E. Find the approximate distance from the point of observation to the two towns.

24. The pipe for a sprinkler system passes under a rectangular house from the northeast corner to a point 12.6 ft south of the northwest corner. If the east-west wall of the house is 45.9 ft long, how long is the portion of pipe that is under the house and what angle does it make with the east-west wall?

25. A weight is suspended from an overhead track and a man is moving the weight by means of a rope attached to it. If the rope is inclined to the horizontal at an angle of 32°24′ and the man is pulling with a force of 150.5 lb, what force is moving the weight along the track?

26. A pilot is flying at an air speed of 200 mph in a wind blowing 30.6 mph from the west. In what direction must he head in order to fly due south? What is his speed relative to the ground?

27. Two observers on an east-west highway note that the angles of elevation of an airplane are 22°31′ and 24°42′. If the plane was 2500 ft directly above a point on the highway east of the observers, how far apart were they?

28. A boy got off a bus at the northwest corner of a section of land whose boundaries run east-west and north-south. He walked along the west boundary to a gate and then along a road that bears S 20°20′ E to a farm house that is 325 yd from the west boundary and 225 yd from the south. How far did he walk? A section is 1760 yd square.

29. In order to measure the width of a river, a stake was driven in the ground on the south bank directly south of a stone on the north bank. Then a distance of 100 ft was measured due west along the south bank and the bearing of the stone from the west extremity of this distance was found to be N 47°50′ E. Find the width of the river.

30. A man built 135 ft of wall that runs S 48°10′ W along one boundary of his triangular lot. If the other boundaries of the lot run north-south and east-west, what is the area of the lot?

31. A pilot flew for 320 mi in the direction 60°10′ and then flew at 150°10′ for 540 mi. From the latter point he returned to the starting point. Find the length and direction of the last flight.

32. Town A is due north of B and due west of C. A straight highway 59.8 mi long runs N 29°20′ E and connects B and C. How many miles out of his way did a salesman go who took the wrong turn and went from B to C by way of A?

33. Two planes left A simultaneously. One flew at 280 mph in the direction 36°40′ and the other in the direction 126°40′ at 400 mph. How far apart were the planes at the end of 1.5 hr? What was the direction of the first plane from the second?

34. Two roommates decided to move their desk. One boy pushed the desk due north with a force of 92 lb while the other pushed due east with a force of 72 lb. In what direction did the desk move?

35. A motor boat crossed a river from west to east at a speed of 20 mph, and there was a 5-mph current flowing toward the north. What was the speed of the boat in still water and in what direction was it headed?

36. A man who was walking along a straight trail headed N 30°37′ E found that his path crossed the corner of a pasture in which there was a dangerous bull. He chose to walk 153.9 ft along the south fence of the pasture and then due north along the east fence to get back on his trail. How much out of the way did he walk for the sake of safety?

37. An airplane was headed due north at 252 mph in a wind blowing from the west at 31.8 mph. Find the ground speed and direction of his flight.

38. With surveyor's instruments an engineer found that the top of the bank of the west side of a river is 5.01 ft below the top of the east bank and that the angle of depression of the former from the latter is 2°20′. How long must a horizontal bridge be to cross the river where these observations were made?

39. From a point on a level plane the angle of elevation of the top of a hill is 23°43′. From a point 1225 ft nearer the hill, the angle of elevation of the top is 30°22′. Find the height of the hill.

40. From an observation balloon 1200 ft above a harbor, an observer found the angle of depression of a boat east of him to be 34°6′ and the angle of depression of a boat south of him to be 31°13′. How far apart are the boats?

3

FUNDAMENTAL
IDENTITIES

3.1. Introduction

In this chapter, we shall use the definitions of the six trigonometric ratios to develop eight relations between the function values. These eight relations are called *fundamental identities* and fall naturally into three groups because of the way in which they are derived. They can be used to change the form of a trigonometric expression. This is very important in trigonometry and calculus. Consequently the mathematician, the physicist, the engineer, and others who are likely to use mathematics should become familiar with the material of this chapter.

3.2. Reciprocal Relations

The trigonometric function values can be grouped in pairs so that the members of each pair are reciprocals. This fact is the basis for the derivation of the three reciprocal relations given below.

By the definition of article 1.15,

$$\sin \theta \csc \theta = \frac{y}{r} \times \frac{r}{y} = 1.$$

Hence, we have

$$\mathbf{\sin \theta \csc \theta = 1.} \tag{3.1}$$

Similarly, we may show that

$$\mathbf{\cos \theta \sec \theta = 1,} \tag{3.2}$$

and

$$\mathbf{\tan \theta \cot \theta = 1.} \tag{3.3}$$

We may solve each of the above relations for either of the two function values involved and obtain the following useful relations:

$$
\begin{cases}
\sin \theta = \dfrac{1}{\csc \theta}, & \csc \theta = \dfrac{1}{\sin \theta}, \\[2ex]
\cos \theta = \dfrac{1}{\sec \theta}, & \sec \theta = \dfrac{1}{\cos \theta}, \\[2ex]
\tan \theta = \dfrac{1}{\cot \theta}, & \cot \theta = \dfrac{1}{\tan \theta}.
\end{cases}
$$

3.3. Ratio Relations

Again referring to the definitions in article 1.15, we have

$$\frac{\sin \theta}{\cos \theta} = \frac{y}{r} \div \frac{x}{r} = \frac{y}{r} \times \frac{r}{x} = \frac{y}{x} = \tan \theta,$$

and

$$\frac{\cos \theta}{\sin \theta} = \frac{x}{r} \div \frac{y}{r} = \frac{x}{r} \times \frac{r}{y} = \frac{x}{y} = \cot \theta.$$

Therefore,

$$\mathbf{\tan \theta = \dfrac{\sin \theta}{\cos \theta},} \tag{3.4}$$

and

$$\mathbf{\cot \theta = \dfrac{\cos \theta}{\sin \theta}.} \tag{3.5}$$

3.4. Pythagorean Identities

Equation (1.5) of article 1.12 states that

$$x^2 + y^2 = r^2.$$

If we divide both members of this equation by r^2, we have

$$\left(\frac{x}{r}\right)^2 + \left(\frac{y}{r}\right)^2 = 1.$$

but, by definition, $x/r = \cos\theta$, and $y/r = \sin\theta$; hence,

$$\cos^2\theta + \sin^2\theta = 1. \tag{3.6}$$

This identity can also be obtained by use of the distance formula and the fact that the coordinates of P in Figure 1.15 are $x = \cos\theta$ and $y = \sin\theta$.

If, on the other hand, we divide both members of the equation $x^2 + y^2 = r^2$ by x^2, we obtain

$$\left(\frac{x}{x}\right)^2 + \left(\frac{y}{x}\right)^2 = \left(\frac{r}{x}\right)^2.$$

By definition, $y/x = \tan\theta$, and $r/x = \sec\theta$. Therefore,

$$1 + \tan^2\theta = \sec^2\theta. \tag{3.7}$$

Furthermore, if both members of $x^2 + y^2 = r^2$ are divided by y^2, it becomes

$$\left(\frac{x}{y}\right)^2 + \left(\frac{y}{y}\right)^2 = \left(\frac{r}{y}\right)^2,$$

which, by use of the definitions of the functions, becomes

$$\cot^2\theta + 1 = \csc^2\theta. \tag{3.8}$$

3.5. Fundamental Identities

For the convenience of the reader, we have collected the fundamental identities just derived. They are very important and should be memorized.

$$\sin\theta \csc\theta = 1. \tag{3.1}$$

$$\cos\theta \sec\theta = 1. \tag{3.2}$$

$$\tan \theta \cot \theta = 1. \tag{3.3}$$

$$\tan \theta = \frac{\sin \theta}{\cos \theta}. \tag{3.4}$$

$$\cot \theta = \frac{\cos \theta}{\sin \theta}. \tag{3.5}$$

$$\cos^2 \theta + \sin^2 \theta = 1. \tag{3.6}$$

$$1 + \tan^2 \theta = \sec^2 \theta. \tag{3.7}$$

$$1 + \cot^2 \theta = \csc^2 \theta. \tag{3.8}$$

3.6. Trigonometric Reductions

It is desirable to be able to reduce a trigonometric expression to another form, since need for reductions of this type arises quite often in higher mathematics. Unfortunately, there is no one method to be followed in all cases in which a reduction is desired, and the student may make many unsuccessful attempts before finding the correct method. Considerable practice is necessary in order to overcome the difficulties associated with problems of this type. Frequently the required process involves performing the indicated algebraic operations and then applying one or more of the fundamental identities. For example, in order to reduce

$$(\sin \theta + \cos \theta)^2 \qquad \text{to} \qquad 1 + 2 \sin \theta \cos \theta,$$

we perform the following steps:

$$(\sin \theta + \cos \theta)^2 = \sin^2 \theta + 2 \sin \theta \cos \theta + \cos^2 \theta \qquad \text{by squaring the sum of two numbers}$$

$$= \sin^2 \theta + \cos^2 \theta + 2 \sin \theta \cos \theta$$

$$= 1 + 2 \sin \theta \cos \theta \qquad \text{by identity (3.6).}$$

In other cases it is advisable to apply the suitable fundamental relation to certain combinations of terms in the problem and then to perform the algebraic operations.

EXAMPLE

Reduce

$$\frac{\tan \theta(1 + \cot^2 \theta)}{1 + \tan^2 \theta}$$

to $\cot \theta$.

Solution:

$$\frac{\tan \theta(1 + \cot^2 \theta)}{1 + \tan^2 \theta} = \frac{\tan \theta(\csc^2 \theta)}{\sec^2 \theta} \qquad \text{by identities (3.8) and (3.7)}$$

$$= \tan \theta \left[\frac{\dfrac{1}{\sin^2 \theta}}{\dfrac{1}{\cos^2 \theta}} \right] \qquad \text{by identities (3.1) and (3.2)}$$

$$= \tan \theta \left(\frac{\cos^2 \theta}{\sin^2 \theta} \right) \qquad \text{by algebraic simplification}$$

$$= \tan \theta \cot^2 \theta \qquad \text{by identity (3.5)}$$

$$= (\tan \theta \cot \theta) \cot \theta$$

$$= \cot \theta \qquad \text{by relation (3.3).}$$

EXERCISE 3.1

Reduce the first expression to the second in each of problems 1 to 32.

1. $\cos A \tan A$, $\sin A$

2. $\cos A \csc A$, $\cot A$

3. $\sin A \sec A$, $\tan A$

4. $\sin A \cot A$, $\cos A$

5. $\dfrac{\sec A}{\csc A}$, $\tan A$

6. $\dfrac{\cot A}{\cos A}$, $\csc A$

7. $\dfrac{\tan A}{\cot A}$, $\tan^2 A$

8. $\dfrac{\cos A}{\sec A}$, $\cos^2 A$

9. $\csc A - \sin A$, $\cot A \cos A$

10. $\sec A - \cos A$, $\tan A \sin A$

11. $\tan A + \cot A$, $\sec A \csc A$

12. $\cos A + \sin A \tan A$, $\sec A$

13. $\cos A(\sec A + \cos A \csc^2 A)$, $\csc^2 A$

14. $(1 + \sin A)(1 - \sin A)$, $\cos^2 A$

15. $\sin A(\csc A + \sin A \sec^2 A)$, $\sec^2 A$

16. $(\sec A + 1)(\sec A - 1)$, $\tan^2 A$

17. $\dfrac{\sec^2 A - 1}{\sin A}$, $\sec A \tan A$

18. $\dfrac{1 + \cot^2 A}{\sec^2 A}$, $\tan^2 A$

19. $\dfrac{1 - \sin^2 A}{\cot A}$, $\sin A \cos A$

20. $\dfrac{1 - \cos^2 A}{\tan A}$, $\cos A \sin A$

21. $\dfrac{\sin A}{\sec^2 A - 1}$, $\cos A \cot A$

22. $\dfrac{\cos A}{\csc^2 A - 1}$, $\tan A \sin A$

23. $\dfrac{1 + \tan^2 A}{\csc^2 A}$, $\tan^2 A$

24. $\dfrac{\cot A}{1 - \sin^2 A}$, $\sec A \csc A$

25. $\dfrac{\cos^2 A}{\sin A} + \sin A$, $\csc A$

26. $\dfrac{1}{\sin A \cos A} - \dfrac{\cos A}{\sin A}$, $\tan A$

27. $\dfrac{\sec A}{\sin A} - \dfrac{\sin A}{\cos A}$, $\cot A$ 28. $\sec A \csc A - \dfrac{\sin A}{\cos A}$, $\cot A$

29. $\dfrac{1}{1 - \sin A} + \dfrac{1}{1 + \sin A}$, $\sec^2 A$

30. $\dfrac{1}{\sec A - \tan A} + \dfrac{1}{\sec A + \tan A}$, $2 \sec A$

31. $\dfrac{\tan A}{\sin A} + \dfrac{\csc A}{\tan A}$, $\sec A \csc^2 A$ 32. $\dfrac{\csc A}{\cos A} - \dfrac{\cos A}{\sin A}$, $\tan A$

3.7.   Identities

The equation is a fundamental concept and tool of mathematics. We shall now define two types of equations.

> **DEFINITION**
>
> *An equation that is not satisfied by any value of the unknown or is satisfied by some values of it and not by others is called a **conditional equation.***

As an example of a conditional equation, we may consider

$$x^2 - 2x - 3 = 0.$$

This equation is satisfied by $x = 3$ and $x = -1$, but by no other value of x. Hence, the solution set is $\{3, -1\}$.

> **DEFINITION**
>
> *An equation that is true for all values of the unknown for which each member has a defined finite value is called an **identity.***

The equation

$$(x + 3)(x - 3) = x^2 - 9$$

is true for every value of x, and therefore it is an identity. Again, the equation

$$\frac{1}{x} - \frac{1}{x + 1} = \frac{1}{x(x + 1)}$$

is true for every value of x except $x = 0$ and $x = -1$. If x is equal to either of these values, a denominator on the left and the denominator on the right become zero, and the value of neither member exists. Nevertheless, this equation is an identity, since it is a true statement for all replacements for x for which each member has a defined finite value.

The eight fundamental identities collected in article 3.5 may be used to prove less fundamental ones.

The simplest method of proving that an equation is an identity is to reduce one member to the same form as the other. As in the previous article, there is no one method for performing this reduction, but the following suggestions will serve as a guide in the work:

1. *It is usually better to work with the more complicated member of the identity.*

2. *If one member contains one or more indicated operations, perform these operations first.*

3. *If one member contains more than one function value, while the other contains only one, express the ones in the first member, by means of the fundamental identities, in terms of the one in the second.*

4. *If the numerator of one member contains several terms and the denominator contains only one term, the desired reduction can sometimes be accomplished by expressing the member as the algebraic sum of several fractions with each term of the numerator appearing as a numerator of one of the fractions and with the original denominator appearing as the denominator of each fraction. Then apply the fundamental identities.*

5. *If either member can be factored, factor it. Then possibly the next step can be seen.*

6. *It is sometimes necessary to multiply the numerator and denominator of a member by the same factor in order to obtain the desired reduction. This is essentially multiplying the fraction by unity.*

7. *If none of the above steps seem applicable, express the function values in the more complicated member in terms of sines and cosines, and simplify.*

EXAMPLE 1

Prove that the following equation is an identity:

$$\frac{\tan \theta}{1 + \sec \theta} - \frac{\tan \theta}{1 - \sec \theta} = \frac{2}{\sin \theta}.$$

Solution: The left member is the more complicated, so we shall work with it. The indicated algebraic operation is subtraction, and, if we perform it, we get

$$\frac{\tan \theta}{1 + \sec \theta} - \frac{\tan \theta}{1 - \sec \theta} = \frac{\tan \theta - \tan \theta \sec \theta - \tan \theta - \tan \theta \sec \theta}{(1 - \sec \theta)(1 + \sec \theta)}$$

$$= \frac{-2 \tan \theta \sec \theta}{1 - \sec^2 \theta}$$

$$= \frac{2 \tan \theta \sec \theta}{\tan^2 \theta} \qquad \text{by (3.7)}$$

$$= \frac{2 \sec \theta}{\tan \theta} \qquad \begin{array}{l}\text{dividing numerator and}\\ \text{denominator by } \tan \theta.\end{array}$$

This completes the algebraic simplification, but we notice that the only function in the right member of the original identity is $\sin \theta$. Hence, we transform all of the functions in the simplified form above into sines and cosines by use of identities (3.2) and (3.4). We thus obtain

$$\frac{2 \sec \theta}{\tan \theta} = \frac{2\left(\dfrac{1}{\cos \theta}\right)}{\dfrac{\sin \theta}{\cos \theta}} = \frac{2}{\sin \theta}.$$

Thus, the proof is completed.

EXAMPLE 2

Prove that

$$\frac{\sin \theta + \cos \theta \tan \theta}{\cos \theta} = 2 \tan \theta.$$

Solution: Since the denominator of the left member of this equation is a single term and the numerator is the sum of terms, we shall try writing the expression as the sum of separate fractions. Thus, we obtain

$$\frac{\sin \theta + \cos \theta \tan \theta}{\cos \theta} = \frac{\sin \theta}{\cos \theta} + \frac{\cos \theta \tan \theta}{\cos \theta}$$

$$= \tan \theta + \tan \theta \qquad \text{by (3.4)}$$

$$= 2 \tan \theta.$$

EXAMPLE 3

Prove that the equation

$$\frac{\cos \theta}{\cos \theta - \sin \theta} = \frac{\cos \theta(\cos \theta - \sin \theta)}{1 - 2 \sin \theta \cos \theta}$$

an identity.

Solution: In this equation we notice that the numerator on the right may be obtained by multiplying the numerator on the left by $\cos \theta - \sin \theta$. This suggests multiplying the numerator and denominator on the left by this factor. If this is done, we have

$$\frac{\cos \theta}{\cos \theta - \sin \theta} = \frac{\cos \theta(\cos \theta - \sin \theta)}{(\cos \theta - \sin \theta)(\cos \theta - \sin \theta)}$$

$$= \frac{\cos \theta(\cos \theta - \sin \theta)}{\cos^2 \theta - 2 \sin \theta \cos \theta + \sin^2 \theta}$$

$$= \frac{\cos \theta(\cos \theta - \sin \theta)}{1 - 2 \sin \theta \cos \theta} \qquad \text{by (3.6).}$$

✓**EXAMPLE 4**

Prove that

$$\sin^2 \theta - \cos^2 \theta = \sin^4 \theta - \cos^4 \theta$$

is an identity.

Solution: The right member of the above equation is factorable, and if we factor it, we get

$$\sin^4 \theta - \cos^4 \theta = (\sin^2 \theta - \cos^2 \theta)(\sin^2 \theta + \cos^2 \theta)$$
$$= \sin^2 \theta - \cos^2 \theta \qquad \text{by (3.6)}.$$

EXERCISE 3.2

Prove that each of the following equations is an identity.

1. $\sin A(\csc A - \sin A) = \cos^2 A$ 2. $\tan A(\cot A + \tan A) = \sec^2 A$

3. $\sec A(\sec A - \cos A) = \tan^2 A$ 4. $\cos A(\sec A - \cos A) = \sin^2 A$

5. $(\sec A - \tan A)(\sec A + \tan A) = 1$

6. $(\csc A - \cot A)(\csc A + \cot A) = 1$

7. $(\sec A + 1)(\sec A - 1) = \tan^2 A$

8. $(1 + \cos A)(1 - \cos A) = \sin^2 A$

9. $\sec^2 A + \csc^2 A = \sec^2 A \csc^2 A$

10. $\tan^2 A \sec^2 A + \sec^2 A = \sec^4 A$

11. $\sec^2 A - \csc^2 A = \tan^2 A - \cot^2 A$

12. $\cos^4 A - \cos^2 A + \sin^2 A = \sin^4 A$

13. $(\sec^2 A + \tan^2 A + 1) \cos^2 A = 2$

14. $2 \cos^2 A - \sin^2 A + 1 = 3 \cos^2 A$

15. $\sin A + \cos A \tan A = 2 \sin A$

16. $\sec^2 A \tan^2 A - \tan^2 A = \tan^4 A$

17. $\dfrac{\csc A - \sin A}{\cot A} - \dfrac{\cot A}{\csc A} = 0$

18. $\dfrac{\cos A}{\sin A} + \dfrac{\sin A}{\cos A} = \sec A \csc A$

19. $\dfrac{\sin A}{\csc A} + \dfrac{\cos A}{\sec A} = \sin A \csc A$

20. $\dfrac{\sin A}{1 - \cos A} - \dfrac{1 + \cos A}{\sin A} = 0$

21. $\dfrac{1 + \cos A}{\sin A} + \dfrac{\sin A}{\cos A} = \dfrac{1 + \cos A}{\sin A \cos A}$

22. $\dfrac{1 + \sec A}{\tan A} - \dfrac{\tan A}{\sec A} = \dfrac{1 + \sec A}{\sec A \tan A}$

23. $\dfrac{1 - \tan A}{\sec A} + \dfrac{\sec A}{\tan A} = \dfrac{1 + \tan A}{\tan A \sec A}$

24. $\dfrac{1 - 2 \sin A}{\cos^2 A} - \dfrac{1 - 3 \sin A}{1 - \sin A} = 3 \tan^2 A$

25. $\dfrac{\sec^2 A - \tan^2 A + \tan A}{\sec A} = \sin A + \cos A$

26. $\dfrac{\sin A \cot A + \cos A}{\cot A} = 2 \sin A$

27. $\dfrac{\sin A + \cos A \tan A}{\tan A} = 2 \cos A$

28. $\dfrac{\cos^3 A - \cos A + \sin A}{\cos A} = \tan A - \sin^2 A$

29. $\dfrac{\tan A + \sec^3 A - \sec A}{\sec A} = \tan^2 A + \sin A$

30. $\dfrac{\cos A + \sin^3 A - \sin A}{\sin A} = \cot A - \cos^2 A$

31. $\dfrac{1 - \csc A \sec^3 A + \sec A \csc A}{\csc A} = \sin A - \tan^2 A \sec A$

32. $\dfrac{1 + \tan A \sin^2 A - \tan A}{\sin A} = \csc A - \cos A$

33. $\dfrac{\sec A}{1 - \cos A} = \dfrac{\sec A + 1}{\sin^2 A}$

34. $\dfrac{\cos A}{\sec A - 1} = \dfrac{1 + \cos A}{\tan^2 A}$

35. $\dfrac{\cos A \cot A}{\sec A - \tan A} = \cot A + \cos A$

36. $\dfrac{\cos A}{1 + \sin A} = \sec A - \tan A$

37. $\dfrac{1}{1 + \sin A} = \dfrac{1 - \sin A}{\cos^2 A}$

38. $\dfrac{\tan A \sin A}{1 - \cos A} = \dfrac{1 + \cos A}{\cos A}$

39. $\dfrac{\sin A}{1 - \cos A} = \dfrac{1 + \cos A}{\sin A}$

40. $\dfrac{\sin A}{\csc A - 1} = \dfrac{1 + \sin A}{\cot^2 A}$

41. $\dfrac{\cot A \cos A}{1 - \sin A} = \csc A + 1$

42. $\dfrac{\sin^2 A}{\sec A - 1} - \cos A = \cos^2 A$

43. $\dfrac{1}{\sec A - \tan A} - \sec A = \tan A$

44. $\dfrac{\sin A}{1 - \cos A} - \cot A = \csc A$

45. $\dfrac{2 \cos^4 A + \sin^2 A \cos^2 A - \sin^4 A}{3 \cos^2 A - 1} = 1$

46. $\dfrac{1 + \tan^2 A}{\sec^2 A (\sin A + \cos A)} = \dfrac{\sec A}{1 + \tan A}$

47. $\dfrac{\sin^3 A + \cos^3 A}{2 \sin^2 A - 1} = \dfrac{\sec A - \sin A}{\tan A - 1}$

48. $\dfrac{1 - \tan^4 A}{\sec^2 A (\sin A + \cos A)} = \sec A (1 - \tan A)$

49. $\dfrac{2 \sin A + 3 - 2 \csc A}{1 + 2 \csc A} = 2 \sin A - 1$

50. $\dfrac{2 \sin^4 A + \sin^2 A \cos^2 A - \cos^4 A}{3 \sin^2 A - 1} = 1$

51. $\dfrac{\tan^3 A + \sin A \sec A - \sin A \cos A}{\sec A - \cos A} = \sin A + \tan A \sec A$

52. $\dfrac{\sin A \sec^2 A + \sin A - 2 \tan A}{(\cos A - 1)^2} = \sec A \tan A$

53. $\dfrac{1 + \cos A}{1 - \cos A} - \dfrac{\csc A - \cot A}{\csc A + \cot A} = 4 \cot A \csc A$

54. $\dfrac{1 + \sin A}{1 - \sin A} - \dfrac{\sec A - \tan A}{\sec A + \tan A} = \dfrac{4 \csc A}{\csc^2 A - 1}$

55. $\dfrac{\sec A \tan^2 A + \sec^2 A - \sec^3 A}{(\sec A - 1)^2} = \dfrac{1}{1 - \cos A}$

56. $\dfrac{\sin^3 A - \cos^3 A}{\sin A - \cos A} = 1 + \sin A \cos A$

57. $\dfrac{\cot B (\cos A - \sin A) + \cos B (1 + \sin A \csc B)}{\sin B + \cos A} = \cot B$

58. $\dfrac{\cot A + \cot B}{\tan A + \tan B} = \cot A \cot B$

59. $\dfrac{(\sin A + \cos B)^2 + (\cos A + \sin B)(\cos A - \sin B)}{\sin A + \cos B} = 2 \cos B$

60. $\dfrac{\tan A (\sin B - \cos A) + \sin A (1 + \cos B \sec A)}{\sin B + \cos B} = \tan A$

61. $\pm\sqrt{\dfrac{1 - \cos A}{1 + \cos A}} = \dfrac{\sin A}{1 + \cos A}.$ *Note.* The plus sign is used if A is in the first

or second quadrant, and the minus sign if A is in the third or fourth quadrant.

62. $\pm\sqrt{\dfrac{\sec A + \tan A}{\sec A - \tan A}} = \sec A + \tan A.$ *Note.* The plus sign is used if A is

in the first or fourth quadrant, and the minus sign if it is in the second or third quadrant.

63. $\pm\sqrt{\dfrac{\sec A - 1}{\sec A + 1}} = \dfrac{\sec A - 1}{\tan A}.$ *Note.* The plus sign is used if A is in the first

or second quadrant, and the minus sign if it is in the third or fourth quadrant.

64. $\pm\sqrt{\dfrac{1 + \sin A}{1 - \sin A}} = \sec A + \tan A.$ *Note.* The plus sign is used if A is in the

first or fourth quadrant, and the minus sign if it is in the second or third quadrant.

4

COMPOSITE ANGLES
AND
TRIGONOMETRIC REDUCTIONS

4.1. Introduction

Heretofore, we have dealt with function values of a single angle. There are, however, conditions under which it is desirable to express function values of the sum or difference of two angles in terms of function values of the separate angles and to express function values of a multiple or submultiple of an angle in terms of function values of the angle. In order to derive such formulas for any two angles, we shall make use of the distance formula as developed in article 1.5.

4.2. The Cosine of the Difference of Two Angles

We shall represent two angles by A and B and find a formula for $\cos (A - B)$. We begin by drawing a circle of radius 1 with center at the origin.

We then draw angles A and B in standard position with their terminal sides intersecting the unit circle at P_2 and P_3, respectively; hence, angle P_2OP_3 is equal to $B - A$. We now designate the intersection of the unit circle and the positive X-axis by P_1 and construct an angle P_1OP_4 that is the same size as $B - A$ but measured in the opposite direction; hence, angle P_1OP_4 is $A - B$, as shown in Figure 4.1. Since chords P_2P_3 and P_1P_4 are intercepted

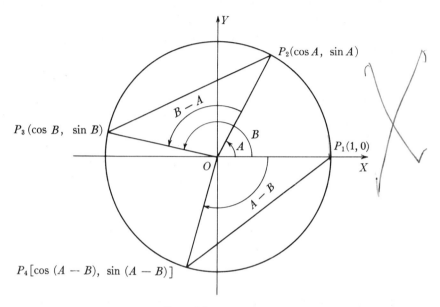

Figure 4.1

by equal central angles, they are equal. We express the coordinates of P_2, P_3, and P_4 in terms of function values as justified by article 1.14, evaluate $(P_1P_4)^2$ and $(P_2P_3)^2$, and then equate the expressions found. By use of the distance formulas, we have

$$(P_1P_4)^2 = [\cos (A - B) - 1]^2 + [\sin (A - B) - 0]^2,$$
$$= \cos^2 (A - B) - 2 \cos (A - B) + 1 + \sin^2 (A - B),$$
$$= 2 - 2 \cos (A - B),$$
$$\text{since } \cos^2 (A - B) + \sin^2 (A - B) = 1; \qquad (1)$$
$$(P_2P_3)^2 = (\cos A - \cos B)^2 + (\sin A - \sin B)^2$$
$$= \cos^2 A - 2 \cos A \cos B + \cos^2 B + \sin^2 A$$
$$\qquad - 2 \sin A \sin B + \sin^2 B$$
$$= 2 - 2 (\cos A \cos B + \sin A \sin B) \qquad \text{by (3.6)} \qquad (2)$$

We can equate the right-hand members of (1) and (2) since the left members are equal. If we do this and solve for cos $(A - B)$, we find that

$$\cos (A - B) = \cos A \cos B + \sin A \sin B. \tag{4.1}$$

key formula $\cos (A + B) = \cos A \cos B - \sin A \sin B$

This can be put in words by saying that *the cosine of the difference of two angles is equal to the product of their cosines increased by the product of their sines.* Furthermore, since the angle may be expressed in degrees or radians and a function value of a given number is equal to that same function value of the given number of radians, it follows that (4.1) is also true for any two real numbers A and B.

EXAMPLE 1

Find the exact value of cos 15°.

Solution: We shall use 15° in the form 60° − 45° and have

$$\cos 15° = \cos (60° - 45°)$$
$$= \cos 60° \cos 45° + \sin 60° \sin 45°$$
$$= \frac{1}{2} \frac{\sqrt{2}}{2} + \frac{\sqrt{3}}{2} \frac{\sqrt{2}}{2}$$
$$= \frac{\sqrt{2}(1 + \sqrt{3})}{4}.$$

4.3. Function Values of $\pi/2 - B$ and of $-B$

In order to find an expression for cos $(\pi/2 - B)$, we shall replace A in (4.1) by $\pi/2$. Thus, we have

$$\cos (\pi/2 - B) = \cos \pi/2 \cos B + \sin \pi/2 \sin B$$
$$= \sin B$$

since cos $\pi/2 = 0$ and sin $\pi/2 = 1$. Therefore,

$$\cos (\pi/2 - B) = \sin B \tag{4.2}$$

for any angle or any real number B.

If B is replaced by $\pi/2 - B$ in (4.2), we get

$$\cos [\pi/2 - (\pi/2 - B)] = \sin (\pi/2 - B)$$

and since

$$[\pi/2 - (\pi/2 - B)] = B, \quad \text{we have}$$
$$\sin (\pi/2 - B) = \cos B \tag{4.3}$$

for any angle or real number B.

The identities for the other function values of $\pi/2 - B$ can now be obtained by use of the ratio and reciprocal relations of articles 3.3 and 3.2, but the mechanics of doing it will be left as an exercise for the reader. When this is done, it is proved that _any trigonometric function value of $\pi/2$, minus any angle or real number, is equal to the corresponding cofunction value of the angle or real number._

We shall now express each function value of the negative of an angle or real number in terms of the same function value of the angle or real number by use of (4.1). If we replace A by zero in (4.1), it becomes

$$\cos (0 - B) = \cos 0 \cos B + \sin 0 \sin B.$$

Hence, since $0 - B = -B$, $\cos 0 = 1$, and $\sin 0 = 0$, we find that

$$\mathbf{\cos (-B) = \cos B.} \qquad (4.4)$$

If we replace B by $-B$ in (4.2), we get $\cos (\pi/2 + B) = \sin (-B)$. Now, since $\pi/2 + B = B + \pi/2 = B - (-\pi/2)$, we have

$$\sin (-B) = \cos [B - (-\pi/2)]$$
$$= \cos B \cos (-\pi/2) + \sin B \sin (-\pi/2).$$

Consequently, since $\cos (-\pi/2) = 0$ and $\sin (-\pi/2) = -1$, we have shown that

$$\mathbf{\sin (-B) = -\sin B.} \qquad (4.5)$$

The identities for the other function values of the negative of an angle can now be obtained by use of the ratio and reciprocal relations of articles 3.3 and 3.2, but doing so will be left as an exercise. When this has been done, it has been proved that _the cosine and secant are even functions and the other four trigonometric functions are odd functions_, since a function f is even if $f(-x) = f(x)$, and is odd if $f(-x) = -f(x)$.

EXAMPLES

1. $\cos 49° \cos (90° - 41°) = \sin 41°$,   by use of (4.2)

2. $\sin 37° \sin (90° - 53°) = \cos 53°$,   by use of (4.3)

3. $\tan (90° - 25°) = \dfrac{\sin (90° - 25°)}{\cos (90° - 25°)} = \dfrac{\cos 25°}{\sin 25°} = \cot 25°$

4. $\cos (-72°) = \cos 72°$,   by use of (4.4)

5. $\sin (-34°) = -\sin 34°$,   by use of (4.5)

6. $\sec (-59°) = \dfrac{1}{\cos (-59°)} = \dfrac{1}{\cos 59°} = \sec 59°$

4.4.   The Cosine of a Sum

Since $[A - (-B)] = A + B$, we see, by use of (4.1), that

$$\cos(A + B) = \cos[A - (-B)],$$
$$= \cos A \cos(-B) + \sin A \sin(-B),$$
$$= \cos A \cos B - \sin A \sin B, \qquad \text{by (4.4) and (4.5).}$$

Hence, we have shown that

$$\cos(A + B) = \cos A \cos B - \sin A \sin B. \tag{4.6}$$

This can be put in words by saying that *the cosine of the sum of two angles or numbers is the product of their cosines decreased by the product of their sines.*

EXAMPLE

Find the exact value of $\cos 75°$.

Solution: We shall use $75° = 45° + 30°$, in (4.6), and get

$$\cos 75° = \cos(45° + 30°)$$
$$= \cos 45° \cos 30° - \sin 45° \sin 30°$$
$$= \frac{\sqrt{2}}{2} \frac{\sqrt{3}}{2} - \frac{\sqrt{2}}{2} \frac{1}{2}$$
$$= \frac{\sqrt{2}(\sqrt{3} - 1)}{4}.$$

4.5.   The Cosine of Twice an Angle or Number

If in (4.6), we let the angle B equal the angle A, we have

$$\cos(A + A) = \cos A \cos A - \sin A \sin A.$$

Therefore,

$$\cos 2A = \cos^2 A - \sin^2 A \tag{4.7a}$$
$$= 1 - 2\sin^2 A, \qquad \text{since } \cos^2 A = 1 - \sin^2 A \tag{4.7b}$$
$$= 2\cos^2 A - 1, \qquad \text{since } \sin^2 A = 1 - \cos^2 A. \tag{4.7c}$$

4.6. The Cosine of $\theta/2$

We shall consider (4.7c), which may be written

$$2 \cos^2 A - 1 = \cos 2A.$$

Adding 1 to each member of this equation and dividing by 2, we have

$$\cos^2 A = \frac{1 + \cos 2A}{2}.$$

Hence,

$$\cos A = \pm \sqrt{\frac{1 + \cos 2A}{2}}.$$

We shall let $2A = \theta$, then $A = \frac{1}{2}\theta$, and the last equation above becomes

$$\cos \tfrac{1}{2}\theta = \pm \sqrt{\frac{1 + \cos \theta}{2}}. \qquad (4.8)$$

This is known as the formula for the cosine of half an angle or number, and by use of it we are able to find the cosine of half an angle or number provided we know the cosine of the angle. The algebraic sign to be used with the radical depends upon the magnitude of $\frac{1}{2}\theta$, and it may be determined by the methods of article 1.11.

4.7. Identities

Most of the problems in the next exercise and in the other exercises of this chapter involve the technique of proving that an equation is an identity. The method is the same as that discussed in article 3.7. The formulas developed in Chapter 3 and those developed in this chapter are used. The reader should review the above article and the formulas of article 3.5 before proceeding further.

EXAMPLE

Prove that

$$\frac{1 - \tan^2 \theta}{1 + \tan^2 \theta} = \cos 2\theta$$

is an identity.

Solution: We shall prove that the above equation is an identity by reducing the left member to the right.

$$\frac{1 - \tan^2 \theta}{1 + \tan^2 \theta} = \frac{1 - \dfrac{\sin^2 \theta}{\cos^2 \theta}}{1 + \dfrac{\sin^2 \theta}{\cos^2 \theta}} \qquad \text{by (3.4)}$$

$$= \frac{\cos^2 \theta - \sin^2 \theta}{\cos^2 \theta + \sin^2 \theta} \qquad \text{multiplying by } \frac{\cos^2 \theta}{\cos^2 \theta}$$

$$= \cos^2 \theta - \sin^2 \theta \qquad \text{by (3.6)}$$

$$= \cos 2\theta \qquad \text{by (4.7a)}.$$

EXERCISE 4.1

Identify the statement in each of problems 1 to 12 as true or false and justify your decision.

1. $\cos 60° + \cos 30° = \cos 90°$
2. $\cos 300° + \cos 60° = \cos 360°$
3. $\cos 270° - \cos 45° = \cos 225°$
4. $\cos 90° - \cos 60° = \cos 150°$
5. $2 \cos \theta - \cos \theta = \cos \theta$
6. $\cos 6\theta = 2 \cos 3\theta$
7. $\cos 2\theta - \cos \theta = \cos \theta$
8. $\cos 5\theta + \cos \theta = 2 \cos 3\theta \cos 2\theta$
9. $\dfrac{\cos 4\pi}{\cos 2\pi} = \cos 2\pi$
10. $\dfrac{\cos 4\theta}{\cos 2\theta} = \cos 2\theta$
11. $\cos 11° = \sqrt{\dfrac{1 + \cos 22°}{2}}$
12. $\cos 163° = -\sqrt{\dfrac{1 + \sin 56°}{2}}$

Make use of the function values of 30°, 45°, and 60° to evaluate each of the following.

13. $\cos 105°$
14. $\cos 22°30'$
15. $\cos 15°$
16. $\cos 120°$

Find $\cos (A + B)$ and $\cos (A - B)$ if A, B, and their function values are as given in problems 17 to 20.

17. $\cos A = 4/5,\ 0 < A < \pi/2;\ \cos B = -3/5,\ \pi/2 < B < \pi$
18. $\cos A = 5/13,\ 3\pi/2 < A < 2\pi;\ \sin B = 4/5,\ \pi/2 < B < \pi$
19. $\sin A = 15/17,\ \pi/2 < A < \pi;\ \cos B = -3/5,\ \pi < B < 3\pi/2$
20. $\sin A = -3/5,\ \pi < A < 3\pi/2;\ \sin B = -7/25,\ 3\pi/2 < B < 2\pi$

If θ and one of its function values are as given, find $\cos 2\theta$ and $\cos \frac{1}{2}\theta$ in each of problems 21 to 24.

21. $\sin \theta = 15/17,\ \pi/2 < \theta < \pi$
22. $\cos \theta = -3/5,\ \pi < \theta < 3\pi/2$
23. $\cos \theta = 5/13,\ 3\pi/2 < \theta < 2\pi$
24. $\sin \theta = -12/13,\ \pi < \theta < 3\pi/2$

Use (4.1) and (4.6) to prove that the equation in each of problems 25 to 28 is an identity.

25. $\cos(\pi - \theta) = -\cos\theta$ **26.** $\cos(2\pi - \theta) = \cos\theta$

27. $\cos(\pi + \theta) = -\cos\theta$ **28.** $\cos(2\pi + \theta) = \cos\theta$

Prove that the equation in each of problems 29 to 44 is an identity.

29. $\cos 3\theta \cos\theta + \sin 3\theta \sin\theta = \cos 2\theta$

30. $\cos 4\theta \cos\theta - \sin 4\theta \sin\theta = \cos 5\theta$

31. $\cos 5\theta \cos 3\theta + \sin 5\theta \sin 3\theta = \cos 2\theta$

32. $\cos 4\theta \cos 2\theta + \sin 4\theta \sin 2\theta = \sin^2 6\theta + \cos^2 6\theta - 2\sin^2\theta$

33. $(1 - 2\sin^2\theta)^2 = 1 - \sin^2 2\theta$ **34.** $\cos^2\theta = \dfrac{1}{2} + \dfrac{1}{2}\cos 2\theta$

35. $\cos 2\theta + 2\sin^2\theta = 1$ **36.** $\dfrac{1 - \cos 2\theta}{2\sin\theta\cos\theta} = \tan\theta$

37. $\cos 5\theta \cos 3\theta - \sin 5\theta \sin 3\theta = \cos^2 4\theta - \sin^2 4\theta$

38. $\cos 3\theta \cos\theta - \sin 3\theta \sin\theta = 2\cos^2 2\theta - 1$

39. $\cos 5\theta \cos\theta + \sin 5\theta \sin\theta = 1 - 4\sin^2\theta\cos^2\theta - \sin^2 2\theta$

40. $\cos^2 2\theta + 4\sin^2\theta\cos^2\theta = 1$

41. $\sin 2\theta + \cos 4\theta = (1 - \sin 2\theta)(1 + 2\sin 2\theta)$

42. $\cos 4\theta + \cos 2\theta = 2\cos^2\theta(4\cos^2\theta - 3)$

43. $\cos 2\theta - \cos\theta = (2\cos\theta + 1)(\cos\theta - 1)$

44. $\cos\theta + \cos 2\theta = (2\cos\theta - 1)(\cos\theta + 1)$

4.8.   The Sine of a Sum

By use of (4.2), we have

$$\sin(A + B) = \cos\left[\frac{\pi}{2} - (A + B)\right]$$

$$= \cos\left[\left(\frac{\pi}{2} - A\right) - B\right]$$

$$= \cos\left(\frac{\pi}{2} - A\right)\cos B + \sin\left(\frac{\pi}{2} - A\right)\sin B$$

$$= \sin A \cos B + \cos A \sin B$$

by use of (4.2) and (4.3). Thus we see that

$$\sin(A + B) = \sin A \cos B + \cos A \sin B. \qquad (4.9)$$

This should be remembered in words as

The sine of the sum of two angles or numbers is equal to the sine of the first times the cosine of the second plus the cosine of the first times the sine of the second.

4.9.   The Sine of 2A

If in (4.9) we let $B = A$, it becomes

$$\sin (A + A) = \sin A \cos A + \cos A \sin A.$$

Hence,

$$\mathbf{\sin 2A = 2 \sin A \cos A.} \tag{4.10}$$

4.10.   The Sine of θ/2

If we solve (4.7b) for $\sin A$, we have

$$\sin A = \pm \sqrt{\frac{1 - \cos 2A}{2}}.$$

If we let $A = \theta/2$, this becomes

$$\mathbf{\sin \tfrac{1}{2}\theta = \pm \sqrt{\frac{1 - \cos \theta}{2}}} \tag{4.11}$$

and is known as *the formula for the sine of half an angle or number*. The algebraic sign to be used is determined by the method of article 1.11.

4.11.   The Sine of a Difference

If in formula (4.9) we replace B by $-C$, we have

$$\sin (A - C) = \sin [A + (-C)]$$
$$= \sin A \cos (-C) + \cos A \sin (-C).$$

Therefore,

$$\mathbf{\sin (A - C) = \sin A \cos C - \cos A \sin C,} \tag{4.12}$$

since $\cos (-C) = \cos C$ and $\sin (-C) = -\sin C$.

This relation should be remembered in the form:

The sine of the difference of two angles or numbers is equal to the sine of the first times the cosine of the second minus the cosine of the first times the sine of the second.

EXAMPLE

Express $\sin 4A$ in terms of $\sin A$ and $\cos A$.

Solution: In order to do this, we need only to use the formulas for the sine and the cosine of twice an angle. Thus,

$$\sin 4A = \sin 2(2A),$$
$$= 2(\sin 2A)(\cos 2A) \qquad \text{by (4.10)},$$
$$= 2(2 \sin A \cos A)(\cos^2 A - \sin^2 A) \qquad \text{by (4.10) and (4.7a)},$$
$$= 4(\sin A \cos^3 A - \sin^3 A \cos A).$$

EXERCISE 4.2

Identify the statement in each of problems 1 to 12 as true or false and justify your decision.

1. $\sin 30° + \sin 60° = \sin 90°$

2. $\sin 150° + \sin 30° = \sin 180°$

3. $\sin 30° + \sin 330° = \sin 360°$

4. $\sin 360° - \sin 60° = \sin 240°$

5. $2 \sin \theta - \sin \theta = \sin \theta$

6. $\sin 6\theta = 2 \sin 3\theta$

7. $\sin 2\theta - \sin \theta = \sin \theta$

8. $\sin 5\theta + \sin \theta = 2 \sin 3\theta \cos 2\theta$

9. $\dfrac{\sin 1.5\pi}{\sin .5\pi} = \sin \pi$

10. $\dfrac{\sin 3\theta}{\sin \theta} = \sin 2\theta$

11. $\sin 13° = \sqrt{\dfrac{1 - \cos 26°}{2}}$

12. $\sin 42° = \sqrt{\dfrac{1 - \cos 84°}{2}}$

Make use of the function values of 30°, 45°, and 60° to evaluate each of the following.

13. $\sin 105°$

14. $\sin 67°30'$

15. $\sin 15°$

16. $\sin 195°$

Find $\sin (A + B)$ and $\sin (A - B)$ if A, B, and their function values are as given in problems 17 to 20.

17. $\sin A = 3/5,\ \pi/2 < A < \pi;\ \sin B = 5/13,\ 0 < B < \pi/2$

18. $\sin A = 12/13,\ 0 < A < \pi/2;\ \cos B = 3/5,\ 3\pi/2 < B < 2\pi$

19. $\cos A = -3/5,\ \pi < A < 3\pi/2;\ \sin B = -4/5,\ \pi < B < 3\pi/2$

20. $\cos A = -5/13,\ \pi/2 < A < \pi;\ \cos B = -3/5,\ \pi < B < 3\pi/2$

If θ and one of its function values are as given, find $\sin 2\theta$ and $\sin \frac{1}{2}\theta$ in each of problems 21 to 24.

21. $\cos \theta = -4/5,\ \pi/2 < \theta < \pi$

22. $\cos \theta = 12/13,\ 3\pi/2 < \theta < 2\pi$

23. $\sin \theta = 5/13,\ \pi < \theta < 3\pi/2$

24. $\sin \theta = -8/17,\ \pi < \theta < 3\pi/2$

Use (4.9) and (4.10) to prove that the equation in each of problems 25 to 28 is an identity.

25. $\sin (\pi - \theta) = \sin \theta$

26. $\sin (2\pi - \theta) = -\sin \theta$

27. $\sin (\pi + \theta) = -\sin \theta$

28. $\sin (2\pi + \theta) = \sin \theta$

Prove that the equation in each of problems 29 to 44 is an identity.

29. $\sin 2\theta \cos \theta + \cos 2\theta \sin \theta = \sin 3\theta$

30. $\sin 5\theta \cos 3\theta - \cos 5\theta \sin 3\theta = \sin 2\theta$

31. $\sin 4\theta \cos 2\theta - \cos 4\theta \sin 2\theta = 2 \sin \theta \cos \theta$

32. $\sin (90° + \theta) \cos (90° - \theta) + \cos (90° + \theta) \sin (90° - \theta) = 0$

33. $\dfrac{\sin^2 2\theta}{1 + \cos 2\theta} = 2 \sin^2 \theta$

34. $\sin 4\theta = 4 \sin \theta \cos \theta (\cos^2 \theta - \sin^2 \theta)$

35. $2 \cot 2\theta = \cot \theta - \tan \theta$

36. $1 - \cos 3\theta \cos \theta + \sin 3\theta \sin \theta = 2 \sin^2 2\theta$

37. $\dfrac{\sin 2\theta}{\sin \theta} - \dfrac{\cos 2\theta}{\cos \theta} = \sec \theta$

38. $\dfrac{\sin 6\theta}{\cos 3\theta} + \dfrac{\cos 6\theta}{\sin 3\theta} = \csc 3\theta$

39. $\sin 5\theta + \sin 3\theta = 8 \sin \theta \cos^2 \theta \cos 2\theta$

40. $\sin 6\theta + \sin 2\theta = 8 \sin \theta \cos \theta \cos^2 2\theta$

41. $\tan 3\theta - \tan \theta = 2 \sin \theta \sec 3\theta$

42. $\sin 5\theta + \sin \theta = 2 \sin 3\theta (\cos^2 \theta - \sin^2 \theta)$

43. $\cot \theta + \tan \theta = 2 \csc 2\theta$

44. $\cot 2\theta - \cot 4\theta = \csc 4\theta$

don't
do

test

4.12.   The Tangent of a Sum

A relation between the tangent of the sum of two angles or numbers and the tangents of the separate angles or numbers can be obtained by making use of three of the formulas already derived. By (3.4)

$$\tan (A + B) = \frac{\sin (A + B)}{\cos (A + B)}$$

$$= \frac{\sin A \cos B + \cos A \sin B}{\cos A \cos B - \sin A \sin B} \qquad \text{by (4.1) and (4.9).}$$

Dividing both numerator and denominator of the right member of this equation by $\cos A \cos B$, we have

$$\tan (A + B) = \frac{\dfrac{\sin A \cos B}{\cos A \cos B} + \dfrac{\cos A \sin B}{\cos A \cos B}}{\dfrac{\cos A \cos B}{\cos A \cos B} - \dfrac{\sin A \sin B}{\cos A \cos B}} = \frac{\dfrac{\sin A}{\cos A} + \dfrac{\sin B}{\cos B}}{1 - \dfrac{\sin A \sin B}{\cos A \cos B}}.$$

Therefore,

$$\tan (A + B) = \frac{\tan A + \tan B}{1 - \tan A \tan B}. \qquad (4.13)$$

Hence,

The tangent of a sum is equal to a fraction whose numerator is the sum of the tangents of the separate angles or numbers and whose denominator is one minus the product of their tangents.

4.13. The Tangent of Twice an Angle or Number

If we let $B = A$ in identity (4.13), we get

$$\tan (A + A) = \frac{\tan A + \tan A}{1 - \tan A \tan A}.$$

Consequently,

$$\tan 2A = \frac{2 \tan A}{1 - \tan^2 A}. \qquad (4.14)$$

4.14. The Tangent of $\theta/2$

In order to derive a formula for $\tan (\theta/2)$ in terms of function values of θ, we begin by making use of identity (3.4). Thus we obtain

$$\tan \tfrac{1}{2}\theta = \frac{\sin \tfrac{1}{2}\theta}{\cos \tfrac{1}{2}\theta}$$

$$= \frac{2 \sin \tfrac{1}{2}\theta \cos \tfrac{1}{2}\theta}{2 \cos^2 \tfrac{1}{2}\theta}, \qquad (1)$$

by multiplying by $(2 \cos \tfrac{1}{2}\theta)/(2 \cos \tfrac{1}{2}\theta)$. We now use (4.10) to replace the numerator by $\sin \theta$, and (4.7c) to replace the denominator by $1 + \cos \theta$. We then have

$$\tan \frac{1}{2}\theta = \frac{\sin \theta}{1 + \cos \theta} \qquad (4.15a)$$

If we multiply the right number of (1) by $(2 \sin \tfrac{1}{2}\theta)/(2 \sin \tfrac{1}{2}\theta)$, and then use (4.7b) and (4.10), we find that

$$\tan \frac{1}{2}\theta = \frac{1 - \cos \theta}{\sin \theta} \qquad (4.15\text{b})$$

4.15. The Tangent of a Difference

By use of (3.4), we have

$$\tan (A - B) = \frac{\sin (A - B)}{\cos (A - B)}.$$

And, by use of (4.12) and (4.1),

$$= \frac{\sin A \cos B - \cos A \sin B}{\cos A \cos B + \sin A \sin B}.$$

Dividing both numerator and denominator of the right member of the above equation by $\cos A \cos B$, we have

$$\tan (A - B) = \frac{\dfrac{\sin A \cos B}{\cos A \cos B} - \dfrac{\cos A \sin B}{\cos A \cos B}}{\dfrac{\cos A \cos B}{\cos A \cos B} + \dfrac{\sin A \sin B}{\cos A \cos B}}.$$

Hence,

$$\tan (A - B) = \frac{\tan A - \tan B}{1 + \tan A \tan B}. \qquad (4.16)$$

Therefore,

> *The tangent of a difference is equal to a fraction whose numerator is the difference of the tangents of the separate angles or numbers and whose denominator is one plus the product of their tangents.*

4.16. An Application Needed in Analytic Geometry

In article 1.16, we saw how to determine the other function values of an angle or number in a given quadrant when one is given. We shall now consider the problem of finding the sine and cosine of an acute angle when the tangent of twice the angle is known. This problem arises in connection with transformation of coordinates in analytic geometry and can be solved by using either of several sets of identities. We shall use

$$\tan 2A = \frac{2 \tan A}{1 - \tan^2 A} \qquad \text{and} \qquad \tan A = \frac{\sin A}{\cos A},$$

as well as $\sin^2 A + \cos^2 A = 1$.

In order to find tan A if tan $2A$ is given, we need only solve the quadratic in tan A that is obtained by replacing tan $2A$ by its value. This gives two values of tan A, but we shall use only the positive one, since we are interested in determining the sine and cosine of an acute angle. Finally, to find sin A and cos A, we make use of the fact that, if tan $A = a/b$, then, dividing numerator and denominator by $\sqrt{a^2 + b^2}$,

$$\tan A = \frac{a/\sqrt{a^2 + b^2}}{b/\sqrt{a^2 + b^2}},$$

and we have

$$\sin A = \frac{a}{\sqrt{a^2 + b^2}} \quad \text{and} \quad \cos A = \frac{b}{\sqrt{a^2 + b^2}}$$

where $\sin^2 A + \cos^2 A = 1$, as must be the case.

EXAMPLE

Find sin A and cos A if tan $2A = -\frac{24}{7}$.

Solution: If we replace tan $2A$ in (4.14) by $-\frac{24}{7}$, we have

$$\frac{-24}{7} = \frac{2 \tan A}{1 - \tan^2 A},$$

$-24 + 24 \tan^2 A = 14 \tan A$,   multiplying by $7(1 - \tan^2 A)$;

$12 \tan^2 A - 7 \tan A - 12 = 0$,   rearranging and dividing by 2;

$(3 \tan A - 4)(4 \tan A + 3) = 0$,   factoring.

Hence,

$$\tan A = \tfrac{4}{3}, -\tfrac{3}{4},$$

and we shall use the positive value since we want an acute angle A. Now,

$$\tan A = \frac{\sin A}{\cos A} = \frac{4}{3} = \frac{4/5}{3/5}.$$

Therefore,

$$\sin A = \tfrac{4}{5} \quad \text{and} \quad \cos A = \tfrac{3}{5}.$$

EXERCISE 4.3

Identify the statement in each of problems 1 to 12 as true or false and give your reason in each case.

1. $\tan 30° + \tan 60° = \tan 90°$
2. $\tan 135° + \tan 45° = \tan 180°$
3. $\tan 120° + \tan 60° = \tan 180°$
4. $\tan 60° + \tan 240° = \tan 300°$

5. $3 \tan \theta - \tan \theta = 2 \tan \theta$

6. $\tan 2\theta = 2 \tan \theta$

7. $\tan 2\theta - \tan \theta = \tan \theta$

8. $\tan 3\theta + \tan \theta = 2 \tan 2\theta \tan \theta$

9. $\dfrac{\tan (5\pi/4)}{\tan \pi/4} = \sin \pi/2$

10. $\dfrac{\tan 2\theta}{\tan \theta} = \sin \theta$

11. $\tan 27° = \dfrac{\sin 126°}{1 + \cos 306°}$

12. $\tan 32° = \dfrac{1 - \cos 64°}{\sin 116°}$

Make use of the function values of 30°, 45°, and 60° to evaluate each of the following.

13. $\tan 75°$; use $75° = 45° + 30°$

14. $\tan 15°$; use $15° = 60° - 45°$

15. $\tan 15°$; use $15° = \frac{1}{2}(30°)$

16. $\tan 120°$; use $120° = 2(60°)$

Find $\tan (A + B)$ and $\tan (A - B)$ if A, B and their function values are as given in problems 17 to 20.

17. $\tan A = -3/4$, $\tan B = 5/12$

18. $\tan A = 15/8$, $\cos B = -4/5$, $\pi/2 < B < \pi$

19. $\sin A = 7/25$, $\pi/2 < A < \pi$, $\tan B = 3/4$

20. $\sin A = -3/5$, $\pi < A < 3\pi/2$, $\cos B = 4/5$, $3\pi/2 < B < 2\pi$

If θ and one of its function values are as given, find $\tan 2\theta$ and $\tan \frac{1}{2}\theta$ in each of problems 21 to 24.

21. $\sin \theta = 3/5$, $\pi/2 < \theta < \pi$

22. $\sin \theta = -5/13$, $3\pi/2 < \theta < 2\pi$

23. $\cos \theta = -12/13$, $\pi < \theta < 3\pi/2$

24. $\cos \theta = 15/17$, $0 < \theta < \pi/2$

Use (4.13) and (4.16) to prove that the equation in each of problems 25 to 28 is an identity.

25. $\tan (\pi - \theta) = -\tan \theta$

26. $\tan (2\pi - \theta) = -\tan \theta$

27. $\tan (\pi + \theta) = \tan \theta$

28. $\tan (2\pi + \theta) = \tan \theta$

Prove that each of the following is an identity.

29. $\tan (\theta + 60°) + \tan (\theta - 60°) = \dfrac{8 \tan \theta}{1 - 3 \tan^2 \theta}$

30. $\tan (\theta + 30°) + \tan (\theta - 30°) = \dfrac{8 \tan \theta}{3 - \tan^2 \theta}$

31. $\tan (\theta + 45°) + \tan (\theta - 45°) = 2 \tan 2\theta$

32. $\tan (\theta + 45°) - \tan (\theta - 45°) = 2 \sec 2\theta$

33. $\tan \left(\dfrac{1}{2}\theta + \dfrac{1}{4}\pi\right) = \dfrac{1 + \sin \theta}{\cos \theta}$

34. $\tan \theta = \dfrac{2}{\cot \frac{1}{2}\theta - \tan \frac{1}{2}\theta}$

35. $\dfrac{(1 + \tan \theta)^2}{1 + \tan^2 \theta} = 1 + \sin 2\theta$

36. $\tan 2\theta - \tan \theta = \tan \theta \sec 2\theta$

37. $\tan 3\theta = \dfrac{3 \tan \theta - \tan^3 \theta}{1 - 3 \tan^2 \theta}$

38. $\tan 4\theta = \dfrac{4 \tan \theta - 4 \tan^3 \theta}{1 - 6 \tan^2 \theta + \tan^4 \theta}$

39. $\tan 2\theta = \dfrac{2}{\cot \theta - \tan \theta}$

40. $\tan \theta \cot \tfrac{1}{2}\theta = 1 + \sec \theta$

41. $\dfrac{\tan 3\theta - \tan \theta}{1 + \tan 3\theta \tan \theta} = \dfrac{2 \tan \theta}{1 - \tan^2 \theta}$

42. $\dfrac{\tan 3\theta + \tan \theta}{1 - \tan 3\theta \tan \theta} = \dfrac{2 \tan 2\theta}{1 - \tan^2 2\theta}$

43. $\cot 4\theta + \tan 2\theta = \csc 4\theta$

44. $1 + \tan \theta \tan 2\theta = \sec 2\theta$

Find $\sin A$ and $\cos A$ if A is a positive acute angle and if $\tan 2A$ has the value given in each of problems 45 to 48.

45. $\tan 2A = 24/7$ **46.** $\tan 2A = -15/8$

47. $\tan 2A = 5/12$ **48.** $\tan 2A = 3/4$

4.17. Reduction Formulas

In this article, we shall express each function value of an angle that differs from π or 2π by a positive acute angle θ in terms of the same function value of θ. We shall give this angle θ a name by saying that *if a given angle other than an integral multiple of 90° is in standard position, then the positive acute angle between its terminal side and the X-axis is called the* **related** *or* **reference angle.** Furthermore, if the given angle is a multiple of 90° and is in standard position, then its related angle is 0 or 90°, according as the terminal side is coincident with the X or Y-axis.

We shall restate problems 25 to 28 of Exercises 4.1, 4.2, and 4.3 here for ready reference. Thus,

$$\left\{ \begin{aligned} \cos (\pi \pm \theta) &= - \cos \theta, & \cos (2\pi \pm \theta) &= \cos \theta, \\ \sin (\pi \pm \theta) &= \mp \sin \theta, & \sin (2\pi \pm \theta) &= \pm \sin \theta, \\ \tan (\pi \pm \theta) &= \pm \tan \theta, & \tan (2\pi \pm \theta) &= \pm \tan \theta. \end{aligned} \right.$$

If θ is a positive acute angle, then the angle on the right in each of these equations is the related or reference angle of the one on the left. Furthermore, the left and right members are equal, or differ only in sign. By use of the fact that the other function values are the reciprocals of these, we can write similar identities for the other function values. Consequently, we know that

> *Each trigonometric function value of a given angle is numerically equal to the same function value of the related or reference angle.*

The algebraic sign is determined as stated in article 1.11.

EXAMPLES

1. $\sin 258° = -\sin 78°$
2. $\cos 258° = -\cos 78°$
3. $\tan 258° = \tan 78°$

The function values of negative angles have been considered in article 4.3 but could be treated by use of the related angle.

EXAMPLES

4. The reference angle of $-73°$ is $73°$; furthermore, $-73°$ is a fourth-quadrant angle. Consequently,

$$\tan(-73°) = -\tan 73° \qquad \text{and} \qquad \cos(-73°) = \cos 73°,$$

as would be obtained by (3.4), (4.4), and (4.5).

5. The reference angle of $-197°$ is $17°$; furthermore, $-197°$ is a second-quadrant angle. Consequently,

$$\sin(-197°) = \sin 17° \qquad \text{and} \qquad \cos(-197°) = -\cos 17°.$$

4.18.   The Product and Factor Formulas

We use relations (4.9), (4.12), (4.6), and (4.1) of this chapter to derive the *product formulas*. These formulas are so named because each of them expresses a product of two sines, of two cosines, or of a sine and a cosine as the sum or difference of two trigonometric function values.

$$\sin(A + B) = \sin A \cos B + \cos A \sin B, \tag{4.9}$$

$$\sin(A - B) = \sin A \cos B - \cos A \sin B, \tag{4.12}$$

$$\cos(A + B) = \cos A \cos B - \sin A \sin B, \tag{4.6}$$

$$\cos(A - B) = \cos A \cos B + \sin A \sin B. \tag{4.1}$$

Adding corresponding members of (4.9) and (4.12), we have

$$2 \sin A \cos B = \sin (A + B) + \sin (A - B). \qquad (4.17)$$

In a similar manner we obtain

$$2 \cos A \sin B = \sin (A + B) - \sin (A - B), \qquad (4.18)$$

$$2 \cos A \cos B = \cos (A + B) + \cos (A - B), \qquad (4.19)$$

$$2 \sin A \sin B = -\cos (A + B) + \cos (A - B). \qquad (4.20)$$

If we let

$$A + B = C \qquad \text{and} \qquad A - B = D,$$

then,

$$A = \tfrac{1}{2}(C + D) \qquad \text{and} \qquad B = \tfrac{1}{2}(C - D).$$

Substituting these values of A and B in formulas (4.17) to (4.20), we have

$$\sin C + \sin D = 2 \sin \tfrac{1}{2}(C + D) \cos \tfrac{1}{2}(C - D), \qquad (4.21)$$

$$\sin C - \sin D = 2 \cos \tfrac{1}{2}(C + D) \sin \tfrac{1}{2}(C - D), \qquad (4.22)$$

$$\cos C + \cos D = 2 \cos \tfrac{1}{2}(C + D) \cos \tfrac{1}{2}(C - D), \qquad (4.23)$$

$$\cos C - \cos D = -2 \sin \tfrac{1}{2}(C + D) \sin \tfrac{1}{2}(C - D). \qquad (4.24)$$

These formulas are called the *factor formulas* because each expresses either a sum or a difference of two function values as a product.

EXAMPLE 1

If we apply (4.20) to $2 \sin 3x \sin x$, we have

$$2 \sin 3x \sin x = -\cos (3x + x) + \cos (3x - x)$$
$$= -\cos 4x + \cos 2x.$$

EXAMPLE 2

If we apply (4.22) to the numerator and (4.23) to the denominator, we see that

$$\frac{\sin 4x - \sin 2x}{\cos 4x + \cos 2x} = \frac{2 \cos \tfrac{1}{2}(4x + 2x) \sin \tfrac{1}{2}(4x - 2x)}{2 \cos \tfrac{1}{2}(4x + 2x) \cos \tfrac{1}{2}(4x - 2x)}$$

$$= \frac{2 \cos 3x \sin x}{2 \cos 3x \cos x}$$

$$= \frac{\sin x}{\cos x}$$

$$= \tan x.$$

EXERCISE 4.4

Draw the angle and the related angle in each of problems 1 to 24, and express the given function value in terms of the same function value of the related angle.

1. sin 134°

2. cos 163°

3. tan 148°

4. cot 97°

5. cos 201°

6. tan 217°

7. cot 184°

8. sin 264°

9. tan 283°

10. cot 309°

11. sin 355°

12. cos 326°

13. cot 444°

14. sin 471°

15. cos 562°

16. tan 713°

17. sin (−108°)

18. cos (−133°)

19. tan (−222°)

20. cot (−282°)

21. cos (−377°)

22. tan (−443°)

23. cot (−503°)

24. sin (−606°)

Express the function value in each of problems 25 to 32 in terms of a function value of a positive angle less than 45°. Use the cofunction complement theorem as needed.

25. cos 76°

26. tan 130°

27. cot 247°

28. sin 308°

29. tan 413°

30. cot 502°

31. sin 617°

32. cos 568°

By use of the related angle theorem and Table III, find each function value called for in problems 33 to 44.

33. tan 227°

34. cot 291°

35. sin 367°

36. cos 111°

37. cos 98°10′

38. tan 196°20′

39. cot 222°30′

40. sin 276°40′

41. cot (−56°)

42. sin (−138°)

43. cos (−234°)

44. tan (−303°)

Express the product in each of problems 45 to 56 as a sum or difference and each sum or difference in problems 57 to 60 as a product.

45. 2 sin 26° cos 19°

46. 2 cos 47° sin 23°

47. 2 sin 71° sin 11°

48. 2 cos 33° cos 12°

49. 2 cos 135° sin 15°

50. 2 sin 67°30′ sin 22°30′

51. 2 cos 45° cos 15°

52. 2 sin 75° sin 45°

53. 2 sin 2θ sin θ

54. 2 cos 6θ cos 2θ

55. 2 cos 3θ sin θ

56. 2 sin 5θ cos 3θ

57. sin 17° + sin 13°

58. sin 43° − sin 47°

59. cos 82° − cos 22°

60. cos 37° + cos 23°

Prove that the equation in each of problems 61 to 76 is an identity.

61. $2 \sin 5x \cos x = \sin 2x(4 \cos^2 2x + 2 \cos 2x - 1)$

62. $2 \sin 3x \sin x = (1 - \cos 2x)(1 + 2 \cos 2x)$

63. $2 \cos 6x \cos 2x = (2 \cos 4x - 1)(\cos 4x + 1)$

64. $2 \cos 6x \sin 2x = 2 \sin 2x \cos 2x(4 \cos^2 2x - 3)$

65. $\sin 3x + \sin x = 2 \cos x \sin 2x$

66. $\sin 4x - \sin 2x = \sec x \cos 3x \sin 2x$

67. $\dfrac{\cos 3x - \cos 5x}{\sin 3x + \sin 5x} = \tan x$

68. $\dfrac{\cos 3x + \cos 5x}{\cos^2 2x - \sin^2 2x} = 2 \cos x$

69. $\dfrac{\sin 5x + \sin 3x}{\sin 4x} = 2 \cos x$ 4.21

70. $\dfrac{\cos x + \cos 5x}{\sin (3x + 90°) - \sin (3x - 90°)} = \cos 2x$

71. $\dfrac{\sin 6x - \sin 4x}{\cos 3x \cos 2x - \sin 3x \sin 2x} = 2 \sin x$

72. $\dfrac{\sin 3x - \sin 2x}{4 \cos^2 x - 2 \cos x - 1} = \sin x$

73. $\sin (x + y) \sin (x - y) = \sin^2 x - \sin^2 y$

74. $\sin (x + y) \sin (x - y) = \cos^2 y - \cos^2 x$

75. $\cos (x + y) \cos (x - y) = \cos^2 x - \sin^2 y$

76. $\sin (x + y) \cos (x - y) = \sin x \cos x + \sin y \cos y$

4.19. Summary of Identities

For the convenience of the reader, we shall collect the fundamental identities that are derived in this chapter and shall then furnish a set of identities that are to be proved.

Sum and difference formulas:

$$\cos (A + B) = \cos A \cos B - \sin A \sin B. \tag{4.6}$$

$$\sin (A + B) = \sin A \cos B + \cos A \sin B. \tag{4.9}$$

$$\cos (A - B) = \cos A \cos B + \sin A \sin B. \tag{4.1}$$

$$\sin (A - B) = \sin A \cos B - \cos A \sin B. \tag{4.12}$$

$$\tan (A + B) = \frac{\tan A + \tan B}{1 - \tan A \tan B}. \tag{4.13}$$

$$\tan (A - B) = \frac{\tan A - \tan B}{1 + \tan A \tan B}. \tag{4.16}$$

Double-angle formulas:

$$\cos 2A = \cos^2 A - \sin^2 A \tag{4.7a}$$
$$= 1 - 2 \sin^2 A \tag{4.7b}$$
$$= 2 \cos^2 A - 1. \tag{4.7c}$$
$$\sin 2A = 2 \sin A \cos A. \tag{4.10}$$
$$\tan 2A = \frac{2 \tan A}{1 - \tan^2 A}. \tag{4.14}$$

Half-angle formulas:

$$\cos \tfrac{1}{2}\theta = \pm \sqrt{\frac{1 + \cos \theta}{2}}. \tag{4.8}$$

$$\sin \tfrac{1}{2}\theta = \pm \sqrt{\frac{1 - \cos \theta}{2}}. \tag{4.11}$$

$$\tan \tfrac{1}{2}\theta = \frac{1 - \cos \theta}{\sin \theta}. \tag{4.15b}$$

$$\tan \tfrac{1}{2}\theta = \frac{\sin \theta}{1 + \cos \theta}. \tag{4.15a}$$

Product and factor formulas:

$$2 \sin A \cos B = \sin (A + B) + \sin (A - B). \tag{4.17}$$
$$2 \cos A \sin B = \sin (A + B) - \sin (A - B). \tag{4.18}$$
$$2 \cos A \cos B = \cos (A + B) + \cos (A - B). \tag{4.19}$$
$$2 \sin A \sin B = - \cos (A + B) + \cos (A - B). \tag{4.20}$$
$$\sin C + \sin D = 2 \sin \tfrac{1}{2}(C + D) \cos \tfrac{1}{2}(C - D). \tag{4.21}$$
$$\sin C - \sin D = 2 \cos \tfrac{1}{2}(C + D) \sin \tfrac{1}{2}(C - D). \tag{4.22}$$
$$\cos C + \cos D = 2 \cos \tfrac{1}{2}(C + D) \cos \tfrac{1}{2}(C - D). \tag{4.23}$$
$$\cos C - \cos D = -2 \sin \tfrac{1}{2}(C + D) \sin \tfrac{1}{2}(C - D). \tag{4.24}$$

EXERCISE 4.5

Prove that the equation in each of problems 1 to 60 is an identity.

1. $\cos 5\theta \cos 2\theta + \sin 5\theta \sin 2\theta = \cos 3\theta$

2. $\cos 6\theta \cos 4\theta + \sin 6\theta \sin 4\theta = 2 \cos^2 \theta - 1$

3. $\dfrac{\cos 4\theta}{\sin \theta} - \dfrac{\sin 4\theta}{\cos \theta} = 2 \cos 5\theta \csc 2\theta$

4. $\dfrac{\sin 5\theta}{\sin \theta} - \dfrac{\cos 5\theta}{\cos \theta} = 4 \cos 2\theta$

5. $\sin 3\theta \cos \theta + \cos 3\theta \sin \theta = 4 \sin \theta \cos \theta \cos 2\theta$

6. $\sin 5\theta \cos \theta - \cos 5\theta \sin \theta = 2 \sin 2\theta(\cos^2 \theta - \sin^2 \theta)$

7. $\dfrac{\sin 3\theta}{\cos \theta} + \dfrac{\cos 3\theta}{\sin \theta} = 2 \cos 2\theta \csc 2\theta$

8. $\dfrac{\sin 3\theta}{\sin 2\theta} + \dfrac{\cos 3\theta}{\cos 2\theta} = 2 \sin 5\theta \csc 4\theta$

9. $\dfrac{\tan 5\theta + \tan 2\theta}{1 - \tan 5\theta \tan 2\theta} = \tan 7\theta$

10. $\tan (\theta + 45°) + \tan (\theta - 45°) = 2 \tan 2\theta$

11. $\tan (\theta + 45°) - \tan (\theta - 45°) = 2 \sec 2\theta$

12. $\dfrac{\tan 3\theta + \tan \theta}{1 - \tan 3\theta \tan \theta} = \dfrac{2 \tan 2\theta}{1 - \tan^2 2\theta}$

13. $\cos A + \sin A \tan B = \sec B \cos (A - B)$

14. $\sin 2A - \cos 2A \tan A = \tan A$

15. $\dfrac{\cot A + \tan B}{\cot A - \tan B} = \dfrac{\sec (A + B)}{\sec (A - B)}$

16. $\cos 3\theta \cos \theta + \sin 3\theta \sin \theta = 1 - 2 \sin^2 \theta$

17. $\sin 4\theta \cos \theta - \cos 4\theta \sin \theta = \sin 3\theta$

18. $\tan 3\theta - \tan \theta = 2 \sin \theta \sec 3\theta$

19. $\tan 5\theta - \tan 3\theta = \sin 2\theta \sec 3\theta \sec 5\theta$

20. $\sin 2\theta - \tan \theta \cos 2\theta = \dfrac{\sin 2\theta}{2 \cos^2 \theta}$

21. $\dfrac{\tan 2\theta - \tan \theta}{1 + \tan 2\theta \tan \theta} = \tan \theta$

22. $\tan (A - B) + \tan B = \sin A \sec B \sec (A - B)$

23. $\tan (\theta - 45°) = \dfrac{\tan \theta - 1}{\tan \theta + 1}$

24. $\dfrac{\tan 4\theta - \tan 2\theta}{1 + \tan 4\theta \tan 2\theta} = \dfrac{2 \tan \theta}{2 - \sec^2 \theta}$

25. $\cos 2\theta - \cos \theta = (2 \cos \theta + 1)(\cos \theta - 1)$

26. $\cos 2\theta + \cos \theta = (2 \cos \theta - 1)(\cos \theta + 1)$

27. $\cos 4\theta = 8 \cos^4 \theta - 8 \cos^2 \theta + 1$

28. $\dfrac{\cos 2\theta - \sin \theta}{\cos^2 \theta} = \dfrac{1 - 2 \sin \theta}{1 - \sin \theta}$

29. $\sin 4\theta = 4 \sin \theta \cos \theta(1 - 2 \sin^2 \theta)$

30. $2 \sin \theta \cos^3 \theta + 2 \sin^3 \theta \cos \theta = \sin 2\theta$

31. $\dfrac{2 + 2 \cos 2\theta}{\sin^2 2\theta} = \csc^2 \theta$

32. $\dfrac{2 \sin 2\theta}{(1 + \cos 2\theta)^2} = \sec^2 \theta \tan \theta$

33. $\tan 4\theta - \tan 2\theta = \tan 2\theta \sec 4\theta$

34. $\tan\left(\dfrac{1}{2}\theta + \dfrac{\pi}{4}\right) = \dfrac{1 + \sin\theta}{\cos\theta}$

35. $\tan 2\theta = \dfrac{2\cos\theta}{\csc\theta - 2\sin\theta}$

36. $2\cot 2\theta = \cot\theta - \tan\theta$

37. $8\cos^4\theta = 3 + 4\cos 2\theta + \cos 4\theta$

38. $1 + \cos 3\theta \cos\theta - \sin 3\theta \sin\theta = 2\cos^2 2\theta$

39. $1 + \cos\dfrac{1}{2}\theta = \dfrac{1 - \cos\theta}{2(1 - \cos\frac{1}{2}\theta)}$

40. $1 + \cos\dfrac{1}{2}\theta + \cos\theta = \dfrac{1 + \cos\theta - 2\cos\theta \cos\frac{1}{2}\theta}{2(1 - \cos\frac{1}{2}\theta)}$

41. $\sin\theta \cos\dfrac{1}{2}\theta + \cos\theta \sin\dfrac{1}{2}\theta = \pm\sqrt{\dfrac{1 - \cos 3\theta}{2}}$

42. $2\sin^2\theta - \sin^2 2\theta = \cos^2 2\theta - \cos 2\theta$

43. $8\sin^4\theta = 3 - 4\cos 2\theta + \cos 4\theta$

44. $4\cos 2\theta + 4\sin^2\theta = \sin^2 2\theta \csc^2\theta$

45. $\dfrac{\tan\frac{1}{2}\theta}{\tan 2\theta} = \dfrac{2\cos^2\theta - 1}{2\cos\theta(\cos\theta + 1)}$

46. $\tan\theta \cot\frac{1}{2}\theta = 1 + \sec\theta$

47. $\csc 2\theta - \cot 2\theta = \tan\theta$

48. $\csc A + \cot A = \cot\frac{1}{2}A$

49. $\tan(-A)\cot(-A) = \sec(-A)\cos A$

50. $\sin(-A)\csc A = \tan(-A)\cot A$

51. $[1 - \cos(-A)][1 + \cos A] = [\sin(-A)]^2$

52. $\sec A \sec(-A) - 1 = [\tan(-A)]^2$

53. $\dfrac{4\sin 3A \cos 2A}{\cos 6A - \cos 4A} = -\csc A - \csc 5A$

54. $\dfrac{\cos 7A - \cos 3A}{\cos 5A \sin 2A} = -2\tan 5A$

55. $\dfrac{\sin 2A + \sin \pi/3}{\cos 2A + \cos \pi/3} = \tan(A + \pi/6)$

56. $\dfrac{\cos 2A + \cos \pi/2}{\sin 2A - \sin \pi/2} = \cot(A - \pi/4)$

57. $2\cos 3A \cos A = \cos^2 2A - \sin^2 2A + \cos^2 A - \sin^2 A$

58. $2\sin 5A \sin A = 8\cos^4 A - 8\cos^2 A - \cos 6A$

59. $2\sin 7A \cos 3A = \sin 10A + 8\sin A \cos^3 A - 4\sin A \cos A$

60. $2\cos 5A \sin A = 2\sin 3A \cos 3A - 4\sin A \cos A + 8\sin^3 A \cos A$

5

GRAPHS OF THE
TRIGONOMETRIC FUNCTIONS

5.1. Introduction

In this chapter, we shall investigate the manner in which the trigonometric function values change as the angle runs through a set of values. We shall then sketch the graphs of the functions and, finally, determine the effect on the graph of multiplying the function value by a constant, of multiplying the angle by a constant, and of adding a constant to a multiple of an angle.

5.2. Periodic Functions

There is a property of the trigonometric function values that enables us to get a complete sample of the graph by examining only a limited portion of it. This property is called *periodicity*, and we shall now define it.

If f is a function whose domain consists of all numbers θ in a certain set, and if p is a number such that

$$f(\theta + p) = f(\theta) \tag{5.1}$$

*for all admissible values of θ, then f is said to be **periodic** and to have **p** as a period. Furthermore, if **p** is the smallest positive number for which (5.1) is true, then **p** is called **the period**.*

If we make use of the sine, cosine, and tangent of the sum of two numbers or angles and replace the second one by 2π, we get

$$\sin (x + 2\pi) = \sin x \cos 2\pi + \cos x \sin 2\pi$$
$$= \sin x,$$
$$\cos (x + 2\pi) = \cos x \cos 2\pi - \sin x \sin 2\pi$$
$$= \cos x,$$

and

$$\tan (x + 2\pi) = \frac{\tan x + \tan 2\pi}{1 - \tan x \tan 2\pi}$$
$$= \tan x,$$

as stated in problems 26 of exercises 4.1, 4.2, and 4.3. Since the other three function values are the reciprocals of these three, we know that *each of the trigonometric functions is periodic and has 2π as a period*. Furthermore, *the tangent and cotangent have π as a period* since

$$\tan (x + \pi) = \frac{\tan x + \tan \pi}{1 - \tan x \tan \pi}$$
$$= \tan x.$$

Because of the periodicity of the sine, cosine, secant, and cosecant functions, we know that the function values are repeated at intervals of 2π and that we can get a complete sample of the graph of each of these functions by drawing the graph over an interval of horizontal length 2π. Furthermore, we have a complete sample of the graphs of the tangent and cotangent functions over an interval of horizontal length π.

5.3. Range of the Sine and Cosine

In this article, we shall determine the variation of $\sin \theta$ and of $\cos \theta$ as θ varies from 0 to 2π, and thereby find the range of the sine and cosine functions. For this purpose we shall construct a circle of unit radius with its center at the origin, as in Figure 5.1, and draw an angle θ in standard posi-

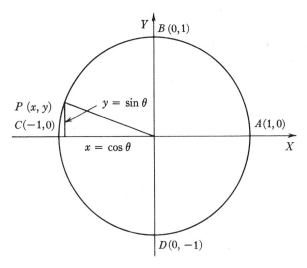

Figure 5.1

tion. If $P(x, y)$ is the point of intersection of the terminal side of θ and the unit circle, then

$$\sin \theta = \frac{y}{r} = \frac{y}{1} = y \quad \text{and} \quad \cos \theta = \frac{x}{r} = \frac{x}{1} = x.$$

Consequently, making use of the periodicity, we need only see how y and x vary as θ changes continuously from 0 to 2π, in order to determine the range of the sine and of the cosine. If θ varies from 0 to $\pi/2$, then $P(x, y)$ varies from $A(1, 0)$ to $B(0, 1)$; hence, $y = \sin \theta$ varies from 0 to 1 and $x = \cos \theta$ varies from 1 to 0. This, and the variation of P, $\sin \theta$, and $\cos \theta$ as θ changes from $\pi/2$ to π to $3\pi/2$ to 2π, are shown below in tabular form.

Variation of θ	Variation of $P(x, y)$	Variation of $y = \sin \theta$	Variation of $x = \cos \theta$
0 to $\pi/2$	$A(1, 0)$ to $B(0, 1)$	0 to 1	1 to 0
$\pi/2$ to π	$B(0, 1)$ to $C(-1, 0)$	1 to 0	0 to -1
π to $3\pi/2$	$C(-1, 0)$ to $D(0, -1)$	0 to -1	-1 to 0
$3\pi/2$ to 2π	$D(0, -1)$ to $A(1, 0)$	-1 to 0	0 to 1

Consequently, *the range of the sine function and of the cosine function is* -1 *to 1 inclusive.* This can be expressed symbolically by writing

$$-1 \le \sin \theta \le 1 \qquad \text{and} \qquad -1 \le \cos \theta \le 1$$

5.4. Bounds and Amplitude

We found in the last article that the function values $\sin \theta$ and $\cos \theta$ are always between -1 and 1, inclusive. Thus, they are hemmed in. A function with this property is said to be *bounded,* and is characterized by the statement that *a function is* **bounded above** *if a number* **M** *exists such that* $f(x) \le M$ *for all* x *in the domain of* f *and is* **bounded below** *if a number* m *exists such that* $f(x) \ge m$ *for all* x *in the domain of* f.

> *is*
> *between*
> *2 nos.,*
> *inclusive*

If a function f is bounded above, the smallest M for which $f(x) \le M$ for all x in the domain of f is called the *least upper bound of* f; furthermore, if f is bounded below, the largest m for which $f(x) \ge m$ for all x in the domain of f is called the *greatest lower bound of* f. The least upper and greatest lower bounds of $\sin \theta$ and $\cos \theta$ are 1 and -1.

If M and m are the least upper and greatest lower bounds of a periodic function f, then $\frac{1}{2}(M - m)$ is called the *amplitude of* f. Consequently, the amplitude of $\cos \theta$ and of $\sin \theta$ is $\frac{1}{2}[1 - (-1)] = 1$.

5.5. The Graph of $y = \sin x$

In sketching the graph of $\{(x, y) \mid y = \sin x\}$, we shall make use of the variation, periodicity, and bounds of $\sin x$ as developed in this chapter, along with a set of corresponding pairs of values of x and $y = \sin x$. We shall assign the multiples of $\pi/6$ and $\pi/4$ from 0 to 2π to x and find each corresponding value of $\sin x$ to one significant digit, as shown in the table below, by using the values of the sine of multiples of $\pi/6$ and $\pi/4$ and the related angle theorem.

x	0	$\pi/6$	$\pi/4$	$\pi/3$	$\pi/2$	$2\pi/3$	$3\pi/4$	$5\pi/6$
$y = \sin x$	0	.5	.7	.9	1	.9	.7	.5

x	π	$7\pi/6$	$5\pi/4$	$4\pi/3$	$3\pi/2$	$5\pi/3$	$7\pi/4$	$11\pi/6$	2π
$y = \sin x$	0	$-.5$	$-.7$	$-.9$	-1	$-.9$	$-.7$	$-.5$	0

We now continue by drawing the X-axis as a horizontal line and the Y-axis as a vertical line, and by choosing any convenient unit on the latter. We then use approximately 3.14 times that unit as π on the X-axis. We can now locate the points that correspond to the number pairs (x, y) that are indicated in the table. If we do that and draw a smooth curve through them, we obtain the solid part of the curve shown in Figure 5.2. Since the

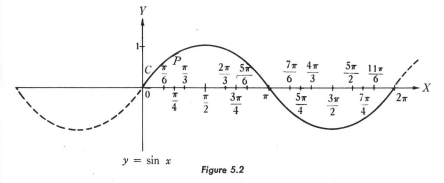

$y = \sin x$

Figure 5.2

bounds of $\sin x$ are 1 and -1 the curve is never above $y = 1$ nor below $y = -1$. If we make use of the periodicity, we can extend the graph as far as we wish in either direction.

5.6. The Graph of $y = \cos x$

The graph of $\{(x, y) \mid y = \cos x\}$, constructed by the methods of the previous article, is shown in Figure 5.3. If the values 0, $\pi/6$, $\pi/4$, $\pi/3$, $\pi/2$,

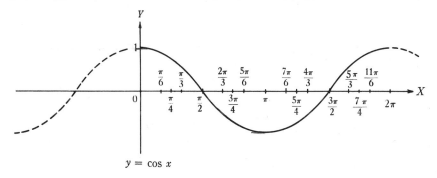

$y = \cos x$

Figure 5.3

$2\pi/3$, and so on up to 2π, are assigned to x, the corresponding values of $\cos x$ are 1, .9, .7, .5, 0, $-.5$, and so forth. By inspecting the table of values in article 5.5, it can be seen that these are the same as the succession of values for sin x starting with $x = \pi/2$ and proceeding to the right. Hence the graph of $y = \cos x$ is the same shape as the sine curve, but it is shifted horizontally to the left a distance $\pi/2$. This can also be seen by observing that $\sin(x + \pi/2) = \sin x \cos \pi/2 + \cos x \sin \pi/2 = \cos x$.

5.7. Range of Tangent

Since $\tan \theta = y/x$, we have the ratio of two numbers each of which changes as θ varies. If $\theta = 0$, then $y = 0$, and $x = 1$ as in Figure 5.1, and $\tan 0 = 0/1 = 0$. Now, as θ increases to a value very near $\pi/2$ rad, y increases from zero to a number very near to one, and x decreases from one to a number very near zero. Hence, $\tan \theta = y/x$ increases from zero to a very large number. In fact, if we take θ sufficiently near $\pi/2$, we can make $\tan \theta$ larger than any given number. However, if $\theta = \pi/2$, $\tan \theta = \frac{1}{0}$ does not exist. This state of affairs is described in mathematical language by saying that, as the angle increases toward $\pi/2$, $\tan \theta$ approaches infinity. This is expressed symbolically by writing

$$\lim_{\theta \to \pi/2^-} \tan \theta = \infty.$$

If, however, $\pi/2 < \theta < \pi$, x is negative and y is positive; hence, $\tan \theta$ is negative. Furthermore, if θ is taken sufficiently near $\pi/2$, then $\tan \theta$ is not only negative but numerically greater than any preassigned number. Consequently, in terms of symbolical notation, we write

$$\lim_{\theta \to \pi/2^+} \tan \theta = -\infty.$$

If θ is in the second quadrant and approaches π, then $\tan \theta$ approaches zero and becomes zero for $\theta = \pi$.

An argument similar to the one above can be used to show that $\tan \theta$ approaches infinity as θ increases toward $3\pi/2$ and that $\tan \theta$ approaches minus infinity as θ decreases toward $3\pi/2$. Consequently, the variation of $\tan \theta$ is as shown in the following table:

θ varies from	0 to $\pi/2$	$\pi/2$ to π	π to $3\pi/2$	$3\pi/2$ to 2π
$\tan \theta$ varies from	0 to ∞	$-\infty$ to 0	0 to ∞	$-\infty$ to 0

Consequently, *the range of the tangent function is all real numbers*, and we write $-\infty < \tan \theta < \infty$.

5.8.   The Graph of $y = \tan x$

In order to obtain a rapid sketch of $\{(x, y)|\ y = \tan x\}$, we shall assign the values shown in the following table to x and then find the corresponding values of $\tan x$.

x	$-\pi/2$	$-\pi/3$	$-\pi/4$	$-\pi/6$	0	$\pi/6$	$\pi/4$	$\pi/3$	$\pi/2$
$\tan x$	$-\infty$	-1.7	-1	$-.6$	0	.6	1	1.7	∞

The symbol ∞ entered below $\pi/2$ indicates that $\tan \pi/2$ does not exist, but that the nearer x approaches $\pi/2$ from the left, the greater the value of $\tan x$ becomes. Similarly, the symbol $-\infty$ under $-\pi/2$ indicates that $\tan x$ is negative, and that the nearer x approaches $-\pi/2$ from the right, the greater the numerical value of $\tan x$ becomes.

Hence, to construct the part of the graph for the values shown in the above table, we plot the points representing the pairs of corresponding values in the table from $x = -\pi/3$ to $x = \pi/3$, then we connect these points with a smooth curve. Finally, we extend the curve upward and to the right, gradually approaching the vertical line through $x = \pi/2$, and downward to the left, approaching the vertical line through $x = -\pi/2$. In this way the solid curve in Figure 5.4 is obtained.

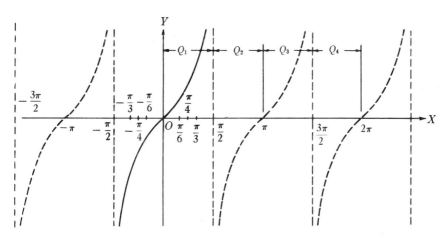

Figure 5.4

Since the tangent of an angle is periodic with π as the period, the branch of the curve that passes through the point $x = \pi$, $y = 0$ is exactly the same shape as the one through $x = 0$, $y = 0$. Furthermore, because of the periodicity, there is a reproduction of this branch in every strip of the plane bounded by vertical lines through $x = \frac{1}{2}(2n - 1)(\pi)$ and $x = \frac{1}{2}(2n + 1)(\pi)$.

From Figure 5.4 we can readily see that, as the angle x moves through the various quadrants, tan x varies as indicated below.

x	$0 \to \pi/2$	$\pi/2 \to \pi$	$\pi \to 3\pi/2$	$3\pi/2 \to 2\pi$
tan θ	$0 \to \infty$	$-\infty \to 0$	$0 \to \infty$	$-\infty \to 0$

5.9. Graphs of the Other Trigonometric Functions

The graph of the function defined by $y = \sec x$ may be obtained by use of the methods discussed in articles 5.5, 5.6, and 5.8. The following table of values is sufficient for sketching the portion of the curve shown in two

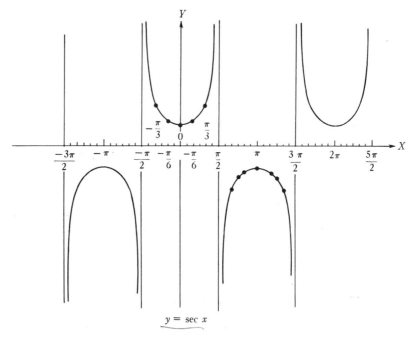

$y = \sec x$

Figure 5.5

parts in Figure 5.5 between the lines $x = -\pi/2$ and $x = 3\pi/2$. This pattern will be reproduced in each strip of the plane between

$$x = \frac{(4n-1)\pi}{2} \quad \text{and} \quad x = \frac{(4n+3)\pi}{2}$$

since the secant is periodic with 2π as a period.

x	$-\pi/2$	$-\pi/3$	$-\pi/4$	$-\pi/6$	0	$\pi/6$	$\pi/4$	$\pi/3$	$\pi/2$	$2\pi/3$
y	∞	2	$\sqrt{2}$	$(2\sqrt{3})/3$	1	$(2\sqrt{3})/3$	$\sqrt{2}$	2	∞	-2

x	$3\pi/4$	$5\pi/6$	π	$7\pi/6$	$5\pi/4$	$4\pi/3$	$3\pi/2$
y	$-\sqrt{2}$	$2\sqrt{3}/(-3)$	-1	$2\sqrt{3}/(-3)$	$-\sqrt{2}$	-2	$-\infty$

The graphs of the function defined by $y = \csc x$ and $y = \cot x$ can be obtained in a similar manner and are shown in Figures 5.6 and 5.7, respectively.

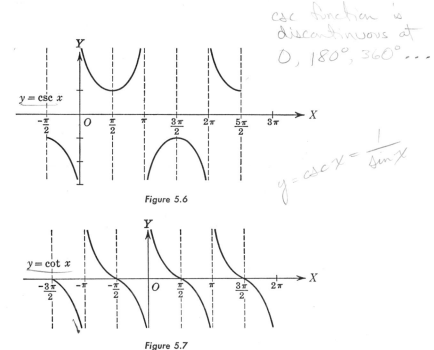

csc function is discontinuous at 0, 180°, 360° ...

$y = \csc x = \frac{1}{\sin x}$

Figure 5.6

Figure 5.7

EXERCISE 5.1

Make a table similar to the one in article 5.5 to show the variation of the function value for the specified interval in each of problems 1 to 8.

1. $\sin x, \ -\pi \leq x \leq \pi$ 2. $\cos x, \ -\pi/2 \leq x \leq 3\pi/2$

3. $\tan x, \ -3\pi/4 \leq x \leq \pi/4$ 4. $\cot x, \ -\pi/3 \leq x \leq 2\pi/3$

5. $\sec x, \ -\pi/2 \leq x \leq 3\pi/2$ 6. $\csc x, \ -\pi \leq x \leq \pi$

7. $\cos x, \ -2\pi \leq x \leq 0$ 8. $\sin x, \ -3\pi/2 \leq x \leq \pi/2$

Make a table of corresponding values of the angle and function value in each of problems 9 to 20 and then sketch the graph of the function defined by the equation.

9. $y = \cos x, \ -\pi/2 \leq x \leq 3\pi/2$ 10. $y = \sin x, \ -\pi \leq x \leq \pi$

11. $y = \sin x, \ -3\pi/2 \leq x \leq \pi/2$ 12. $y = \cos x, \ \pi/2 \leq x \leq 5\pi/2$

13. $y = \tan x, \ \pi/2 \leq x \leq 3\pi/2$ 14. $y = \cot x, \ -3\pi/2 \leq x \leq \pi/2$

15. $y = \cot x, \ 0 \leq x \leq \pi$ 16. $y = \tan x, \ -\pi \leq x \leq \pi$

17. $y = \csc x, \ -\pi/2 \leq x \leq 3\pi/2$ 18. $y = \sec x, \ 0 \leq x \leq 2\pi$

19. $y = \sec x, \ -3\pi/2 \leq x \leq 3\pi/2$ 20. $y = \csc x, \ \pi/2 \leq x \leq 5\pi/2$

5.10. Graphs of Functions of the Type
$\{(x, y) \mid y = af(bx + c)\}$

The graphs of the trigonometric functions are used in physics and many branches of engineering as well as in mathematics. We saw in articles 5.5, 5.6, 5.8, and 5.9 how to sketch the graph of $\{(x, y) \mid y = f(x)\}$ where $f(x)$ is a trigonometric function value. In this article, we shall consider the graph of the function determined by $y = af(bx + c)$, where a, b, and c are constants and f is a trigonometric function.

The function $\{(x, y) \mid y = f(bx)\}$. If $y_1 = f(x_1)$ and $y_2 = f(bx)$, and, in the latter, we replace x by x_1/b, then

$$y_2 = f(bx_1/b) = f(x_1) = y_1.$$

Consequently, if (x, y) is a point on the graph of $y = f(x)$, then $(x/b, y)$ is a point on the graph of $y = f(bx)$.

If we make use of the fact that $b(x + 2\pi/b) = bx + 2\pi$, we see that x must change by $2\pi/b$ in order for bx to change by 2π. Consequently, the period of a trigonometric function value of bx is $1/b$ times that of the same function value of x. Thus, we know that *multiplying the angle or number by a constant divides the period by that constant*. Therefore, the period of $\sin 3x$ is $2\pi/3$, and the period of $\tan 5x$ is $\pi/5$.

The function $\{(x, y) \mid y = f(bx + c)\}$. In order to investigate the effect on the graph of adding a constant to a multiple of an angle or a number, we shall compare the values of x for which $bx + c$ and bx have the same value. If $bx + c = x_1$, then $x = (x_1 - c)/b$; and if $bx = x_1$, then $x = x_1/b$. These values differ by c/b. Consequently, *the graph of a trigonometric function of $bx + c$ is c/b units to the left of that of the same function of bx.* Thus, the graph of $y = \cos (2x + 7)$ is $7/2$ units to the left of the graph of $y = \cos 2x$, and the graph of $y = \tan (3x - \pi)$ is $-\pi/3$ units to the left of that of $y = \tan 3x$. Instead of saying $-\pi/3$ units to the left, some readers may prefer to say $\pi/3$ units to the right.

Some engineers and physicists would say the graphs of the functions determined by $y = f(bx)$ and $y = f(bx + c)$, f being a trigonometric function, differ in phase by c/b.

The function $\{(x, y) \mid y = af(bx + c)\}$. If $y_1 = f(bx + c)$ and $y_2 = af(bx + c)$, then $y_2 = ay_1$. Therefore, the ordinate of each point on $y_2 = af(bx + c)$ is a times the ordinate on the graph of $y_1 = f(bx + c)$, if the same value of x is used in both. Thus, *multiplying a trigonometric function value by a constant multiplies the ordinate of each point on the graph by that constant.* Consequently, we know that multiplying a trigonometric function value by a constant multiplies the amplitude by that constant. Accordingly, for any given value of x, the ordinate of the graph of $y = 5 \cos (2x + \pi)$ is five times the ordinate of $y = \cos (2x + \pi)$.

EXAMPLE 1

Find the period and amplitude of $y = 2 \cos (3x + 5)$ and the position of the graph relative to that of $y = 2 \cos 3x$.

Solution: In arriving at the solution to this problem, we make use of the fact that the period and amplitude of $y = \cos x$ are 2π and 1, respectively, along with the italicized statements in this article. Therefore, the period of $y = 2 \cos (3x + 5)$ is $2\pi/3$ since the angle x is multiplied by 3; the amplitude is 2 since the function value is multiplied by 2; finally, the graph is $5/3$ of a unit to the left of that of $y = 2 \cos 3x$ since 5 is added to the angle $3x$.

EXAMPLE 2

Sketch the graphs of $y = \sin x$, $y = \sin 2x$, $y = \sin (2x + \pi)$, and $y = 3 \sin (2x + \pi)$ about the same axes.

Solution: The first of the graphs called for is the one given in Figure 5.2 and the part of it for $0 \leq x \leq 2\pi$ is repeated in Figure 5.8. The graph of $y = \sin 2x$ is obtained from the reproduced portion by dividing the period by 2. Then the graph of $y = \sin (2x + \pi)$ is obtained by moving the graph of $y = \sin 2x$ to the left $\pi/2$ units. Finally, the graph of $y = 3 \sin (2x + \pi)$ is obtained from that of $y = \sin (2x + \pi)$ by multiplying each ordinate by 3. All four graphs are shown in Figure 5.8.

period of $\sin 5x = \dfrac{360°}{5}$

$= 72°$

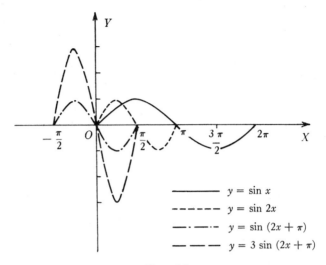

Figure 5.8

5.11.   Composition of Ordinates

If a function is made up of the sum of two or more simpler functions with the same or overlapping domains, it is frequently desirable to get the graph by sketching the graphs of the simpler functions and then adding ordinates, as illustrated in the following example:

EXAMPLE

Sketch the graph of $y = .2x + \cos x$.

Solution: We shall begin by sketching the graph of $y_1 = .2x$ and of $y_2 = \cos x$ for $-\pi \leq x \leq 3\pi$, as shown in Figure 5.9. After this is done, we can obtain as many points as we wish on the graph of $y = .2x + \cos x$ by assigning any desired value to x, computing the corresponding values of y_1 and y_2, and plotting $(x, y_1 + y_2)$. Several such points were used in drawing the sketch in Figure 5.9. For example, if $x = OA$, then $y_1 = AB$, $y_2 = AC$, and $y = AC + AB = AC + CD = AD$.

EXERCISE 5.2

Find the period and the amplitude of the function defined by the equation in each of problems 1 to 16.

1. $2 \cos 3x$	2. $3 \tan 2x$	3. $4 \cot 5x$	4. $3 \sec 2x$
5. $2 \csc 5x$	6. $5 \sin 6x$	7. $2 \tan \frac{1}{2}x$	8. $7 \cot \frac{1}{4}x$
9. $6 \sec \frac{2}{3}x$	10. $4 \csc 1.5x$	11. $4 \sin 3.6x$	12. $3 \cos 2.4x$
13. $6 \sin \pi x$	14. $5 \cos 2\pi x$	15. $2 \cot \frac{\pi}{2}x$	16. $5 \tan \frac{\pi}{3}x$

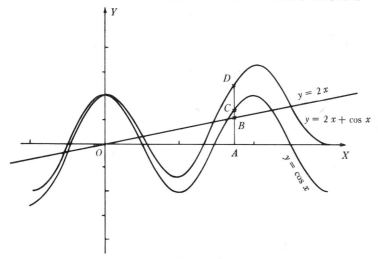

Figure 5.9

Make use of the effect on the period of multiplying the angle by a constant in order to find the value of a in each of problems 17 to 24 so that the period is the number given after the comma.

17. $\tan ax,\ \pi/2$ **18.** $\cot ax,\ \pi/3$ **19.** $\sec ax,\ \pi/2$ **20.** $\csc ax,\ \pi/3$

21. $\sin ax,\ \pi$ **22.** $\cos ax,\ 2\pi$ **23.** $\cos ax,\ 3\pi$ **24.** $\sin ax,\ 4\pi$

Find the period and amplitude of the function defined by the first equation in each of problems 25 to 32 and the position of its graph relative to that of the function defined by the second equation.

25. $y = 2 \sin (3x + 6),\ y = 2 \sin 3x$ **26.** $y = 3 \cos (4x + 12),\ y = 3 \cos 4x$

27. $y = 4 \tan (2x - 3),\ y = 4 \tan 2x$ **28.** $y = 8 \cot (5x - 6),\ y = 8 \cot 5x$

29. $y = 2 \sec (2x - \pi),\ y = 2 \sec 2x$ **30.** $y = 2 \csc (5x - 3\pi),\ y = 2 \csc 5x$

31. $y = 5 \sin (\tfrac{2}{3}x + 3\pi),\ y = 5 \sin \tfrac{2}{3}x$ **32.** $y = 4 \cos (\tfrac{3}{4}x + 5\pi),\ y = 4 \cos \tfrac{3}{4}x$

In each of problems 33 to 40, sketch the graphs of the four functions determined by the given equations about the same axes by use of the facts developed in article 5.10.

33. $y = \cos x,\ y = \cos 2x,\ y = \cos (2x + \pi),\ y = 3 \cos (2x + \pi)$

34. $y = \tan x,\ y = \tan 3x,\ y = \tan (3x + 2\pi),\ y = 2 \tan (3x + 2\pi)$

35. $y = \cot x,\ y = \cot 2x,\ y = \cot (2x - \pi),\ y = 4 \cot (2x - \pi)$

36. $y = \sec x,\ y = \sec 4x,\ y = \sec (4x - 2\pi),\ y = 3 \sec (4x - 2\pi)$

37. $y = \csc x,\ y = \csc \tfrac{1}{2}x,\ y = \csc (\tfrac{1}{2}x - \tfrac{1}{2}),\ y = 2 \csc (\tfrac{1}{2}x - \tfrac{1}{2})$

38. $y = \sin x,\ y = \sin \tfrac{1}{3}x,\ y = \sin (\tfrac{1}{3}x - \tfrac{2}{3}),\ y = 4 \sin (\tfrac{1}{3}x - \tfrac{2}{3})$

39. $y = \sin x,\ y = \sin \tfrac{1}{2}x,\ y = \sin (\tfrac{1}{2}x + 1),\ y = 3 \sin (\tfrac{1}{2}x + 1)$

40. $y = \cos x,\ y = \cos \tfrac{2}{3}x,\ y = \cos (\tfrac{2}{3}x + \tfrac{4}{3}),\ y = 2 \cos (\tfrac{2}{3}x + \tfrac{4}{3})$

Sketch the graph of each of the following by composition of ordinates.

41. $y = x - \sin x$ **42.** $y = x + \cos x$

43. $y = \tan x + x$ **44.** $y = \cot x - x$

45. $y = \sin x + \csc x$ **46.** $y = \sec x - \cos x$

47. $y = \cot x - \sec x$ **48.** $y = \tan x + \csc x$

6

LOGARITHMS

6.1. Introduction

Logarithms were invented in the seventeenth century by Sir John Napier. They are used in arithmetic computation, in the applications of mathematics in chemistry, physics, celestial navigations, and engineering, and are essential for some parts of advanced mathematics. In this chapter, we shall define the logarithm of a number to a base, develop several of the properties of logarithms, and show how to use them in computation and in solving certain types of equations. The definition and theory of logarithms are based on the laws of exponents, and the reader is advised to re-familiarize himself with that topic.

6.2. Laws of Exponents

In the equation

$$b^L = N, \tag{1}$$

the number b is called the *base*, L is the exponent, and N is the Lth power of b. We shall require that b be positive and not equal to 1.

In equation (1) the number L may be an integer, a number of the form q/p where p and q are positive integers, or an irrational number. Furthermore, L may be positive, zero, or negative.

If L is a positive integer, then equation (1) means that

$$b \times b \times b \times \cdots \text{ to } L \text{ factors} = N.$$

If L is of the form q/p, then $b^L = b^{q/p}$ is defined in algebra to mean $\sqrt[p]{b^q}$, and the value of b^L can be computed. If L is an irrational number, the meaning of b^L is defined in more advanced mathematics, and methods are presented for computing the approximate value of b^L if definite values are assigned to b and L.

If $L = 0$, then we have $b^L = b^0$, and b^0 is defined to be equal to 1. If $L = -n$, $n > 0$, then, by definition, $b^L = b^{-n} = 1/b^n$.

Consequently, if in equation (1), L is a real number, a unique value of N exists for any value assigned to b.

We shall conclude this article by repeating three of the above definitions, and by stating five laws of exponents.

$$b^0 = 1 \qquad b \neq 0. \tag{2}$$

$$b^{q/p} = \sqrt[p]{b^q} \qquad \textbf{\textit{p and q positive integers, b > 0.}} \tag{3}$$

$$b^{-n} = \frac{1}{b^n} \qquad b \neq 0. \tag{4}$$

$$b^m b^n = b^{m+n}. \tag{5}$$

$$\frac{b^m}{b^n} = b^{m-n} \qquad b \neq 0. \tag{6}$$

$$(ab)^n = a^n b^n. \tag{7}$$

$$\left(\frac{a}{b}\right)^n = \frac{a^n}{b^n} \qquad b \neq 0. \tag{8}$$

$$(b^m)^n = b^{mn}. \tag{9}$$

6.3.   Definition of a Logarithm

In equation (1) of article 6.2, we say that L is the *exponent* of b. The number L, however, also is related to the number N, and this relationship is described by saying that L *is the logarithm of N to the base b.* Thus, if $b^L = N$, we say that the *logarithm* to the base b of N is L. We state this formally by saying that *the* **logarithm** *to the base b of a positive number N is the exponent that indicates the power to which b must be raised in order to obtain N.* The phrase *the logarithm to the base b of N* is abbreviated by writing $\log_b N$.

We may now write the above definition more concisely in the form

$$\log_b N = L \quad \text{if, and only if,} \quad b^L = N. \tag{6.1}$$

EXAMPLES

1. $\log_3 9 = 2$, since $3^2 = 9$.
2. $\log_2 8 = 3$, since $2^3 = 8$.
3. $\log_9 3 = \frac{1}{2}$, since $9^{1/2} = 3$.
4. $\log_8 4 = \frac{2}{3}$, since $8^{2/3} = (8^{1/3})^2 = 4$.

If the values of any two of the three letters b, L, and N in (6.1) are known, the third often can be found by inspection.

EXAMPLE 5

Find N if $\log_3 N = 4$.

Solution: By (6.1), $\log_3 N = 4$ means that $3^4 = N$. Hence, $N = 81$.

EXAMPLE 6

Find L if $\log_8 64 = L$.

Solution: By (6.1),

$$8^L = 64.$$

Hence, since

$$8^2 = 64,$$
$$8^L = 8^2,$$

and

$$L = 2.$$

EXAMPLE 7

Find b if $\log_b 125 = 3$.

Solution: By (6.1),

$$b^3 = 125.$$

Hence,

$$b = \sqrt[3]{125}$$
$$= 5.$$

EXERCISE 6.1

Express the equation in each of problems 1 to 8 in logarithmic form and in each of problems 9 to 16 in exponential form.

1. $8^2 = 64$ **2.** $4^3 = 64$ **3.** $2^6 = 64$ **4.** $64^1 = 64$

5. $25^{1/2} = 5$ **6.** $16^{1/4} = 2$ **7.** $81^{3/4} = 27$ **8.** $27^{2/3} = 9$

9. $\log_6 36 = 2$ **10.** $\log_3 81 = 4$

11. $\log_4 4096 = 6$ **12.** $\log_7 343 = 3$

13. $\log_{25} 125 = 3/2$ **14.** $\log_8 4 = 2/3$

15. $\log_{1296} 36 = 1/2$ **16.** $\log_{27} 81 = 4/3$

Find the value of the logarithm to the specified base of the given number in each of problems 17 to 32.

17. $\log_{10} 100$ **18.** $\log_6 216$ **19.** $\log_2 32$ **20.** $\log_3 81$

21. $\log_7 \frac{1}{7}$ **22.** $\log_2 \frac{1}{8}$ **23.** $\log_3 \frac{1}{27}$ **24.** $\log_6 \frac{1}{1296}$

25. $\log_8 32$ **26.** $\log_{32} 8$ **27.** $\log_9 27$ **28.** $\log_{27} 243$

29. $\log_{1/2} \frac{1}{4}$ **30.** $\log_{2/3} \frac{4}{9}$ **31.** $\log_{3/4} \frac{27}{64}$ **32.** $\log_{5/2} \frac{625}{16}$

In each of problems 33 to 68, find the replacement for N, b, or L so that the equation is a true statement.

33. $\log_2 N = 3$ **34.** $\log_3 N = -2$ **35.** $\log_5 N = -3$

36. $\log_7 N = 4$ **37.** $\log_8 N = 2/3$ **38.** $\log_{32} N = 3/5$

39. $\log_{27} N = 2/3$ **40.** $\log_{81} N = 5/4$ **41.** $\log_{1/2} N = -2$

42. $\log_{1/3} N = 3$ **43.** $\log_{2/5} N = 3$ **44.** $\log_{3/2} N = -4$

45. $\log_b 36 = 2$ **46.** $\log_b 8 = 3$ **47.** $\log_b 27 = 3$

48. $\log_b 64 = 6$ **49.** $\log_b 36 = -2$ **50.** $\log_b 8 = -3$

51. $\log_b 27 = -3$ **52.** $\log_b 64 = -6$ **53.** $\log_b \frac{4}{9} = 2$

54. $\log_b \frac{27}{125} = 3$ **55.** $\log_b \frac{125}{64} = -3$ **56.** $\log_b \frac{16}{81} = -4$

57. $\log_4 64 = L$ **58.** $\log_3 81 = L$ **59.** $\log_2 32 = L$

60. $\log_5 625 = L$ **61.** $\log_{16} 8 = L$ **62.** $\log_{27} 9 = L$

63. $\log_{25} 125 = L$ **64.** $\log_{36} 216 = L$ **65.** $\log_8 \frac{1}{64} = L$

66. $\log_2 \frac{1}{32} = L$ **67.** $\log_6 \frac{1}{216} = L$ **68.** $\log_5 \frac{1}{25} = L$

6.4.   Properties of Logarithms

Many of the useful properties of logarithms may be derived by an application of the laws of exponents. For example, by (5), article 6.2, the product of the two numbers

$$M = b^x \qquad \text{and} \qquad N = b^y$$

is

$$MN = b^x b^y = b^{x+y}.$$

Therefore,

$$\log_b MN = \log_b b^{x+y} = x + y,$$

since $x + y$ is the exponent that indicates the power to which b must be raised in order to obtain b^{x+y}. However, by the definition of a logarithm,

$$x = \log_b M \qquad \text{and} \qquad y = \log_b N.$$

Hence,

$$\log_b MN = \log_b M + \log_b N, \qquad\qquad (6.2)$$

and we have the following theorem:

THEOREM 1

The logarithm of the product of two numbers is equal to the sum of their logarithms.

Let us again consider the numbers

$$M = b^x \qquad \text{and} \qquad N = b^y.$$

Their quotient is

$$\frac{M}{N} = \frac{b^x}{b^y} = b^{x-y} \qquad \text{by (6), article 6.2.}$$

Hence,

$$\log_b \frac{M}{N} = x - y,$$

and therefore,

$$\log_b \frac{M}{N} = \log_b M - \log_b N. \qquad\qquad (6.3)$$

Consequently, we have the theorem below:

THEOREM 2

The logarithm of a quotient is the logarithm of the dividend minus the logarithm of the divisor.

If $N = b^x$, then

$$N^p = (b^x)^p = b^{px} \qquad \text{by (9), article 6.2,}$$

and

$$\log_b N^p = \log_b b^{px}$$
$$= px \qquad \text{by (6.1), article 6.3.}$$

Hence,

$$\log_b N^p = p \log_b N, \tag{6.4}$$

since x is the exponent that indicates the power to which b must be raised in order to obtain N. Thus we have the following theorem:

THEOREM 3

The logarithm of a power of a positive number is the exponent of the power multiplied by the logarithm of the number.

By (3), article 6.2,

$$\sqrt[q]{N^p} = N^{p/q} \qquad N > 0.$$

Hence, in order to obtain the logarithm of a root of a power of a positive number, we need only to change from the radical to the exponential form and apply the theorem above.

6.5. The Common, or Briggs, System of Logarithms

We stated in article 6.1 that logarithms are used in computation as well as in theoretical work. The remainder of this chapter will be devoted to a study of how to use logarithms in computation.

The set of logarithms obtained for a specified base is called a *system of logarithms*. The base used for most computational work is 10, and the system, when 10 is the base, is called *the Common, or Briggs, system*. It is customary to omit the base in using the common logarithm of a number. Thus log N is used to indicate the common logarithm of the number N.

If c is any real number, then 10^c is positive; hence, the common logarithm of zero and a negative number do not exist as real numbers.

6.6. Characteristic and Mantissa

At this time, we shall review the concept of scientific notation discussed in article 2.2. If a positive number is in scientific notation, it is expressed as the product of an integral power of 10 and a number greater than or equal to 1 and less than 10. Thus,

$$N = N'10^m,$$

when $1 \leq N' < 10$ and m is an integer. If all digits in N are significant or are used for placing the decimal point, the value of m is equal numerically to the number of digits over which the decimal point in N was shifted in order to obtain N'. Furthermore, m is positive or negative according as the decimal point was shifted to the left or right.

The position immediately to the right of the first nonzero digit in N is called the *reference position* of the decimal point. Thus, the decimal point is in reference position in $N = 3.501$.

We can now say that in expressing N in scientific notation, the value of N' is obtained by shifting the decimal point in $N = N'10^m$ to reference position. Also, m is numerically equal to the number of digits between the decimal point and reference position, and is positive or negative according as the decimal point in N is to the right or left of reference position.

If we take the common logarithm of each member of $N = N'10^m$, we obtain

$$\begin{aligned}
\log N &= \log N'10^m \\
&= \log N' + \log 10^m \qquad \text{by (6.2)} \\
&= \log N' + m \qquad\quad\ \text{since } \log 10^m = m.
\end{aligned}$$

Therefore, by the commutative axiom, we have

$$\log N = m + \log N'. \tag{1}$$

Since $1 \leq N' < 10$, it follows that $\log_{10} 1 \leq \log_{10} N' < \log_{10} 10$. Also, since $10^0 = 1$ and $10^1 = 10$, we have $0 \leq \log_{10} N < 1$. Therefore $\log_{10} N'$ can be expressed as a non-negative decimal fraction. Furthermore, since m is an integer, $\log_{10} N$ is equal to an integer plus a non-negative fraction. The integer is called the *characteristic* of the logarithm, and the non-negative decimal fraction is the *mantissa*.

If we refer to (1) in the preceding explanation of the method for obtaining the value of m, we see that the following rule for obtaining the characteristic of the logarithm of a number applies:

> *The characteristic of the common logarithm of the number N is numerically equal to the number of digits between the decimal point and the reference position. The characteristic is positive or negative according as the decimal point is to the right or to the left of the reference position.*

Thus, the characteristic of log 86.3 is 1, since the decimal point is one place to the right of the reference position. Furthermore, the characteristic of log .0863 is -2, since the decimal point is two places to the left of the reference position.

If the mantissa of log 86.3 is .93601, then log 86.3 = 1.93601, and $86.3 = 10^{1.93601}$.

Furthermore,

$$863 = 86.3 \times 10 = 10^{1.93601} \times 10^1 = 10^{2.93601},$$
$$8630 = 86.3 \times 100 = 10^{1.93601} \times 10^2 = 10^{3.93601},$$

and

$$8.63 = 86.3 \div 10 = 10^{1.93601} \div 10^1 = 10^{.93601}.$$

Hence, we have

$$\log 86.3 = 1.93601,$$
$$\log 863 = 2.93601,$$
$$\log 8630 = 3.93601,$$

and

$$\log 8.63 = 0.93601.$$

The above example illustrates the fact that shifting the decimal point to the right or left in a number affects only the characteristic of the logarithm. Furthermore, when the characteristic is positive or zero, the logarithm can be written as a single number. However, in the case of 0.863, we have $0.863 = 86.3 \div 100 = 10^{1.93601} \div 10^2 = 10^{-1+.93601}$. Hence, log 0.863 = $-1 + .93601$. The only way to express this logarithm as a single number is to perform the addition indicated by $-1 + .93601$, obtaining -0.06399. However, in this form, the fractional part of the logarithm is negative and is therefore not a mantissa. We avoid this difficulty by writing the logarithm $-1 + .93601$ in the form $9.93601 - 10$. This illustrates the general rule, *if the characteristic of the logarithm of a number is $-c$, we write $-c$ in the form $n - 10$ and then write the mantissa, preceded by a decimal point, to the right of n.*

6.7. Use of Tables to Obtain the Mantissa

In order to obtain the mantissa of the logarithm of a four-digit number, we look in the column of Table I which is headed by N until we find the first three digits of the number. In line with this, and in the column headed by the fourth digit of the number, we find the desired mantissa. For example, in order to find the mantissa of the logarithm of 1762, we first find 176 in the column headed by N. In line with this, and in the column headed by 2, we find 601. We must use the 24 which is opposite 174 and under 0. This number is printed only once, but it is needed with every mantissa until an asterisk (*) or 25 appears. We must supply a decimal point to the left of each mantissa; hence, the mantissa of log 1762 is .24601.

EXERCISE 6.2

Apply the computation theorems to determine the logarithm of the number in each of problems 1 to 12.

1. ab	**2.** bcd	**3.** $5pq$	**4.** $amrt$
5. a/b	**6.** bc/d	**7.** $5/pq$	**8.** am/rt
9. a^b	**10.** $(bc)^d$	**11.** 5^{pq}	**12.** $(am/r)^t$

Make use of the computation theorems and the fact that log 2 = .30, log 3 = .48, and log 7 = .85 to find the common logarithm of the number or combination of numbers in each of problems 13 to 24.

13. 6	**14.** 21	**15.** 14	**16.** 42
17. 7/3	**18.** 14/3	**19.** 7/6	**20.** 21/2
21. 2^4	**22.** 3^2	**23.** $2^3 3^2$	**24.** $7^2/2^5$

Determine the characteristic of the logarithm of the number in each of problems 25 to 40.

25. 34	**26.** 507	**27.** 2714	**28.** 34672
29. 48.2	**30.** 719.3	**31.** 2.61	**32.** 353.8
33. .015	**34.** .105	**35.** .003	**36.** .265
37. .0029	**38.** .00027	**39.** .0259	**40.** .0016

Find the common logarithm of the number in each of problems 41 to 68.

41. 6	**42.** 7	**43.** .9	**44.** .03
45. .073	**46.** .81	**47.** 3.4	**48.** 63
49. 25.9	**50.** 3.27	**51.** .0879	**52.** .914

53. 814.2	**54.** 59.18	**55.** 6.123	**56.** 7654
57. 2.934	**58.** 776.6	**59.** 2801	**60.** 48.92
61. .5032	**62.** .01257	**63.** .002468	**64.** .0004002
65. .05174	**66.** .007007	**67.** .0001101	**68.** .8527

6.8. Interpolation

If there are more than four digits in a number, the mantissa of the logarithm of the number cannot be found directly from Table I. It is possible, however, to find the mantissa of the logarithm of such a number from Table I by resorting to linear interpolation. We shall omit all discussion of interpolation and merely give an example, since interpolation was presented in article 2.7 and the procedure here is the same.

EXAMPLE

Find log 27548 by use of interpolation.

Solution: Since $27540 < 27548 < 27550$, it follows that $\log 27540 < \log 27548 < \log 27550$; furthermore, the mantissa of log 2754 is equal to the mantissa of log 27540 and manlog 2755 = manlog 27550. We shall now give the usual diagram.

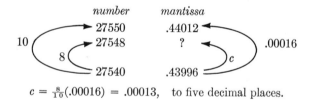

$c = \frac{8}{10}(.00016) = .00013$,   to five decimal places.

Consequently, the mantissa of log 27548 is $.43996 + .00013 = .44009$. Therefore, log 27548 = 4.44009 since the decimal point is four places to the right of reference position.

If a number contains more than five digits, we round it off to five places and then obtain the logarithm of the resulting number by interpolating as above.

6.9. Given log N: To Find N

The same general method of procedure may be used here as was used in article 2.6. For example, if we are given log $N = 1.90639$ and wish to find N, we look in the body of Table I until we find 90639. It is in the same line as 806 and in the same column as 1; hence, the sequence of digits is 8061.

The position of the decimal point in N may be determined by means of the rule for determining the characteristic in article 6.6. The decimal point is one place to the right of the reference position, since the characteristic of the logarithm of the number is 1. Hence, $N = 80.61$.

The value of log N given above is one of the entries in the table. However, it is often the case that the given mantissa is not listed. In such cases we must again resort to interpolation.

EXAMPLE

Find the value of N if $\log N = 8.76375 - 10$.

Solution: The mantissa is not listed in the table; hence, we find the two entries that are nearer 76375 than any others. These entries and the corresponding sequences of digits are shown below.

$$c = \frac{.00002}{.00007} \text{ of } 1 = \frac{2}{7} \text{ of } 1 = .3 \text{ correct to one significant figure}$$

$5804 + .3 = 5804.3$

Hence $.76375 =$ mantissa of log 58043

The characteristic of log N is -2. Hence, the decimal point is two places to the left of the reference position, and $N = .058043$.

EXERCISE 6.3

Find the logarithm of the number in each of problems 1 to 12.

1. 72.361   **2.** 7.6206   **3.** 13131   **4.** 453.17

5. 2.7438   **6.** 62.971   **7.** 8032.9   **8.** 18326

9. 123.456   **10.** .423455   **11.** 3.45723   **12.** 468.134

Find the value of N if $\log N$ is the number in each of problems 13 to 32. Express N in scientific notation in each of problems 21 to 32.

13. 0.21272   **14.** 1.57807   **15.** 3.93922

16. 2.64326   **17.** 8.35122 − 10   **18.** 9.99996 − 10

19. 1.76380   **20.** 3.94905   **21.** 6.86864

22. 5.90020   **23.** 6.52061   **24.** 8.91892

25. 6.81351   **26.** 5.28240   **27.** 5.70200   **28.** 6.44154

29. 7.69020   **30.** 7.83544   **31.** 7.91566   **32.** 6.95119

If log N is the number given in each of problems 33 to 44, find the value of N to four digits by using the number that corresponds to the mantissa in the table that is nearest the one given in the problem.

33. 1.47213	**34.** 3.28618	**35.** 0.37605
36. 2.53376	**37.** 1.62384	**38.** 2.80372
39. 3.77809	**40.** 2.57883	**41.** 8.71536 − 10
42. 7.91918 − 10	**43.** 7.85559 − 10	**44.** 9.38752 − 10

By use of interpolation, find the value of N to five digits if log N is the number given in each of problems 45 to 60. Express N in scientific notation in each of problems 57 to 60.

45. 2.47823	**46.** 3.58902	**47.** 1.26262
48. 0.91807	**49.** 8.72359 − 10	**50.** 9.92378 − 10
51. 7.77777 − 10	**52.** 9.80357 − 10	**53.** 6.48332 − 10
54. 5.42293 − 10	**55.** 8.73629 − 10	**56.** 7.73866 − 10

57. 6.38247	**58.** 5.41838	**59.** 6.31148	**60.** 5.66543

6.10. Logarithmic Computation

The following examples will illustrate the method to be used in the application of logarithms to computation:

EXAMPLE 1

By use of logarithms, find the value of N where

$$N = (21.36)(.01245)(4162).$$

Solution: Since N is the product of three numbers, log N is equal to the sum of the logarithms of the three factors. Hence, we shall obtain the logarithm of each of the factors, add them together, and thus have log N. Then we can find N by use of the tables. Before turning to the tables, it is advisable to make an outline of the solution, leaving blanks in which to enter the logarithms as they are found. We suggest the following plan. Note that the characteristic of the logarithm of each number is entered when the number is listed.

$$
\begin{array}{ll}
\log 21.36 = 1. & \\
\log .01245 = 8. & \quad -10 \\
\underline{\log 4162 = 3.} & \\
\log N = \underline{\hspace{4cm}} & \\
\qquad\qquad \text{(enter sum here)} & \\
N = \underline{\hspace{4cm}} &
\end{array}
$$

Now we turn to the tables, get the mantissa of each of the logarithms, and enter it in the proper place in the outline. Next, we perform the addition and then find N from the tables by the method of article 6.9. The completed solution then appears as follows:

$$\begin{aligned}
\log 21.36 &= 1.32960 \\
\log .01245 &= 8.09517 - 10 \\
\underline{\log 4162} &= \underline{3.61930} \\
\log N &= 13.04407 - 10 \text{ (sum)} \\
N &= 1107.
\end{aligned}$$

Note. The value of N to five places obtained by interpolation is 1106.8. However, since each of the numbers in the problem contains only four significant figures, we round the product off to four significant figures. Note also that the characteristic of $\log N$ is $13 - 10 = 3$. Hence, the decimal point in N is three places to the right of the reference position.

In this example, we have written the outline twice in order to show its appearance after each step. In practice, it is necessary to write it only once, since each operation requires the filling of a separate blank.

EXAMPLE 2

By use of logarithms, find the value of

$$N = \frac{(4.120)(.6412)}{(.2671)(3.134)}.$$

Solution: In this problem, N is the quotient of two numbers, each of which is a product. Hence, we find the sum of the logarithms of the two factors of the dividend and also the sum of the logarithms of the two factors of the divisor. Then we subtract the latter sum from the former and obtain $\log N$. We suggest the following outline for the solution:

$$\begin{aligned}
\log 4.120 &= 0.61490 \\
\underline{\log .6412} &= \underline{9.80699 - 10} \\
\log \text{dividend} &= \qquad\qquad 10.42189 - 10 \text{ (sum)} \\
\log .2671 &= 9.42667 - 10 \\
\underline{\log 3.134} &= \underline{0.49610} \\
\log \text{divisor} &= \qquad\qquad 9.92277 - 10 \text{ (sum)} \\
\log N &= \qquad\qquad 0.49912 \qquad \text{(difference)} \\
N &= 3.156
\end{aligned}$$

The value of N to five digits by interpolation is 3.1559. However, since the data involves numbers of four digits only, we round the value of N off to four digits.

EXAMPLE 3

By use of logarithms, find the value of $N = \sqrt[3]{.23678}$.

Solution: Since

$$\sqrt[3]{.23678} = (.23678)^{1/3} \qquad \text{by (3), article 6.2,}$$

we have

$$\log N = \tfrac{1}{3}(\log .23678)$$
$$= \tfrac{1}{3}(9.37434 - 10).$$

If we perform the indicated operation in the last line above, we obtain $\log N = 3.12478 - 3.33333$. This is the correct value of $\log N$, but we cannot use the tables to find N when $\log N$ is in this form. We avoid this difficulty by adding and subtracting 20 to the characteristic of $\log .23678$, thus obtaining

$$\log .23678 = 29.37434 - 30.$$

Hence,

$$\log N = \tfrac{1}{3}(29.37434 - 30)$$
$$= 9.79145 - 10.$$

Therefore,

$$N = .61866.$$

EXAMPLE 4

By use of logarithms, find the value of

$$N = \frac{13672}{46891}.$$

Solution: By (6.3), we have

$$\begin{array}{l} \log 13672 = 4.13583 \\ \underline{\log 46891 = 4.67109} \\ \quad\log N = \qquad\quad \text{(difference)} \end{array}$$

where $\log N$ is obtained by subtracting the second logarithm from the first. This operation yields $\log N = -.53526$. However, since this fraction is negative, the value of N cannot be obtained from the tables. We avoid this difficulty by adding $10 - 10$ to the characteristic of the first logarithm, and then we complete the solution as below.

$$\begin{array}{l} \log 13672 = 14.13583 - 10 \\ \underline{\log 46891 = \;\;4.67109} \\ \quad\log N = \;\;9.46474 - 10 \text{ (difference)} \\ \qquad N = \;\;.29157. \end{array}$$

EXAMPLE 5

Find the value of

$$N = \sqrt[3]{\frac{(82.13)\sqrt{1.236}}{(1.621)^2(3.274)}}.$$

Solution: Since each of the numbers in the problem contains four digits, we shall find the value of N to four significant figures. The steps in the solution follow:

$$
\begin{aligned}
\log 82.13 \ \ &= & & 1.91450 \\
\log \sqrt{1.236} &= \tfrac{1}{2}(\log 1.236) = \tfrac{1}{2}(0.09202) = 0.04601 \\
& & \text{log dividend } = & \ 1.96051 \\
\log (1.621)^2 &= 2(\log 1.621) = 2(0.20978) = 0.41956 \\
\log 3.274 \ \ &= & & 0.51508 \\
\hline
& & \text{log divisor } = & \quad .93464 \\
& & & 3\,|\,1.02587 \\
& & \log N = & \quad .34196 \\
& & N = 2.198. &
\end{aligned}
$$

EXERCISE 6.4

By use of the logarithmic computation theorems, perform the indicated operations in problems 1 to 36. Carry each result out to four significant figures.

1. $(274.1)(.0632)$ 2. $(234.7)(1.562)$

3. $(.5678)(392.8)$ 4. $(6.197)(.7118)$

5. $\dfrac{1.982}{87.26}$ 6. $\dfrac{8924}{38.17}$

7. $\dfrac{283.1}{543.0}$ 8. $\dfrac{8.423}{29.67}$

9. $(23.61)(1.263)(.01681)$ 10. $(.3627)(.01324)(4620)$

11. $(.02679)(.001321)(8670)$ 12. $(9.173)(598.6)(87.61)$

13. $\dfrac{(286.3)(3.714)}{67.93}$ 14. $\dfrac{(.06917)(13.62)}{781.9}$

15. $\dfrac{9363}{(406.2)(63.41)}$ 16. $\dfrac{.4936}{(64.63)(.5016)}$

17. $(.06217)^4$ 18. $(3.261)^5$

19. $(.9675)^3$ 20. $(.5741)^8$

21. $\sqrt[3]{24.73}$ 22. $\sqrt[3]{597.3}/\sqrt[4]{33.17}$

23. $\dfrac{\sqrt[3]{875.9}}{28.47}$ 24. $\sqrt[6]{564.8}$

25. $(.007961)^{3/7}$ 26. $(.7215)^{1/7}$

27. $(.5136)^{1/5}$ 28. $(.05728)^{1/8}$

29. $(802.9)\sqrt[3]{7.183}$ 30. $\dfrac{(5.421)^3}{\sqrt[4]{.8174}}$

31. $\dfrac{\sqrt[3]{69.71}}{(3.146)^2}$ 32. $(100.6)^2\sqrt[3]{19.01}$

33. $\sqrt{\dfrac{(62.71)(.5324)^2}{8964}}$ **34.** $\sqrt[3]{\dfrac{\sqrt{.2364}(61.43)^2}{534.2}}$

35. $\sqrt{\dfrac{(13.43)(6.706)^2}{1013}}$ **36.** $\sqrt[4]{\dfrac{\sqrt{(1.663)}(987.6)}{26.72}}$

Obtain the value of the combination of numbers in each of problems 37 to 48 to five figures by use of logarithms.

37. $\dfrac{(.13672)(46.213)}{.92614}$ **38.** $\dfrac{(25.311)(3.9245)}{.76873}$

39. $\dfrac{(36.117)(2.0054)}{.91763}$ **40.** $\dfrac{(43.332)(561.14)}{3.2287}$

41. $\dfrac{\sqrt[3]{847}}{\sqrt[5]{3800}\sqrt{47.8}} = \dfrac{847^{1/3}}{(3800)^{1/5}(47.8)^{1/2}}$ **42.** $(357.26)\sqrt{27.810}$

43. $(84.628)\sqrt[3]{568.30}$ **44.** $(2693.8)\sqrt{76.225}$

45. $(71.963)^{2/5}(\sqrt{7.8634})^5$ **46.** $\dfrac{45.766}{847.63\sqrt[3]{362.50}}$

47. $\dfrac{(68.471)^2\sqrt{68.742}}{86.744}$ **48.** $\dfrac{(711.63)^2}{(5.0135)\sqrt[3]{8.1139}}$

Obtain the value of the combination of numbers in problems 49 to 64 to as many significant digits as are justified under the assumption that the given numbers are approximations. Use logarithms.

49. $(38.234)(2.713)$ **50.** $(2.6173)^2(52.764)$

51. $(931.01)(2.363)$ **52.** $(76.468)(5.426)$

53. $\dfrac{598.34}{43.89}$ **54.** $\dfrac{(.12631)(56.82)}{(763.24)(42.35)}$

55. $\dfrac{7318.2}{1.985}$ **56.** $\dfrac{658.17}{48.28}$

57. $\dfrac{(28.573)\sqrt{627.50}}{384.26}$ **58.** $(92.672)^2(2.7764)$

59. $(1.3087)^2(26.382)$ **60.** $\dfrac{(.127003)^{3/2}}{(1.9437)^{4/3}}$

61. $\sqrt{\dfrac{(216.8)(37.14)}{(382.51)(.76803)}}$ **62.** $\dfrac{\sqrt{(584.73)(.20961)}}{\sqrt[3]{(9475.3)(.37625)}}$

63. $\dfrac{(284.763)^{2/3}}{(548.24)^{3/2}}$ **64.** $\dfrac{(265.951)^{3/4}}{(55.953)^{2/3}}$

$= (\tfrac{1}{2}\log 584.73 + \tfrac{1}{2}\log .2096$
$- (\tfrac{1}{3}\log 9475.3 +$
$\tfrac{1}{3}\log .37625)$

7

LOGARITHMIC SOLUTION
OF TRIANGLES

7.1. Logarithms of the Trigonometric Function Values

The logarithm of a trigonometric function value of an angle can be obtained
by first finding the function value from Table III and then getting its
logarithm by use of Table I. This, however, is not the procedure generally
followed. The work required to obtain the logarithm of a trigonometric
function value is materially lessened by the use of Table II. Each integral
multiple of $1'$ from $0°$ to $90°$ is listed in this table, and the logarithm of the
function value is given for the sine, cosine, tangent, and cotangent of each
angle listed is given. The abbreviation L. Sin x is used for the logarithm
of the sine of the angle x. Corresponding abbreviations are used in the
other cases.

The arrangement of the material is the same as in Table III and the
process of interpolation is carried out here as in all previous cases. On each

page of Table II there is a table of proportionate parts, headed by **P.P.**, and columns, headed by **d** and **c.d.**, that give the significant digits in the difference between consecutive entries in the table. The use of these columns and the table of proportional parts will be explained at the end of this article. If the angle is between 0° and 45°, we use the column heading and the angle at the left. However, if the angle is between 45° and 90°, we must use the function named at the foot of the column and the angles at the right of the page.

In using Table II we must recall that the sine and cosine of an angle are never greater than 1; furthermore, the tangent of any angle between 0° and 45° and the cotangent of any angle between 45° and 90° are each less than 1. Therefore, the logarithmic sine and the logarithmic cosine of any angle, the logarithmic tangent of any angle between 0° and 45°, and the logarithmic cotangent of any angle between 45° and 90° have negative characteristics. A negative characteristic is generally written in the form $n - 10$. The positive integer n is found, in each case, in the table; but the negative 10 must be supplied by the reader. In order that there may be a uniform procedure, the table is so constructed that 10 must be subtracted from each entry.

Hence, in order to find log cos 88°20′, we must look on the right of the page for the angle, and we must use the column with "L. Cos." at its foot, since the angle is greater than 45°. In this column and in line with 88°20′, we find 8.46366. Supplying the necessary −10, we see that log cos 88°20′ = 8.46366 − 10.

By means of interpolation, we can use Table II to obtain the logarithms of function values of angles that involve tenths of a minute. As an example, we shall find log tan 17°34.6′. The angle is less than 45°; hence, we must use the minutes as listed on the left of the page and the column headings at the top. Obviously, log tan 17°34.6′ is between log tan 17°34′ and log tan 17°35′. The entry in the column headed by **c.d.** that is opposite these two values in Table II is 44. Hence these two values differ by .00044. Therefore,

$$\log \tan 17°34.6' = \log \tan 17°34' + .6(.00044)$$
$$= (9.50048 - 10) + .00026$$
$$= 9.50074 - 10.$$

The product .6(.00044) can be obtained from the table of proportional parts by first locating the column headed by 44. Then, in this column in line with 6, we find 26.4. We drop the .4 and then prefix three zeros and a decimal point, thus obtaining .00026.

7.2. Logarithmic Solution of Right Triangles

In article 2.9 we discussed the solution of right triangles, and the only way in which this discussion will differ from the previous one is that we shall use logarithms to facilitate the numerical computation. The essential step in the procedure is to select a trigonometric function value so that two known parts and the desired unknown part of the given triangle are involved.

We shall illustrate the process by solving two illustrative examples.

EXAMPLE 1

Given the right triangle in which $a = 17.02$ and $b = 12.01$, find the value of each of the unknown parts (see Figure 7.1).

Solution: By definition,

$$\tan B = \frac{12.01}{17.02}.$$

Therefore,

$$\log \tan B = \log 12.01 - \log 17.02.$$
$$\log 12.01 = 11.07954 - 10$$
$$\log 17.02 = 1.23096$$
$$\overline{\log \tan B = 9.84858 - 10.}$$

Hence,

$$B = 35°12'.$$

Since angles A and B are complementary, $A = 90° - 35°12' = 54°48'$. In order to calculate the side c, we shall use the sine of A, since this involves two known parts and the desired unknown part.

$$\text{Sin } A = \frac{17.02}{c}.$$

Hence,

$$c = \frac{17.02}{\sin 54°48'},$$

and

$$\log c = \log 17.02 - \log \sin 54°48'.$$
$$\log 17.02 = 11.23096 - 10$$
$$\log \sin 54°48' = 9.91230 - 10$$
$$\overline{\log c = 1.31866.}$$

Therefore,

$$c = 20.83.$$

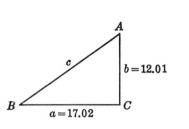

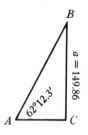

Figure 7.1   Figure 7.2

EXAMPLE 2

Solve the right triangle in which $A = 62°12.3'$ and $a = 149.86$ (see Figure 7.2).

Solution: Since A and B are complementary, we have

$$B = 90° - 62°12.3' = 27°47.7'.$$

By definition,

$$\tan 62°12.3' = \frac{149.86}{b}.$$

Hence,

$$b = \frac{149.86}{\tan 62°12.3'}.$$

The value of b can now be found by use of logarithms as shown below.

$$\log b = \log 149.86 - \log \tan 62°12.3'.$$
$$\log 149.86 = 2.17568$$
$$\underline{\log \tan 62°12.3' = 0.27808}$$
$$\log b = 1.89760.$$

Therefore, by interpolation, we have

$$b = 78.995.$$

In order to find c, we shall use

$$\sin 62°12.3' = \frac{149.86}{c},$$

$$c = \frac{149.86}{\sin 62°12.3'},$$

and, by use of logarithms, we complete the solution as below.

$$\log 149.86 = 12.17568 - 10$$
$$\underline{\log \sin 62°12.3' = \ \ 9.94676 - 10}$$
$$\log c = \ \ 2.22892.$$

Therefore, by interpolation,

$$c = 169.40.$$

It should be noted that the use of interpolation in Tables I and II enables us to obtain the value of a side of a triangle to five significant digits and the value of an angle to the nearest tenth of a minute. Hence, the accuracy of five significant digits in the sides of a triangle corresponds to an accuracy of tenths of a minute in the angles. This information should be used in connection with that given in article 2.7.

7.3. Oblique Triangles

A very common application of trigonometry is in the solution of oblique triangles. The physicist and engineer often meet with problems of this type. We say that a triangle is solved if its sides, angles, and area are known. We can determine the unknown quantities provided we know one side and two other parts, and we shall derive several formulas that are useful for this purpose. Three parts of a triangle including at least one side may be given in the following ways:

Case A.   Two angles and a side.
Case B.   Two sides and the angle opposite one of them.
Case C.   Two sides and the included angle.
Case D.   Three sides.

It should be noted that the data in case A are equivalent to three angles and a side, since the third angle may be obtained by subtracting the sum of the two given angles from 180°. In order to solve for an unknown part of a triangle, we must select a formula that involves three known parts and the desired unknown.

7.4. The Law of Sines

We shall consider a triangle whose angles are A, B, and C, while the sides opposite them are a, b, and c, respectively. We shall drop a perpendicular from either vertex to the opposite side, or to the opposite side produced, and call its foot D and its length h. Now, using either Figure 7.3 or Figure 7.4, we have

$$\sin C = \frac{h}{a}.$$

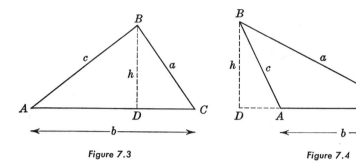

Figure 7.3   Figure 7.4

Hence,

$$h = a \sin C. \tag{1}$$

Furthermore, using Figure 7.3,

$$\sin A = \frac{h}{c},$$

and

$$h = c \sin A. \tag{2}$$

Angle DAB in Figure 7.4 is the related angle of CAB. Consequently, $\sin A$ is numerically equal to $\sin DAB$. They have the same sign because each is positive and less than 180°. Therefore, using Figure 7.4, we also have

$$h = c \sin A$$

since

$$\sin A = \sin DAB$$

$$= \frac{h}{c}.$$

Equating the two forms of the value of h as given by (1) and (2), we have

$$a \sin C = c \sin A.$$

Dividing by $\sin A \sin C$, we have

$$\frac{a}{\sin A} = \frac{c}{\sin C}.$$

In a similar manner, we may prove that

$$\frac{a}{\sin A} = \frac{b}{\sin B}.$$

This relation between the sides of a triangle and the sines of the angles may be expressed conveniently in the form

$$\frac{a}{\sin A} = \frac{b}{\sin B} = \frac{c}{\sin C}. \qquad (7.1)$$

This is a symbolic statement of the law of sines that appears in words below.

LAW OF SINES

In any triangle, the three ratios obtained by dividing a side by the sine of the opposite angle are equal.

The law of sines may be used when three of the quantities involved in any two of the ratios are known, and we use only the two ratios which involve the three known parts.

It should be noticed that any two of the ratios of formula (7.1) involve two sides and the angles opposite them. Hence, we can solve a triangle by means of this formula if we know either

Case A: two angles and a side; or

Case B: two sides and the angle opposite one of them.

7.5. The Area of a Triangle

It is a familiar fact of plane geometry that the area of a triangle is equal to one-half the product of the base and the altitude. The altitude in Figures 7.3 and 7.4 is h, and the base is b. Hence, one expression for the area K of a triangle is

$$K = \tfrac{1}{2}bh = \tfrac{1}{2}ab \sin C. \qquad (7.2)$$

Since the particular lettering of the triangle is immaterial, we have

THEOREM 1

The area of a triangle is equal to one-half the product of any two of its sides and the sine of the included angle.

Solving the equation composed of the first two members of (7.1) for a we have

$$a = \frac{b \sin A}{\sin B}.$$

Substituting this in (7.2), we have

$$K = \frac{\frac{1}{2}b^2 \sin A \sin C}{\sin B}. \tag{7.3}$$

This relation may be expressed in words in

THEOREM 2

The area of any triangle is equal to one-half the product of the square of any side and the sines of the adjacent angles divided by the sine of the opposite angle.

7.6. Case A

In order to illustrate the use of the last two articles, we shall solve a triangle in which $a = 12.30$, $A = 48°10'$, and $C = 84°17'$. The law of sines is applicable here, since we know three of the four quantities involved in it. Hence, we have

$$\frac{c}{\sin 84°17'} = \frac{12.30}{\sin 48°10'}.$$

Therefore,

$$c = \frac{(12.30)(\sin 84°17')}{\sin 48°10'},$$

and

$$\log c = \log 12.30 + \log \sin 84°17' - \log \sin 48°10'.$$

Performing the operations indicated above, we have

$$
\begin{array}{rl}
\log 12.30 = & 1.08991 \\
\log \sin 84°17' = & 9.99783 - 10 \\
\hline
 & 11.08774 - 10 \\
\log \sin 48°10' = & 9.87221 - 10 \\
\hline
\log c = & 1.21553 \\
c = & 16.43.
\end{array}
$$

Furthermore,

$$
\begin{aligned}
B &= 180° - (48°10' + 84°17') \\
&= 47°33'.
\end{aligned}
$$

Now, using the law of sines again, we have

$$\frac{b}{\sin 47°33'} = \frac{12.30}{\sin 48°10'}.$$

Hence,

$$b = \frac{(12.30)(\sin 47°33')}{\sin 48°10'},$$

and

$$\log b = \log 12.30 + \log \sin 47°33' - \log \sin 48°10'.$$

$$\begin{array}{rl}
\log 12.30 = & 1.08991 \\
\log \sin 47°33' = & \underline{9.86798 - 10} \\
 & 10.95789 - 10 \\
\log \sin 48°10' = & \underline{9.87221 - 10} \\
\log b = & 1.08568.
\end{array}$$

Hence,

$$b = 12.18.$$

These values of a, b, c, A, B, and C can be checked by use of either of *Mollweide's equations*, which are proved in Appendix A. Equation (M₂) is

$$\frac{a + b}{c} = \frac{\cos \dfrac{(A - B)}{2}}{\sin \dfrac{C}{2}}.$$

Substituting into this, we have

$$\frac{12.30 + 12.18}{16.43} = \frac{\cos \dfrac{(48°10' - 47°33')}{2}}{\sin \dfrac{84°17'}{2}},$$

$$\frac{24.48}{16.43} = \frac{\cos 19'}{\sin 42°9'}.$$

$$\begin{array}{rl}
\log 24.48 = & 1.38881 \\
\log 16.43 = & \underline{1.21564} \\
\log \text{ quotient} = & 0.17317.
\end{array}$$
$$\begin{array}{rl}
\log \cos 19' = & 9.99999 - 10 \\
\log \sin 42°9' = & \underline{9.82677 - 10} \\
\log \text{ quotient} = & 0.17322.
\end{array}$$

Using either value of the logarithm of the quotient, we find the quotient to be 1.490. Hence, the two values are the same to four significant figures, and that is the degree of accuracy to which the sides are given.

In order to find the area of the triangle, we shall use Theorem 1 of article 7.5. Choosing b and c as the sides, we have

$$K = \tfrac{1}{2}bc \sin A$$
$$= \frac{(12.18)(16.43)(\sin 48°10')}{2}.$$

Therefore,

$$\log K = \log 12.18 + \log 16.43 + \log \sin 48°10' - \log 2.$$

$$
\begin{aligned}
\log 12.18 &= 1.08568 \\
\log 16.43 &= 1.21553 \\
\log \sin 48°10' &= \underline{9.87221 - 10} \\
&\ 12.17342 - 10 \\
\log 2 &= \underline{.30103} \\
\log K &= 1.87239.
\end{aligned}
$$

$$K = 74.54.$$

EXERCISE 7.1

Solve each triangle in problems 1 to 24 by finding the missing sides and angles to the justified degree of accuracy under the assumption that the given parts are approximations. Find the area of each triangle.

1. $A = 72°10'$, $B = 43°20'$, $b = 276$
2. $A = 39°40'$, $B = 68°10'$, $a = 213$
3. $A = 48°20'$, $C = 69°30'$, $b = 7.09$
4. $B = 81°10'$, $C = 72°20'$, $a = 14.7$
5. $A = 69°26'$, $B = 57°29'$, $b = .6243$
6. $A = 25°16'$, $B = 57°47'$, $a = 75.58$
7. $B = 106°14'$, $C = 31°5'$, $b = 9.119$
8. $B = 28°42'$, $C = 84°27'$, $b = 10.76$
9. $A = 26°24'$, $C = 79°38'$, $c = 123.4$
10. $B = 26°42'$, $C = 15°31'$, $c = 46.13$
11. $A = 13°31'$, $B = 76°3'$, $a = 6119$
12. $A = 111°43'$, $B = 26°19'$, $a = .9054$
13. $B = 121°5'$, $C = 17°24'$, $b = .4273$
14. $A = 67°51'$, $B = 37°48'$, $a = 4891$
15. $A = 76°14'$, $C = 88°51'$, $c = 917.3$
16. $C = 84°48'$, $A = 62°26'$, $c = 94.49$

17. $B = 26°53.2'$, $C = 83°17.6'$, $c = .026351$

18. $A = 29°32.6'$, $C = 112°41.3'$, $c = .36125$

19. $A = 77°25.8'$, $B = 59°36.2'$, $a = 935.06$

20. $A = 51°59.4'$, $C = 10°13.8'$, $a = 2311.2$

21. $A = 136°26.3'$, $C = 24°35.2'$, $a = 5.2763$

22. $B = 68°38.7'$, $C = 73°27.3'$, $b = .071624$

23. $A = 28°15.9'$, $B = 64°4.3'$, $b = 21.996$

24. $B = 67°22.4'$, $C = 12°7.6'$, $b = .11851$

25. One ranger station was 1.850 mi due east of another. They sighted a fire which was N 63°50′ W of the first and N 65°9′ E of the second. Which one was nearer the fire and how far was he from it?

26. The angle of elevation of the top of a tree is 58° at one point and is 84° at a point on a horizontal ground and 25 ft nearer the base of the tree than the first point. How tall is the tree?

27. An observer in Paulsen saw an unidentified flying object in the direction N 68°40′ E. At the same time, an observer in Charleston found it to be N 10°20′ W. What distance was the object from Paulsen if Charleston is 2.34 mi N 75°20′ E of Paulsen?

28. A gorge is crossed by a street with a sidewalk on each side. An observer on one of the sidewalks measured the angle of depression of a rock on the bottom of the gorge to be 89°30′. He then walked 30.0 ft along the horizontal sidewalk and away from the rock and found the angle of depression of the rock to be 78°30′. How far was the rock from the observer's first position?

29. A motorist, traveling N 40° E on a highway, intended to turn off on a farm-to-market road that proceeded N 52° E to a town 25 mi from the intersection. He failed to turn as intended and took a second road that ran S 60° E to the town. How many extra miles did he travel?

30. A surveyor at the top of a hill whose surface makes an angle of 15°10′ with the horizontal found that the angle of depression of a small tree at the base of the hill was 3°20′. Find the height of the tree if its base was 72.5 ft from the observation point.

31. After walking 98.8 ft along horizontal ground from the base of a hill, an observer found that the angle of elevation of a marker on the top of the hill had changed from 33°10′ to 15°30′. How far from the observer's second position was the marker?

32. The ground slopes upward from a street at an angle of 6°10′. A builder wishes to dig a level lot 50.0 ft long in the slope. For the sake of safety, the cut base at the back of the lot must be cut to have an angle of elevation of 26°40′. How far from the street, measured along the present slope, will the excavation extend?

7.7. Case B: The Ambiguous Case

If two sides and an angle opposite one of them are given, we have the *ambiguous case* of the law of sines. It is so named because there may be no solution, one solution, or two solutions. If the given parts are a, b, and A, we can solve $a/\sin A = b/\sin B$ for $\sin B$, and have

$$\sin B = \frac{b \sin A}{a}.$$

The value of $\sin B$ is determined by the given parts and is greater than, equal to, or less than 1.

If $\sin B > 1$, then there is no angle B and no triangle.

If $\sin B = 1$, then $B = 90°$ and there is no triangle if A is obtuse or a right angle. If A is an acute angle, it fits into a triangle with $B = 90°$, and there is one triangle.

If $\sin B < 1$, then a value of B can be found from a table of natural function values, and $B' = 180° - B$ is another angle whose sine is equal to $\sin B$ since the sine of an angle and of its supplement are equal. Therefore, there are two triangles if $B + A < 180°$ and $B' + A < 180°$, there is one triangle if $B + A < 180°$ and $B' + A \geq 180°$, and there is no solution if both $B + A$ and $B' + A$ are equal to or greater than $180°$.

It is not necessary to memorize the tests for the number of solutions since all that is needed is to *evaluate sin B, find all positive values of B that are less than 180°, and use each of them that fits into a triangle with the given angle.* We can evaluate $\sin B$ with or without the use of logarithms.

EXAMPLE 1

We shall solve the triangle in which $b = 246$, $a = 197$, and $A = 48°10'$.

Solution: Using the law of sines, we have

$$\frac{\sin B}{246} = \frac{\sin 48°10'}{197}.$$

Therefore,

$$\sin B = \frac{246 \sin 48°10'}{197},$$

and

$$\log \sin B = \log 246 + \log \sin 48°10' - \log 197.$$

$$
\begin{aligned}
\log 246 &= 2.39094 \\
\log \sin 48°10' &= \underline{9.87221 - 10} \\
&\; 12.26315 - 10 \\
\log 197 &= \underline{2.29447} \\
\log \sin B &= \;\;9.96868 - 10.
\end{aligned}
$$

Hence,

$$B = 68°30'.$$

Furthermore,

$$B' = 180° - 68°30' = 111°30'.$$

Since

$$B + A = 68°30' + 48°10' = 116°40' < 180°$$

and

$$B' + A = 111°30' + 48°10' = 159°40' < 180°,$$

there are two triangles that satisfy the given conditions. We shall call the first triangle ABC and the second $AB'C'$. The third side of the first will be designated

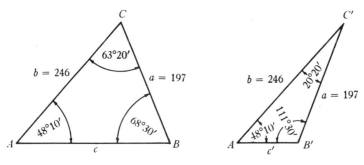

Figure 7.5

by c, and the third side of the second by c'. These two triangles are shown in Figure 7.5. The angle

$$C = 180° - (68°30' + 48°10') = 63°20',$$

and the angle

$$C' = 180° - (111°30' + 48°10') = 20°20'.$$

We may now find the sides c and c' by the law of sines.

$$\frac{c}{\sin 63°20'} = \frac{197}{\sin 48°10'}.$$

Therefore,

$$c = \frac{197 \sin 63°20'}{\sin 48°10'},$$

and

$$\log c = \log 197 + \log \sin 63°20' - \log \sin 48°10'.$$

$$
\begin{aligned}
\log 197 &= 2.29447 \\
\log \sin 63°20' &= \underline{9.95116 - 10} \\
& 12.24563 - 10 \\
\log \sin 48°10' &= \underline{9.87221 - 10} \\
\log c &= 2.37342.
\end{aligned}
$$

Therefore,

$$c = 236.$$

Similarly,

$$c' = \frac{197 \sin 20°20'}{\sin 48°10'}.$$

Therefore,

$$\log c' = \log 197 + \log \sin 20°20' - \log \sin 48°10'.$$

$$
\begin{aligned}
\log 197 &= 2.29447 \\
\log \sin 20°20' &= \underline{9.54093 - 10} \\
& 11.83540 - 10 \\
\log \sin 48°10' &= \underline{9.87221 - 10} \\
\log c' &= 1.96319,
\end{aligned}
$$

and

$$c' = 91.9.$$

We shall use equation (M$_2$) in order to check these solutions. It is

$$\frac{a+b}{c} = \frac{\cos \dfrac{A-B}{2}}{\sin \dfrac{C}{2}}.$$

Substituting the solution of the triangle $AB'C'$, we have

$$\frac{197 + 246}{91.9} = \frac{\cos \frac{1}{2}(48°10' - 111°30')}{\sin \frac{1}{2}(20°20')},$$

$$\frac{443}{91.9} = \frac{\cos (-31°40')}{\sin 10°10'}$$

$$= \frac{\cos 31°40'}{\sin 10°10'}.$$

$\log 443 = 2.64640$	$\log \cos 31°40' = 9.92999 - 10$
$\log 91.9 = 1.96332$	$\log \sin 10°10' = 9.24677 - 10$
$\log$ quotient $= 0.68308$	$\log$ quotient $= 0.68322$

The number whose logarithm is 0.68308 and the number whose logarithm is 0.68322 is, in each case, 4.82, correct to three significant figures. Hence, this solution is checked and correct.

The solution of the triangle ABC may be checked similarly.
By (7.2), the area is

$$K = \tfrac{1}{2}(246)(197)(\sin 63°20').$$

$$\begin{aligned}
\log 246 &= 2.39094\\
\log 197 &= 2.29447\\
\log \sin 63°20' &= \underline{9.95116 - 10}\\
&\ \ 14.63657 - 10.\\
\log 2 &= \underline{.30103}
\end{aligned}$$

Therefore,

$$\log K = 4.33554,$$

and

$$K = 2.17(10^4).$$

Similarly,

$$K' = 8.42(10^3).$$

EXAMPLE 2

Given $A = 145°20'$, $b = 172$, and $a = 213$, solve the triangle.

Solution: Using the law of sines, we have

$$\frac{\sin B}{172} = \frac{\sin 145°20'}{213}.$$

Therefore,

$$\sin B = \frac{172 \sin 145°20'}{213},$$

and

$$\log \sin B = \log 172 + \log \sin 145°20' - \log 213.$$

$$\begin{aligned}
\log 172 &= 2.23553\\
\log \sin 145°20' = \log \sin 34°40' &= \underline{9.75496 - 10}\\
&\ \ 11.99049 - 10\\
\log 213 &= \underline{2.32838}\\
\log \sin B &= 9.66211 - 10,\\
B &= 27°20'.
\end{aligned}$$

Furthermore,

$$B' = 180° - 27°20' = 152°40'.$$

However, since

$$A + B' = 145°20' + 152°40' = 298° > 180°,$$

the angle B' must be discarded.
The angle

$$C = 180° - (145°20' + 27°20') = 7°20'.$$

Using the law of sines again, we have

$$\frac{c}{\sin 7°20'} = \frac{213}{\sin 145°20'}.$$

Therefore,

$$\log c = \log 213 + \log \sin 7°20' - \log \sin 145°20'.$$

$$\begin{aligned}
\log 213 &= 2.32838 \\
\log \sin 7°20' &= \underline{9.10599 - 10} \\
&\ \ 11.43437 - 10 \\
\log \sin 145°20' &= \underline{9.75496 - 10} \\
\log c &= \ \ 1.67941, \\
c &= 47.8.
\end{aligned}$$

This solution may be checked by use of either of Mollweide's equations. Making use of (7.2) with b and a as the known sides, we have

$$\begin{aligned}
K &= \tfrac{1}{2}(172)(213)(\sin 7°20') \\
&= 2.34(10^3).
\end{aligned}$$

EXAMPLE 3

Given the two line segments $b = 47$ and $c = 40$ and the angle $C = 65°$, solve the triangle.

Solution: An application of the law of sines gives

$$\begin{aligned}
\sin B &= \frac{47 \sin 65°}{40} \\
&= \frac{47(.9063)}{40} \\
&= \frac{42.6}{40} \\
&= 1.07.
\end{aligned}$$

Since this result is greater than 1, there is no angle B and no solution. The calculation above, made by use of logarithms, is given below.

$$\sin B = \frac{47 \sin 65°}{40}.$$

$$\log \sin B = \log 47 + \log \sin 65° - \log 40.$$

$$\begin{aligned}
\log 47 &= 1.67210 \\
\log \sin 65° &= \underline{9.95728 - 10} \\
&\ \ 11.62938 - 10 \\
\log 40 &= \underline{1.60206} \\
\log \sin B &= \ \ .02732.
\end{aligned}$$

Since $\log \sin B$ is greater than 0, it follows that $\sin B$ is greater than 1, and, consequently, there is no angle B and no solution.

EXERCISE 7.2

Solve each triangle in problems 1 to 20 by finding the area and the missing sides and angles to the degree of accuracy justified under the assumption that the given parts are approximations.

1. $b = 876$, $c = 987$, $C = 76°40'$
2. $a = 16.7$, $b = 19.2$, $A = 57°20'$
3. $a = 371$, $b = 812$, $A = 31°20'$
4. $a = 487$, $b = 606$, $A = 126°30'$
5. $a = 30.52$, $b = 24.37$, $A = 56°13'$
6. $a = 8612$, $c = 5234$, $A = 36°18'$
7. $a = .5053$, $c = .7811$, $A = 41°14'$
8. $a = .5183$, $b = .7918$, $B = 65°14'$
9. $a = .3214$, $b = .6217$, $B = 43°12'$
10. $b = 91.37$, $c = 216.3$, $C = 78°16'$
11. $b = 5.881$, $c = 4.976$, $B = 21°14'$
12. $a = 617.6$, $c = 591.1$, $C = 48°56'$
13. $a = .21682$, $c = .062173$, $A = 54°12.8'$
14. $b = 9863.1$, $c = 21364$, $B = 21°16.7'$
15. $b = 1197.6$, $c = 3917.4$, $C = 26°21.8'$
16. $a = 69873$, $b = 62025$, $B = 61°15.6'$
17. $a = 3.2647$, $b = 10.162$, $B = 26°21.7'$
18. $a = 25623$, $b = 37281$, $A = 58°16.3'$
19. $b = 19726$, $c = 42728$, $B = 21°16.7'$
20. $a = .41221$, $b = .21272$, $B = 18°23.5'$

21. A pilot flew in the direction 140° from A to B and then at 240° from B to C. If A is 540 mi from B and 800 mi from C, find the distance BC and the direction from A to C.

22. If one travels N 26° E from A, he arrives at C. If he travels N 72° E from A, he arrives at B. If C is 38.96 mi from B and 21.36 mi from A, find the distance between A and B and the direction of C from B.

23. The owner wished to fence a triangular lot whose vertices are at A, B, and C, but the marker at C has been destroyed. He, however, has a deed that shows the lengths of AC and AB to be 448 ft and 263 ft, respectively, and that the angle at B is 130°10'. What distance from B is the correct position of C?

24. Due to an error in calculation, a pilot flew for 341 mi in a direction that was 19°20′ off his course. He then turned through an acute angle toward his destination and reached it after flying 212 mi. If the speed of the plane was 320 mph, find the extra time in flight due to the error.

7.8. The Law of Cosines

The second fundamental law for solving oblique triangles is the law of cosines. In order to derive this law, we shall consider the triangle ABC in

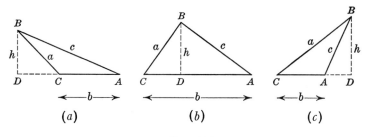

Figure 7.6

Figure 7.6. From any vertex B, we shall drop a perpendicular to the opposite side or the opposite side produced.

Applying the Pythagorean theorem to each triangle in Figure 7.6, we have

$$c^2 = h^2 + (DA)^2. \tag{i}$$

But

$$h = a \sin C, \qquad \text{since } \sin C = \frac{h}{a}.$$

Furthermore,

$$DA = b + DC$$
$$= b - CD, \qquad \text{since } DC = -CD,$$
$$= b - a \cos C, \qquad \text{since } \cos C = \frac{CD}{a}.$$

Substituting these values of h and DA in (i), we have

$$c^2 = (a \sin C)^2 + (b - a \cos C)^2$$
$$= a^2 \sin^2 C + b^2 - 2ab \cos C + a^2 \cos^2 C$$
$$= a^2(\sin^2 C + \cos^2 C) + b^2 - 2ab \cos C.$$

Hence,

$$c^2 = a^2 + b^2 - 2ab \cos C. \qquad (7.4)$$

Since the lettering used is immaterial, we have the following law:

LAW OF COSINES (form 1)

The square of any side of a triangle is equal to the sum of the squares of the other two sides minus twice the product of these sides and the cosine of the angle between them.

If we solve (7.4) for cos C, we obtain

$$\cos C = \frac{a^2 + b^2 - c^2}{2ab}. \qquad (7.5)$$

Since the lettering used is immaterial, we have the second form of the law of cosines.

LAW OF COSINES (form 2)

The cosine of any angle of a triangle is equal to a fraction whose numerator is the sum of the squares of the adjacent sides minus the square of the third side and whose denominator is twice the product of the adjacent sides.

The right member of (7.4) contains two sides and the included angle, and enables us to obtain the third side. Formula (7.5) enables us to obtain an angle when the three sides are given. Hence, the law of cosines suffices for cases C and D. However, since logarithms cannot be conveniently applied to it, the computation is tedious when the numbers involved contain three or more significant figures.

If two sides and an angle opposite one of them are given, we can use the law of cosines. This is case B, and was solved by use of the law of sines.

EXAMPLE 1

Given $a = 20$, $b = 30$, $C = 23°$. Solve the triangle.

Solution: By the law of cosines, we have

$$
\begin{aligned}
c^2 &= a^2 + b^2 - 2ab \cos C \\
&= (20)^2 + (30)^2 - 2(20)(30)(\cos 23°) \\
&= 400 + 900 - 1200(.9205) \\
&= 200 \qquad \text{to two significant digits.}
\end{aligned}
$$

Hence,

$$c = 14 \qquad \text{to two significant figures.}$$

We shall employ the second form of the law of cosines to get angles A and B. Thus,

$$\cos A = \frac{b^2 + c^2 - a^2}{2bc}$$

$$= \frac{900 + 196 - 400}{840}$$

$$= .83 \qquad \text{to two significant figures.}$$

Hence,

$$A = 34° \qquad \text{to the nearest degree.}$$

Furthermore,

$$\cos B = \frac{a^2 + c^2 - b^2}{2ac}$$

$$= \frac{400 + 196 - 900}{560}$$

$$= \frac{-304}{560} = -.54.$$

$$B = 123° \qquad \text{to the nearest degree.}$$

The value of B was obtained from the fact that

$$\cos (\text{related angle of } B) = .54,$$
$$\text{related angle of } B = 57°,$$
$$B = 180° - 57°$$
$$= 123°.$$

As a check, we have

$$A + B + C = 34° + 123° + 23° = 180°.$$

EXAMPLE 2

Determine the angles of a triangle whose sides are $a = 15$, $b = 20$, and $c = 25$.

Solution: Using the second form of the law of cosines, we have

$$\cos C = \frac{15^2 + 20^2 - 25^2}{2(15)(20)}$$

$$= \frac{225 + 400 - 625}{600}$$

$$= 0.$$

Therefore,

$$C = 90°.$$

Furthermore,

$$\cos B = \frac{15^2 + 25^2 - 20^2}{2(15)(25)}$$

$$= \frac{225 + 625 - 400}{750}$$

$$= \frac{450}{750}$$

$$= .6000.$$

Therefore,

$$B = 53°.$$

Similarly,

$$\cos A = \frac{20^2 + 25^2 - 15^2}{2(20)(25)}$$

$$= \frac{800}{1000}$$

$$= .8000,$$

and

$$A = 37°.$$

Since $90° + 53° + 37° = 180°$, the above solution is correct.

If two sides and an angle opposite one of them are given and we use the law of cosines, we obtain a quadratic equation. If the given parts are a, c and C, then the quadratic is equation (7.4). In order to find the values, if any, of b, we substitute the given values in (7.4) and solve for b. The solution set may consist of two distinct real positive numbers, one positive and one negative, two negative, two equal positive, or two imaginary numbers. The only usable values of b are the real positive ones, since the length of a side of a triangle is a real positive number. We can thus find the number of usable values of b and the number of triangles and can continue the solution of each by use of the law of sines.

EXAMPLE 3

How many triangles are there if $a = 2\sqrt{3}$, $c = 3\sqrt{2}$, and $C = 60°$? Find them.

Solution: Since the given parts are a, c, and C, we shall substitute them in $c^2 = a^2 + b^2 - 2ab \cos C$, and have

$$(3\sqrt{2})^2 = (2\sqrt{3})^2 + b^2 - 2(2\sqrt{3})b(\tfrac{1}{2}), \qquad \text{since } \cos 60° = \tfrac{1}{2}$$

$$18 = 12 + b^2 - 2\sqrt{3}\,b$$

$$b^2 - 2\sqrt{3}\,b - 6 = 0$$

$$b = \frac{2\sqrt{3} \pm \sqrt{(-2\sqrt{3})^2 - 4(1)(-6)}}{2(1)}$$

$$= \frac{2\sqrt{3} \pm 6}{2}$$

$$= \sqrt{3} \pm 3.$$

Thus, one member of the solution set of the quadratic is positive and the other is negative. Consequently, there is one triangle with the given parts. We can find readily that $\sin A = 1/\sqrt{2}$; hence, $A = 45°, 135°$. Since $60° + 135° > 180°$, it follows that we cannot use $A = 135°$. Therefore, $B = 180° - (60° + 45°) = 75°$.

EXERCISE 7.3

In each of problems 1 to 8, find the length of the third side.

1. $a = 20, b = 50, C = 60°$
2. $a = 200, c = 300, B = 120°$
3. $b = 81, c = 30, A = 120°$
4. $a = 15, b = 25, C = 45°$
5. $a = 60, c = 30, B = 40°$
6. $b = 8, c = 12, A = 25°$
7. $a = 14, c = 12, B = 100°$
8. $b = 24, c = 48, A = 108°$
9. If $a = 12, b = 13, c = 20$, find A
10. If $a = 40, b = 25, c = 30$, find B
11. If $a = 50, b = 30, c = 22$, find C
12. If $a = 12, b = 13, c = 20$, find C
13. If $a = 7.0, b = 3.0, c = 5.0$, find C
14. If $a = 60, b = 40, c = 70$, find B
15. If $a = 90, b = 60, c = 100$, find C
16. If $a = 90, b = 63, c = 45$, find B

Find each usable value of the third side and the number of triangles possible in each of problems 17 to 20.

17. $a = 6\sqrt{3}, c = 6\sqrt{3}, C = 30°$
18. $a = 8\sqrt{2}, c = 8, C = 45°$
19. $a = 14, c = 7, C = 60°$
20. $a = 10, c = 9, C = 60°$

Find all three angles in each of problems 21 to 24.

21. $a = 15, b = 16, c = 17$
22. $a = 33, b = 40, c = 50$
23. $a = 600, b = 550, c = 625$
24. $a = 300, b = 150, c = 200$

25. Point A is located on the shore of a pond and is accessible to B and C. If $AB = 2000$ ft, $AC = 3000$ ft, and angle BAC is $30°$, find BC.

26. Two forces of 150 lb and 200 lb are acting on the same point in directions that differ by 36°. Find the magnitude of their resultant and the angle its direction makes with the smaller force.

27. A pilot is flying with an air speed of 287 mph and a heading of 200°. Find the ground speed of the plane and the direction of flight if the wind is blowing from the east at 35 mph.

28. To drive from Baker to Taylor, one drives 60 mi east and 25 mi N 23°30′ E. How far apart are the towns?

7.9. The Law of Tangents

We shall use the law of sines as a basis for deriving another formula for use in solving triangles. This new formula will lend itself more readily than the law of cosines to the use of logarithms. If we begin with the law of sines in the form

$$\frac{\sin A}{a} = \frac{\sin B}{b} \tag{1}$$

and multiply each member by $a/\sin B$, we get

$$\frac{\sin A}{\sin B} = \frac{a}{b}. \tag{2}$$

Now subtracting 1 from each member of (2) and simplifying each new member, we have

$$\frac{\sin A - \sin B}{\sin B} = \frac{a - b}{b}. \tag{3}$$

Similarly, adding 1 to each member of (2) and simplifying each new member, we obtain

$$\frac{\sin A + \sin B}{\sin B} = \frac{a + b}{b}. \tag{4}$$

If we divide each member of (3) by the corresponding member of (4), we have

$$\frac{\sin A - \sin B}{\sin A + \sin B} = \frac{a - b}{a + b},$$

and applying (4.22) to the numerator and (4.21) to the denominator leads to

$$\frac{2 \cos \frac{1}{2}(A + B) \sin \frac{1}{2}(A - B)}{2 \sin \frac{1}{2}(A + B) \cos \frac{1}{2}(A - B)} = \frac{a - b}{a + b}.$$

Hence

$$\cot \tfrac{1}{2}(A + B) \tan \tfrac{1}{2}(A - B) = \frac{a - b}{a + b}.$$

Finally, since

$$\cot \tfrac{1}{2}(A + B) = \frac{1}{\tan \tfrac{1}{2}(A + B)},$$

we have

$$\frac{a - b}{a + b} = \frac{\tan \tfrac{1}{2}(A - B)}{\tan \tfrac{1}{2}(A + B)}, \qquad (7.6)$$

which is known as the *law of tangents*.

Since the lettering of a triangle is immaterial, we can state the law as follows:

LAW OF TANGENTS

The ratio of the difference of any two sides of a triangle to their sum is equal to the ratio of the tangent of half the difference of the angles opposite the two sides to the tangent of half their sum, the two differences being taken in the same order.

If we know one angle, we can readily find half the sum of the other two, since the sum of the three angles of a triangle is 180°. In particular, if we know C, we can find $\tfrac{1}{2}(A + B)$. Therefore, we can use the law of tangents to solve a triangle, provided we know two sides and the included angle.

EXAMPLE

Solve the triangle in which $b = 147.0$, $c = 216.0$, and $A = 83°10'$.

Solution: Since the sides b and c are known, we must first find the value of $\tfrac{1}{2}(B + C)$. Furthermore, the length of each of c and b is given to four significant figures. Hence, we shall obtain the length of a to four significant figures and the values of B and C to the nearest minute. Since

$$A + B + C = 180°$$
$$\tfrac{1}{2}(B + C) = \tfrac{1}{2}(180° - A)$$
$$= \tfrac{1}{2}(180° - 83°10') = \tfrac{1}{2}(96°50')$$
$$= 48°25'.$$

Furthermore, $c - b = 69$, and $c + b = 363$. Using the law of tangents, we have

$$\frac{69}{363} = \frac{\tan \tfrac{1}{2}(C - B)}{\tan 48°25'}.$$

Therefore,

$$\tan \tfrac{1}{2}(C - B) = \frac{69 \tan 48°25'}{363},$$

and

$$\log \tan \tfrac{1}{2}(C - B) = \log 69 + \log \tan 48°25' - \log 363.$$

$$
\begin{aligned}
\log 69 &= 1.83885 \\
\log \tan 48°25' &= 10.05192 - 10 \\
\hline
&= 11.89077 - 10 \\
\log 363 &= 2.55991 \\
\hline
\log \tan \tfrac{1}{2}(C - B) &= 9.33086 - 10.
\end{aligned}
$$

Hence,

$$\tfrac{1}{2}(C - B) = 12°5'.$$

But,

$$\tfrac{1}{2}(C + B) = 48°25'.$$

Therefore, adding the last two equations, we have

$$C = 60°30',$$

and subtracting them,

$$B = 36°20'.$$

Using the law of sines in order to calculate a, we have

$$
\begin{aligned}
a &= \frac{b \sin A}{\sin B} \\[2mm]
&= \frac{147 \sin 83°10'}{\sin 36°20'} \\[2mm]
&= 246.3.
\end{aligned}
$$

This solution may be checked by use of either of Mollweide's equations. Substituting in (M_2), we have

$$\frac{246.3 + 147.0}{216.0} = \frac{\cos \dfrac{83°10' - 36°20'}{2}}{\sin \dfrac{60°30'}{2}},$$

$$\frac{393.3}{216.0} = \frac{\cos 23°25'}{\sin 30°15'}.$$

$\log 393.3 = 2.59472$	$\log \cos 23°25' = 9.96267 - 10$
$\log 216.0 = 2.33445$	$\log \sin 30°15' = 9.70224 - 10$
$\log$ quotient $= .26027$	$\log$ quotient $= .26043$

The quotient in each case is 1.821, correct to four significant figures. Hence, the values determined are correct.

We are now in position to use either (7.2) or (7.3) in order to calculate the area of the triangle. If we employ the latter, we obtain

$$K = \frac{\frac{1}{2}(147)^2(\sin 83°10')(\sin 60°30')}{\sin 36°20'}$$

$$= 1.576(10^4).$$

EXERCISE 7.4

Solve the triangle in each of problems 1 to 20.

1. $b = 2.44,\ c = 8.10,\ A = 46°0'$
2. $a = 842,\ b = 638,\ C = 99°20'$
3. $b = 203,\ c = 315,\ A = 48°20'$
4. $a = 606,\ b = 660,\ C = 82°20'$
5. $b = .02163,\ c = .03217,\ A = 38°18'$
6. $a = 463.4,\ c = 326.8,\ B = 72°24'$
7. $a = 25.14,\ b = 36.97,\ C = 43°22'$
8. $a = 443.8,\ c = 211.9,\ B = 57°18'$
9. $b = 7.605,\ c = 9.182,\ A = 28°28'$
10. $b = .09763,\ c = .1194,\ A = 85°44'$
11. $a = .3650,\ c = .2588,\ B = 119°46'$
12. $a = 2543,\ c = 1142,\ B = 22°54'$
13. $b = 31.627,\ c = 43.421,\ A = 36°21.2'$
14. $a = 4.1672,\ c = 4.0234,\ B = 61°12.7'$
15. $a = 263.41,\ b = 128.32,\ C = 61°22.8'$
16. $a = 136.47,\ c = 326.41,\ B = 25°23.2'$
17. $b = 5.6732,\ c = 7.0213,\ A = 31°33.6'$
18. $a = 41915,\ b = 31174,\ C = 27°29.8'$
19. $b = 71.716,\ c = 56.548,\ A = 88°44.4'$
20. $a = .21176,\ c = .47695,\ B = 72°17.6'$

21. Two forces of 258 and 603 lb are acting at the same point and make an angle of 67°40' with one another. Find the magnitude of their resultant and the angle the resultant makes with the smaller force.

22. A triangle is inscribed in a circle, has one angle of 106°28', and the including sides are 26.42 and 31.57 in. Find the other angles.

23. Two ships leave a port simultaneously. One steams S 37° E at 24 knots and the other goes N 62° E at 32 knots. How far apart are they after three hours?

24. Towns A and C are 226 mi N 50° E and 138 mi S 70° E of town B, respectively. A truck driver leaves A with deliveries for B and C and finishes unloading at C at 3:30 P.M. How fast must he drive over a straight highway from C to A in order to keep an opera engagement at 8 P.M. if an hour is needed to get to the opera after arriving at A?

25. An airplane is headed at 56° with an airspeed of 275 mph. If the wind is blowing from due north at 32 mph, find the direction of the course and the ground speed of the plane.

26. A garage adjoins a house at an angle of 136°20′. A triangular patio in the corner extends 19.5 ft along the house and 13.0 ft along the garage. What is the area of the patio?

27. Two streets intersect at an angle of 114°. A vacant lot on the corner has a frontage of 98.3 ft on one street and of 62.8 ft on the other. What distance can be saved by walking diagonally across the lot instead of going around the edges?

28. A designer found that forces of 988 lb and 876 lb would act on a rivet. If the angle between the directions of the forces is 59°40′, what is the resultant force on the rivet?

7.10. The Half-Angle Formulas

We shall now derive two more formulas for use in finding the angles and area of a triangle when the sides are known.

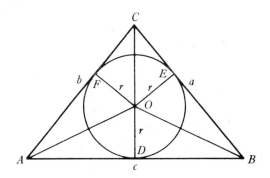

Figure 7.7

In Figure 7.7 we shall let r represent the radius of the inscribed circle and s denote one-half of the perimeter of the triangle. Hence,

$$s = \tfrac{1}{2}(a + b + c).$$

We have the following facts from plane geometry:

1. The angles at D, E, and F are right angles, since the radius of a circle is perpendicular to a tangent at the point of tangency.

2. The line segment OA bisects angle A, OB bisects angle B, and OC bisects angle C, since the line connecting an exterior point to the center of a circle bisects the angle between the tangents from the point to the circle.

3. The area of the triangle is given by Heron's formula

$$K = \sqrt{s(s-a)(s-b)(s-c)}.$$

4. Since the tangents from a point to a circle are equal, we have

$$AD = AF,$$
$$BE = BD,$$
$$CE = CF.$$

In the right triangle ODA, $\tan \frac{1}{2}A = r/AD$, and we shall now prove that $AD = s - a$. Since $s = \frac{1}{2}(a + b + c)$, we have

$$2s = a + b + c = BE + CE + CF + AF + AD + BD.$$

Now if we use the equalities in (4), we may substitute CE for CF, AD for AF, and BE for BD in the above and get

$$
\begin{aligned}
2s &= BE + CE + CE + AD + AD + BE \\
&= 2AD + 2CE + 2BE \\
&= 2AD + 2(CE + BE) \\
&= 2AD + 2a.
\end{aligned}
$$

Hence,

$$2AD = 2s - 2a,$$

and

$$AD = s - a.$$

By a similar argument, we may show that $BE = s - b$ and $CF = s - c$. Therefore, we have

$$\tan \tfrac{1}{2}A = \frac{r}{s-a}, \qquad (7.7a)$$

$$\tan \tfrac{1}{2}B = \frac{r}{s-b}, \qquad (7.7b)$$

$$\tan \tfrac{1}{2}C = \frac{r}{s-c}. \qquad (7.7c)$$

We shall next express r in terms of a, b, c, and s. From Figure 7.7, we see that

K = area of ABC = area of BCO + area of CAO + area of ABO

$\qquad = \frac{1}{2}ra + \frac{1}{2}rb + \frac{1}{2}rc$

$\qquad = \frac{1}{2}r(a + b + c)$

$\qquad = \frac{1}{2}r(2s).$

Hence,

$$K = rs. \tag{7.8}$$

Now, equating this to the area of a triangle as given by Heron's formula, we have

$$rs = \sqrt{s(s - a)(s - b)(s - c)},$$

$$r = \frac{\sqrt{s(s - a)(s - b)(s - c)}}{s}.$$

$$r = \sqrt{\frac{(s - a)(s - b)(s - c)}{s}}.$$

The results of this article are summarized in the following theorems:

THEOREM 1

The radius of the circle inscribed in a triangle with **a**, **b**, *and* **c** *as sides is*

$$r = \sqrt{\frac{(s - a)(s - b)(s - c)}{s}} \tag{7.9}$$

where

$$s = \frac{1}{2}(a + b + c).$$

THEOREM 2

The area of a triangle is **rs.**

THEOREM 3

The tangent of half an angle of a triangle is a fraction whose numerator is **r** *and whose denominator is* **s** *minus the length of the side opposite the angle.*

In order to illustrate the use of the formulas developed in this article we shall solve the following problem:

EXAMPLE

Given the triangle whose sides are $a = 12.96$, $b = 16.92$, and $c = 14.04$, find the values of the angles and the area.

Solution: It is first necessary to compute the values of s, $s - a$, $s - b$, and $s - c$, as indicated below.

$$s = \tfrac{1}{2}(12.96 + 16.92 + 14.04) = 21.96,$$
$$s - a = 9.00,$$
$$s - b = 5.04,$$
$$s - c = 7.92.$$

If we add the differences just obtained, we get 21.96, and this is the value of s. This serves as a check on the previous computation since

$$(s - a) + (s - b) + (s - c) = 3s - (a + b + c) = 3s - 2s = s.$$

The next step is to compute the value of r by use of (7.9). Using logarithms, we have

$$\log r = \tfrac{1}{2}[\log (s - a) + \log (s - b) + \log (s - c) - \log s]$$

and we complete the computation below.

$$
\begin{aligned}
\log (s - a) &= \log 9.00 = & 0.95424 \\
\log (s - b) &= \log 5.04 = & 0.70243 \\
\log (s - c) &= \log 7.92 = & 0.89873 \\
& & \overline{2.55540} \\
\log s &= \log 21.96 = & 1.34163 \\
& & 2\overline{\,|\,1.21377\,} \\
\log r &= & 0.60689.
\end{aligned}
$$

Since only the logarithm of r is needed in the following calculation, we shall not determine the value of r, and shall proceed directly to the computation of the angles.

By (7.7a),

$$\tan \tfrac{1}{2}A = \frac{r}{s - a}.$$

Hence,

$$
\begin{aligned}
\log \tan \tfrac{1}{2}A &= \log r - \log (s - a). \\
\log r &= 10.60689 - 10 \\
\log (s - a) &= \ \ 0.95424 \\
\log \tan \tfrac{1}{2}A &= \ \overline{9.65265} - 10. \\
\tfrac{1}{2}A &= 24°12', \\
A &= 48°24'.
\end{aligned}
$$

Similarly, using (7.7b), we have

$$\tan \tfrac{1}{2}B = \frac{r}{s - b},$$

$$
\begin{aligned}
\log \tan \tfrac{1}{2}B &= \log r - \log (s - b). \\
\log r &= 10.60689 - 10 \\
\log (s - b) &= \ \ 0.70243 \\
\log \tan \tfrac{1}{2}B &= \ \overline{9.90446} - 10. \\
\tfrac{1}{2}B &= 38°45', \\
B &= 77°30'.
\end{aligned}
$$

Likewise, using (7.7c) we get

$$\begin{aligned}
\log r &= 10.60689 - 10 \\
\log (s - c) &= \underline{0.89873} \\
\log \tan \tfrac{1}{2}C &= 9.70816 - 10. \\
\tfrac{1}{2}C &= 27°3', \\
C &= 54°6'.
\end{aligned}$$

As a check, we have

$$A + B + C = 48°24' + 77°30' + 54°6' = 179°60' = 180°.$$

In order to find K, we shall use (7.8):

$$K = rs.$$

Therefore,

$$\begin{aligned}
\log K &= \log r + \log s. \\
\log r &= 0.60689 \\
\log s &= \underline{1.34163} \\
\log K &= 1.94852. \\
K &= 88.82.
\end{aligned}$$

EXERCISE 7.5

Solve each triangle whose parts are given in problems 1 to 20.

1. $a = .563, b = .330, c = .689$ **2.** $a = 312, b = 532, c = 612$

3. $a = 12.6, b = 17.9, c = 18.7$ **4.** $a = 9.82, b = 7.96, c = 8.14$

5. $a = 213.8, b = 312.7, c = 402.9$ **6.** $a = .3178, b = .7289, c = .9613$

7. $a = 7.329, b = 6.517, c = 5.286$ **8.** $a = 3.674, b = .9829, c = 2.931$

9. $a = 9646, b = 8846, c = 7988$ **10.** $a = .6676, b = .6982, c = .4714$

11. $a = .09112, b = .08145, c = .06163$

12. $a = 37.46, b = 43.12, c = 56.14$

13. $a = .27632, b = .42186, c = .35248$

14. $a = 6371.8, b = 2463.5, c = 7293.1$

15. $a = .096314, b = .17831, c = .21876$

16. $a = 26.132, b = 28.218, c = 27.421$

17. $a = 26.285, b = 39.903, c = 40.112$

18. $a = 798.51, b = 615.97, c = 553.92$

19. $a = 1.1177, b = 2.9083, c = 2.6674$

20. $a = .15976, b = .39114, c = .25972$

21. The distances from Santa Fe, New Mexico to Amarillo, Texas and El Paso, Texas are 286 mi and 333 mi, respectively, and Amarillo is 423 mi from El Paso. If the direction from Santa Fe to Amarillo is S 80° E, find the directions from Santa Fe and Amarillo to El Paso.

22. The airspeed and heading of a plane are 253 mph and 212°, whereas the ground speed is 261 mph and the wind velocity is 19 mph. Find the direction of the wind if the course of the plane is north of the heading.

23. Two forces and their resultant are 325 lb, 465 lb, and 512 lb, respectively. Find the angle between the forces.

24. A triangle with sides of 10.5, 12.6, and 15.8 cm in length is circumscribed about a circle. Find the radius.

25. What is the radius of the largest circular water tank that can be placed with its base in a triangle with sides 10.2, 11.8, and 14.0 ft?

26. A triangular breakfast bar with sides of 4.75, 9.50, and 5.67 ft is planned for a home. How far out in the room will it extend if the long side is against the wall?

27. How much soil is needed to fill a triangular planter that is 1.5 ft high and has sides of inside length 7.5, 9.3, and 8.2 ft?

28. After a pole was driven vertically in the ground on a hillside of uniform slope, its top was 7.15 ft above the ground. What was the angle of elevation of the hillside if, from a point on the hill, it was 101.80 ft to the base of the pole and 99.20 ft to the top of the pole?

7.11. Summary

In this chapter, we have derived four formulas for use in solving triangles and three for use in finding areas. We shall now collect these formulas so as to have them readily available for use in solving the following set of review problems.

LAW OF SINES

$$\frac{a}{\sin A} = \frac{b}{\sin B} = \frac{c}{\sin C} \tag{7.1}$$

LAW OF COSINES

$$a^2 = b^2 + c^2 - 2bc \cos A \tag{7.4}$$

LAW OF TANGENTS

$$\frac{\tan \frac{1}{2}(A - B)}{\tan \frac{1}{2}(A + B)} = \frac{a - b}{a + b} \tag{7.6}$$

HALF-ANGLE FORMULAS

$$\tan \frac{A}{2} = \frac{r}{s-a} \tag{7.7a}$$

where

$$r = \frac{\sqrt{(s-a)(s-b)(s-c)}}{s} \tag{7.9}$$

and

$$s = \tfrac{1}{2}(a+b+c)$$

AREA FORMULAS

$$K = \tfrac{1}{2}bc \sin A \tag{7.2}$$

$$K = \frac{c^2 \sin A \sin B}{2 \sin C} \tag{7.3}$$

$$K = rs \tag{7.8}$$

EXERCISE 7.6

Solve each triangle in problems 1 to 32.

1. $A = 52°14'$, $B = 43°38'$, $a = 617.2$
2. $A = 72°37'$, $C = 39°13'$, $b = 3.945$
3. $A = 63°15'$, $C = 27°41'$, $a = 1631$
4. $A = 60°26'$, $C = 57°44'$, $b = 5176$
5. $C = 73°10'$, $b = 3.267$, $c = 3.013$
6. $C = 38°56'$, $a = 599.3$, $b = 718.7$
7. $C = 66°12'$, $b = 1.128$, $c = 2.072$
8. $C = 55°42'$, $a = 9058$, $c = 8762$
9. $B = 79°31'$, $a = 159.8$, $c = 203.1$
10. $A = 48°14'$, $b = .3762$, $c = .5621$
11. $A = 12°14'$, $b = 569.8$, $c = 617.2$
12. $C = 39°52'$, $a = 47.51$, $b = 51.94$
13. $a = 58.62$, $b = 48.23$, $c = 53.37$
14. $a = .03627$, $b = .01867$, $c = .02571$
15. $a = .2074$, $b = .2842$, $c = .3656$
16. $a = .03918$, $b = .04572$, $c = .03214$
17. $A = 47°29.3'$, $C = 64°10.8'$, $b = 68475$
18. $A = 44°16.6'$, $C = 56°27.8'$, $b = 4967.2$

19. $A = 77°44.8'$, $B = 69°58.4'$, $c = 4.1769$

20. $A = 57°18.9'$, $B = 79°39.6'$, $c = .031392$

21. $B = 56°13.8'$, $b = 1.1786$, $c = 1.3624$

22. $A = 65°13.3'$, $a = .10091$, $b = .12315$

23. $A = 66°21.2'$, $a = 6.9174$, $b = 4.8562$

24. $A = 68°51.8'$, $a = 4.3061$, $c = 3.6632$

25. $C = 56°34.8'$, $a = 64.632$, $b = 54.812$

26. $B = 28°11.6'$, $a = 362.41$, $c = 286.79$

27. $A = 76°13.8'$, $b = 8.7163$, $c = 6.6218$

28. $C = 56°36.4'$, $a = 263.44$, $b = 198.98$

29. $a = .037216$, $b = .016372$, $c = .028763$

30. $a = 32.146$, $b = 26.188$, $c = 43.514$

31. $a = 7682.9$, $b = 6134.7$, $c = 5367.8$

32. $a = 34162$, $b = 21176$, $c = 46916$

33. Find the area of a triangle if two of its sides are of length 176 ft and 223 ft and meet at an angle of 79°20'.

34. Find the third side and the area of a triangle if two of its sides meet at an angle of 71°40' and are 225 ft and 276 ft in length.

35. A boy hiked 187 paces due north, then 226 paced N 81°20' E. How many steps would he have saved by taking a short cut straight between the beginning and ending points?

36. If a quart of paint covers 24 sq ft, how much paint is used in painting a triangular section of wall that is 9.3 ft by 7.0 ft by 6.2 ft?

37. If land is worth 26.3 cents per sq ft, find to the nearest $10 the cost of a lot that is 183 ft by 217 ft by 250 ft.

38. Two forces and their resultant are 127 lb, 152 lb, and 209 lb, respectively. Find the angle between the forces.

39. Calculations show a force of 15.2 tons applied vertically to a footing and a force of 10.5 tons applied to the footing at an angle of 68°20' with the horizontal. What is the resultant force on the footing and what direction does it make with the horizontal?

40. An airplane headed at 146°0' is flying with an airspeed of 260 mph. If the wind is blowing from the north at 24.8 mph, find the direction and ground speed of the plane.

41. Two forces of 173 and 228 lb are acting on the same point. If the first acts N 23°10' E and the second S 40°30' E, find the magnitude and direction of their resultant.

42. Find the diameter of the largest circular tent that can be pitched on a triangular lot with sides of 102, 118, and 140 ft.

43. Find the diameter of the largest circular braided rug that will fit into a triangular area that is 9.5 by 8.3 by 6.7 ft.

44. Two tractors are pulling simultaneously on a mired car. If the angle between their directions of pull is 36°20′ and the larger pulls with a force of 1.90 times that of the other, find the direction the car moves with respect to the line of pull of the larger tractor.

45. A tower stands on a hill that is inclined at 17°20′ to the horizontal. Find the height of the tower if the angle of elevation of its top is 28°10′ at a point 592 ft. downhill from the base of the tower.

46. In triangle ABC, $A = 36°15′$, $c = 1626$ ft, and the bisector of angle B is 1215 ft in length. Find the angles and sides.

47. Some miners were trapped 689 ft from the mine entrance in a straight mine shaft that sloped upward at an angle of 28°20′ with the ground surface. A rescue shaft was started 563 ft directly uphill from the entrance. Find the length of the rescue shaft.

48. If a pound of grass seed is needed for each 1200 sq ft, how much seed is required for a triangular tract that is 33 by 31 by 46 ft?

49. The angle between the directions of two forces acting at a point is 61°12′ and their resultant is 513.9 lb. Find the other force if one is 362.8 lb.

50. A battery at B is firing at a hidden target T and is directed by an observer at H that is 6186 yd S 26°31′ W of B. If T is 1672 yd S 27°43′ E from H, find the distance and direction from B to T.

51. Two boys are pulling on a wagon. One is pulling N 52°20′ W and the other S 62°10′ W, while the wagon is moving S 84°50′ W. Find the ratio of the second force to the first.

52. A door is 36.0 in. wide and has a 7.2 in. safety chain attached at its edge and to the wall 3.2 in. from the edge of the door. Through what angle can the door open?

53. The diagonals of a parallelogram are 136 and 178 ft in length and intersect at 69°. Find the area.

54. One side of a flat V-shaped channel has an angle of elevation of 29°30′ and measures 10.5 ft from the channel bottom to the top of the bank. The other side has an angle of elevation of 24°50′ and measures 11.2 ft. What volume of material is needed to fill the channel from bank to bank for a distance of 20 ft?

55. One corner in a fenced backyard measured 87°50′. If the corner is fenced off as a triangular service area by using 25.0 ft of new fencing and if the service area extends 12.8 ft along the back fence, what is the area?

56. From the top and bottom of a 74-ft lighthouse, the angles of depression of a ship are 41° and 37°. Find the height of the top of the lighthouse above sea level and the distance of the base of the lighthouse from the ship.

57. A circle of radius 12.6 ft is inscribed in an isosceles triangle. If the angle between the equal sides is 35°20′, find the sides and other angles.

58. The resultant of two forces is 537 lb and acts N 30°30′ W. If one force is 342 lb and acts N 86°10′ W, find the direction and magnitude of the other.

59. If street parking were restricted for a radius of 500 ft from a stadium entrance, would a place 200 ft north of the entrance and then 350 ft N 51°20′ E be within the restricted zone? How far from the boundary of the restricted zone is it?

60. At a certain point, the bearing of a cypress tree at one end of a lake is N 45°40′ E and that of a tree at the other end is N 20°20′ W. How long is the lake if the observation point is 289 ft from the first tree and 407 ft from the second?

61. Prove that a median of a triangle divides the angle into two parts whose sines are proportional to the lengths of the sides that include the angle from which the median is drawn.

62. A straight line is drawn from a vertex of an isosceles triangle to a point on the base. Prove that the two segments of the base have the same ratio as the sines of the two parts into which the angle is divided.

63. Prove that the radius R of the circumscribed circle of the triangle ABC satisfies the equation

$$R = \frac{a}{2 \sin A} = \frac{b}{2 \sin B} = \frac{c}{2 \sin C}.$$

Hint. Let A be an acute angle of ABC. Draw the triangle $A'BC$ with $A'C$ as a diameter of the circle. Then $A'C = 2R$, and angle $CA'B$ is equal to angle A.

64. Show that the radius R of the circle circumscribed about the triangle ABC is given by

$$R = \frac{abc}{4rs}$$

where r is the radius of the inscribed circle. *Hint.* Use result in Problem 63 and the formulas for the area.

8

TRIGONOMETRIC
EQUATIONS

8.1. Introduction

An equation that involves at least one trigonometric function value is
called a *trigonometric equation*. In order to find the angles or numbers that
constitute the solution set if the equation involves function values of only
one angle or number, we first solve the given equation for a function value
and then find certain of the angles or numbers by use of a table. The entire
solution set can be found by a variety of procedures. We shall find it by
first obtaining those elements between zero and 2π, including zero. Then
the entire solution set can be found by adding $2n\pi$, n being an integer, to
each non-negative element less than 2π.

If a trigonometric equation involves $k\theta$ instead of or in addition to θ,
we shall begin by finding the elements of the solution set that are between

165

zero and $k\,2\pi$ so as to be able to get the elements of the solution set between zero and 2π.

Quite often it will be necessary to use one or more identities in order to change an equation to an equivalent one that is in a more desirable form.

8.2. Equations that Contain Only One Function Value and Only One Angle

If the equation contains only one function value of only one angle, the first step in finding the solution set is to solve the equation for the function value. We can then obtain certain elements of the solution set from a table and then find all elements θ such that $0 \leq \theta < 2\pi$. Finally, we obtain the solution set by adding $2n\pi$, n being an integer, to each of these elements.

EXAMPLE 1

Solve the equation

$$3 \sin \theta = 1 + 2 \sin \theta.$$

Solution: If we add $-2 \sin \theta$ to each member, we get

$$3 \sin \theta - 2 \sin \theta = 1,$$

or

$$\sin \theta = 1.$$

Therefore, the only solution in $0 \leq \theta < 2\pi$ is $\pi/2$. Hence, the solution set is $\{\theta \mid \theta = \pi/2 + 2n\pi\}$, where n is an integer.

EXAMPLE 2

Solve the equation

$$2 \cos^2 \theta - 3 \cos \theta = 2.$$

Solution: This is a quadratic equation in the unknown $\cos \theta$, and we may solve for this unknown by any of the methods for solving quadratics. We shall use the method of completing the square.

We first divide through by the coefficient of $\cos^2 \theta$ and get

$$\cos^2 \theta - \tfrac{3}{2} \cos \theta = 1.$$

Now we add the square of half the coefficient of $\cos \theta$ to each member and obtain

$$\cos^2 \theta - \tfrac{3}{2} \cos \theta + \tfrac{9}{16} = 1 + \tfrac{9}{16},$$
$$(\cos \theta - \tfrac{3}{4})^2 = \tfrac{25}{16}.$$

Next, we equate the square roots of the members and get

$$\cos \theta - \tfrac{3}{4} = \pm \tfrac{5}{4}.$$

Hence,

$$\cos \theta = 2 \text{ or } -\tfrac{1}{2}.$$

Since there is no value of θ for which $\cos \theta = 2$, this number must be excluded. However, if

$$\cos \theta = -\tfrac{1}{2},$$

then the values of θ in $0 \le \theta < 2\pi$ are $2\pi/3$ and $4\pi/3$, and the solution set is

$$\left\{ \theta \mid \theta = \frac{2\pi}{3} + 2n\pi \right\} \cup \left\{ \theta \mid \theta = \frac{4\pi}{3} + 2n\pi \right\}$$

where n is an integer.

8.3. Equations with One Member Factorable and the Other Zero

If one member of a trigonometric equation is zero and the other is factorable with each factor containing only one function value of only one angle, the first step in solving is to factor the non-zero member and set each factor equal to zero. The solution can then be completed by the procedure used in article 8.2.

EXAMPLE 1

Solve the equation

$$2 \sin \theta \cos \theta - \cos \theta = 0.$$

Solution: If we factor the left member, we obtain

$$\cos \theta (2 \sin \theta - 1) = 0.$$

Next, we set each factor equal to zero and get

$$\cos \theta = 0, \tag{1}$$

and

$$2 \sin \theta - 1 = 0,$$

or

$$\sin \theta = \tfrac{1}{2}. \tag{2}$$

Hence, the values of θ in $0 \le \theta < 2\pi$ are $\pi/2$ and $3\pi/2$ from (1), and $\pi/6$ and $5\pi/6$ from (2). Therefore, the solution set is

$$\{\theta \mid \theta = \pi/6 + 2n\pi, \ \pi/2 + 2n\pi, \ 5\pi/6 + 2n\pi, \ 3\pi/2 + 2n\pi\},$$

where n is an integer.

EXAMPLE 2

Solve the equation

$$\tan 2\theta \sin \theta - \sin \theta - \tan 2\theta + 1 = 0.$$

Solution: In this equation, we have two function values and two angles. However, if we factor the left member, we have only one function value of only one angle in each factor. Thus,

$$\sin \theta(\tan 2\theta - 1) - (\tan 2\theta - 1) = 0,$$

or

$$(\sin \theta - 1)(\tan 2\theta - 1) = 0.$$

Since each of the above factors contains only one function value and only one angle, we may set each factor equal to zero and solve. Thus, from the first, we obtain

$$\sin \theta - 1 = 0,$$
$$\sin \theta = 1,$$
$$\theta = \frac{\pi}{2};$$

and from the second,

$$\tan 2\theta - 1 = 0,$$
$$\tan 2\theta = 1,$$
$$2\theta = \frac{\pi}{4}, \frac{5\pi}{4}, \frac{9\pi}{4}, \frac{13\pi}{4}$$
$$\theta = \frac{\pi}{8}, \frac{5\pi}{8}, \frac{9\pi}{8}, \frac{13\pi}{8}.$$

In this case it was necessary to get all values of 2θ between 0 and $2(2\pi)$ in order to obtain all positive values of θ less than 2π.

Hence, the solution set is $\{\pi/8, \pi/2, 5\pi/8, 9\pi/8, 13\pi/8\}$ in the interval between 0 and 2π, and the complete solution set can be obtained as usual by adding $2n\pi$ to each number.

In obtaining the solution set of

$$\tan 2\theta - 1 = 0,$$

instead of using the procedure that we did, we could have found the values of 2θ

such that $0 \leq 2\theta < 2\pi$ and have added $2n\pi$ to each of them and then have solved for θ by dividing by 2. If we had done this, we would have had

$$2\theta = \pi/4 + 2n\pi, \quad 5\pi/4 + 2n\pi$$
$$\theta = \pi/8 + n\pi, \quad 5\pi/8 + n\pi.$$

It can be readily verified that these are the same values as obtained by the other procedure.

EXERCISE 8.1

Find the elements of the solution sets of the following equations that are in $0 \leq \theta < 2\pi$.

1. $\sec \theta - 2 = 0$
2. $2 \sin \theta - 1 = 0$
3. $\tan \theta + \sqrt{3} = 0$
4. $\sqrt{3} \sec \theta - 2 = 0$
5. $4 \sin^2 \theta - 3 = 0$
6. $3 \sec^2 \theta - 4 = 0$
7. $3 \cot^2 \theta - 1 = 0$
8. $4 \cos^2 - 3 = 0$
9. $\tan^2 \theta + \tan \theta = 0$
10. $2 \cos^2 \theta - \sqrt{3} \cos \theta = 0$
11. $\sqrt{3} \tan^2 \theta + \tan \theta = 0$
12. $\sec^2 \theta - 2 \sec \theta = 0$
13. $2 \cos^3 \theta - \cos \theta = 0$
14. $\sec^3 \theta - 4 \sec \theta = 0$
15. $3 \sec^3 \theta - 4 \sec \theta = 0$
16. $\tan^3 \theta - 3 \tan \theta = 0$
17. $2 \sin^2 \theta - \sin \theta - 1 = 0$
18. $2 \sin^2 \theta - 5 \sin \theta - 3 = 0$
19. $2 \cos^2 \theta + 3 \cos \theta - 2 = 0$
20. $\cos^2 \theta + 2 \cos \theta - 3 = 0$
21. $2 \cos^2 \theta - \cos \theta - 1 = 0$
22. $\sqrt{3} \tan^2 \theta - 2 \tan \theta = \sqrt{3}$
23. $2 \sin^2 \theta + 3 \sin \theta - 2 = 0$
24. $2\sqrt{3} \cos^2 \theta + \cos \theta = 2\sqrt{3}$
25. $\tan^2 \theta - (1 + \sqrt{3}) \tan \theta + \sqrt{3} = 0$
26. $\sqrt{3} \cot^2 \theta + (\sqrt{3} - 1) \cot \theta - 1 = 0$
27. $\sqrt{3} \sec^2 \theta - 2(1 + \sqrt{3}) \sec \theta + 4 = 0$
28. $\sqrt{3} \tan^2 \theta - 4 \tan \theta + \sqrt{3} = 0$
29. $2 \sin \theta \cos \theta + \sin \theta = 0$
30. $2 \sin \theta \tan \theta - \tan \theta = 0$
31. $2 \cos \theta \sin \theta + \cos \theta = 0$
32. $\sqrt{3} \cos \theta \tan \theta - \cos \theta = 0$
33. $\sec \theta \tan \theta - 2 \tan \theta + \sec \theta - 2 = 0$
34. $2 \sin \theta \sec \theta - \sec \theta - 4 \sin \theta + 2 = 0$
35. $\sin \theta \sec \theta - \sec \theta - \sin \theta + 1 = 0$

36. $2 \sin \theta \cos \theta + 2 \cos \theta - \sin \theta - 1 = 0$

37. $2 \sin \theta \tan \theta - 2\sqrt{3} \sin \theta - \tan \theta + \sqrt{3} = 0$

38. $\cos \theta \tan \theta - \sqrt{3} \cos \theta + \tan \theta - \sqrt{3} = 0$

39. $\sqrt{3} \tan \theta - 3 \tan \theta \cot \theta - 1 + \sqrt{3} \cot \theta = 0$

40. $\sec \theta \cot \theta + \sqrt{3} \sec \theta - 2 \cot \theta - 2\sqrt{3} = 0$

41. $2 \sin \theta \cos 2\theta - \cos 2\theta + 2 \sin \theta - 1 = 0$

42. $\sin \theta \sin 2\theta + 2 \sin 2\theta + \sin \theta + 2 = 0$

43. $2 \sin \theta \tan 3\theta + \tan 3\theta - 2\sqrt{3} \sin \theta - \sqrt{3} = 0$

44. $\sec \theta \csc 2\theta - 2 \sec \theta + 2 \csc 2\theta - 4 = 0$

8.4. Equations Reducible to a Form Solvable by Factoring

In order to solve an equation of this type, it is necessary to use the identities of Chapters 3 and 4 to transform the equation into an equivalent one that contains only one function value of one angle or into one that can be factored so that each factor contains only one function value of one angle. The resulting equation may then be solved by the methods of 8.2 or 8.3.

EXAMPLE 1

Solve the equation

$$2 \cos \theta - 5 + 2 \sec \theta = 0.$$

Solution: This equation contains two function values of one angle but can be transformed into an equation that contains only one function value of one angle by multiplying through by $\cos \theta$, since $\sec \theta \cos \theta = 1$. Thus we get the quadratic

$$2 \cos^2 \theta - 5 \cos \theta + 2 = 0$$

$$(2 \cos \theta - 1)(\cos \theta - 2) = 0 \qquad \text{factoring.}$$

Therefore, solving for $\cos \theta$, we find that

$$\cos \theta = \tfrac{1}{2}, \ 2.$$

There are no real angles corresponding to $\cos \theta = 2$, since $\cos \theta$ is bounded above by 1. The values of θ in $0 \leq \theta < 2\pi$ for which $\cos \theta = \tfrac{1}{2}$ are $\pi/3$ and $5\pi/3$. Hence, the solution set is

$$\{\theta \mid \theta = \pi/3 + 2n\pi\} \cup \{\theta \mid \theta = 5\pi/3 + 2n\pi\},$$

where n is an integer.

EXAMPLE 2

Solve

$$\cos 3\theta + \cos \theta - 2 \cos 2\theta = 0.$$

Solution: If we apply (4.23) to the first two terms of the given equation, we get

$$2 \cos 2\theta \cos \theta - 2 \cos 2\theta = 0$$
$$2 \cos 2\theta(\cos \theta - 1) = 0, \quad \text{factoring.}$$

If we set the first factor equal to zero and solve for values of 2θ between 0 and $2(2\pi)$, we find that $2\theta = \pi/2, 3\pi/2, 5\pi/2, 7\pi/2$. Therefore, $\theta = \pi/4, 3\pi/4, 5\pi/4, 7\pi/4$. If we set the second factor equal to 0 and solve for $\cos \theta$, we find that

$$\cos \theta = 1.$$

Therefore, $\theta = 0$. Hence, the solution set with non-negative elements less than 2π is $\{0, \pi/4, 3\pi/4, 5\pi/4, 7\pi/4\}$.

8.5. Equations of the Form $A \sin \theta + B \cos \theta = C$

An equation of the form

$$A \sin \theta + B \cos \theta = C \tag{1}$$

can be solved by letting r be a positive number and setting

$$A = r \cos \alpha \tag{2}$$

and

$$B = r \sin \alpha. \tag{3}$$

If this is done, (1) becomes

$$r(\cos \alpha \sin \theta + \sin \alpha \cos \theta) = C,$$
$$r \sin (\theta + \alpha) = C \quad \text{by (4.9)}$$

Hence,

$$\sin (\theta + \alpha) = \frac{C}{r}. \tag{4}$$

We shall now see how to determine r and α since, if r is known, we can obtain the value of $\theta + \alpha$ and then determine θ after α is found. If we add the squares of the corresponding members of (2) and (3), we have

$$A^2 + B^2 = r^2(\cos^2 \alpha + \sin^2 \alpha)$$
$$= r^2 \quad \text{by (3.6).}$$

Therefore, since r is positive, we have

$$r = \sqrt{A^2 + B^2}.$$

In order to determine α, we divide each member of (3) by the corresponding member of (2) and get

$$\frac{B}{A} = \frac{\sin \alpha}{\cos \alpha}$$

$$= \tan \alpha;$$

consequently, α is an angle whose tangent is B/A, but it must be so selected that both (2) and (3) are satisfied.

EXAMPLE

Solve the equation

$$1 \sin \theta - \sqrt{3} \cos \theta = 1.$$

Solution: In this problem, $A = 1$, and $B = -\sqrt{3}$; hence, we let

$$1 = r \cos \alpha, \quad \text{and} \quad -\sqrt{3} = r \sin \alpha. \tag{5}$$

Therefore, the given equation becomes

$$r(\cos \alpha \sin \theta + \sin \alpha \cos \theta) = 1,$$

or, dividing by r and using (4.9),

$$\sin (\theta + \alpha) = \frac{1}{r}. \tag{6}$$

Furthermore,

$$r = \sqrt{A^2 + B^2}$$

becomes

$$r = \sqrt{1^2 + (-\sqrt{3})^2} = 2,$$

and

$$\alpha \text{ is an angle whose tangent is } \frac{B}{A}$$

becomes

$$\alpha \text{ is an angle whose tangent is } \frac{-\sqrt{3}}{1}.$$

From this statement alone we could have many values of α, including $2\pi/3$ and $-\pi/3$, but $\sin 2\pi/3$ is positive, and $\cos 2\pi/3$ is negative; hence, these values do not satisfy (5). We can use $-\pi/3$ since, as required by (5), $\cos \alpha = \cos (-\pi/3)$ is positive, and $\sin \alpha = \sin (-\pi/3)$ is negative. Now, substituting $r = 2$ and $\alpha = -\pi/3$ in (6), we have

$$\sin \left(\theta - \frac{\pi}{3}\right) = \frac{1}{2};$$

hence,

$$\theta - \frac{\pi}{3} = \frac{\pi}{6}, \frac{5\pi}{6},$$

and

$$\theta = \frac{\pi}{2}, \frac{7\pi}{6}$$

are the nonnegative solutions less than 2π. Consequently, the solution set is
$$\{\theta \mid \theta = \pi/2 + 2n\pi\} \cup \{\theta \mid \theta = 7\pi/6 + 2n\pi\}.$$

EXERCISE 8.2

Solve each of the following equations for nonnegative values of θ that are less than 2π.

1. $2 \sin^2 \theta + 3 \cos \theta - 3 = 0$
2. $4 \tan^2 \theta - 3 \sec^2 \theta = 0$
3. $\tan^2 \theta + \sec \theta - 1 = 0$
4. $2 \sin^2 \theta + 5 \cos \theta + 1 = 0$
5. $2 \cos \theta + 1 - \sec \theta = 0$
6. $2 \sin \theta + \sqrt{3} - 2 - \sqrt{3} \csc \theta = 0$
7. $\sqrt{3} \tan \theta + \sqrt{3} - 1 - \cot \theta = 0$
8. $\csc \theta - 3 + 2 \sin \theta = 0$
9. $\cos 2\theta - \sin \theta = 0$
10. $2 \cos^2 \theta + 2 \cos 2\theta = 1$
11. $\tan 2\theta = \cot \theta$
12. $\cos 2\theta = 2 \sin^2 \theta$
13. $\sin 2\theta - \cos^2 \theta + \sin^2 \theta = 0$
14. $2(\cos 2\theta - \sin 2\theta)(\cos 2\theta + \sin 2\theta) = 1$
15. $\tan \frac{1}{2}\theta + \sin 2\theta = \csc \theta$
16. $2 \cos^2 \frac{\theta}{2} + 2 \cos \theta = 1$
17. $\cos^2 \frac{\theta}{2} - \cos \theta = 0$

18. $\cos 2\theta + 2 \cos^2 \dfrac{\theta}{2} = 0$

19. $2 \sin^2 \dfrac{\theta}{2} - 2 = \cos \theta$

20. $\tan \dfrac{\theta}{2} + 1 = \cos \theta$

21. $\sin 2\theta \cos 3\theta - \sin 3\theta \cos 2\theta = 0$

22. $\sin 3\theta \cos \theta - \cos 3\theta \sin \theta = 0$

23. $\sin 2\theta \cos \theta + \cos 2\theta \sin \theta = .5$

24. $\sin 4\theta \cos 2\theta - \cos 4\theta \sin 2\theta = \sqrt{3}/2$

25. $\cos 3\theta \cos \theta + \sin 3\theta \sin \theta = \sin 4\theta$

26. $\cos 2\theta \cos \theta - \sin 2\theta \sin \theta = 0$

27. $\dfrac{\tan 3\theta - \tan \theta}{1 + \tan 3\theta \tan \theta} = \sqrt{3}$

28. $\dfrac{\tan 2\theta + \tan \theta}{1 - \tan 2\theta \tan \theta} = \cot 3\theta$

29. $\cos 3\theta - \cos \theta = 0$

30. $\cos 3\theta + \cos 2\theta = 0$

31. $\sin 2\theta + \sin 4\theta = 0$

32. $\sin 3\theta - \sin \theta = 0$

33. $-\sin \theta + \sqrt{3} \cos \theta = 1$

34. $\sqrt{3} \sin \theta + \cos \theta = 3$

35. $\sin \theta - \sqrt{3} \cos \theta = 1$

36. $-7 \sin \theta + 24 \cos \theta = 12.5\sqrt{3}$

37. $2 \sin \theta + 2 \cos \theta = \sqrt{2}$

38. $\sin \theta + \cos \theta = \sqrt{6}/2$

39. $\sin \theta + \sqrt{3} \cos \theta = 2$

40. $\sin 2\theta \cos \theta + \cos 2\theta \sin \theta + \sin 2\theta \sin \theta - \cos 2\theta \cos \theta = 0$

9

THE INVERSE
TRIGONOMETRIC
FUNCTIONS

9.1. Functions and Inverse Relations

We shall begin this chapter by recalling that a *function* is a set of ordered pairs (x, y) such that no two pairs have the same first element and different second elements. Furthermore, the set of values assumed by the first elements is the *domain*, and the set of values taken on by the second elements is the *range*. Thus,

$$f = \{(1, 4), (2, 5), (3, 6)\} \qquad \text{and} \qquad g = \{(1, 4), (2, 5), (3, 5)\}$$

are functions with domain $\{1, 2, 3\}$. The range of the function f is $\{4, 5, 6\}$ and the range of g is $\{4, 5\}$. We should recall that the notation

$$f = \{(x, y) \mid y = f(x)\}$$

is used to designate a function and that $f(x)$ is called the function value.

If we interchange the first and second elements in the ordered number pairs in f above, we obtain the set of pairs

$$\{(4, 1), (5, 2), (6, 3)\}$$

and they constitute a function with domain $\{4, 5, 6\}$ and range $\{1, 2, 3\}$. If, however, we interchange the first and second elements in g above, we obtain

$$\{(4, 1), (5, 2), (5, 3)\},$$

and this set is not a function since two ordered pairs have the same first element 5 and different second elements 2 and 3.

The set of ordered pairs obtained by interchanging the elements in each ordered pair of a function is called the *inverse relation*. Furthermore, if the inverse relation is a function, it is called the *inverse function*.

Since by interchanging the elements in the ordered pairs of

$$f = \{(1, 4), (2, 5), (3, 6)\}$$

we obtain the function

$$\{(4, 1), (5, 2), (6, 3)\},$$

the latter is called the inverse function of f. It is designated by f^{-1}.

It follows from the definition of a function that *if a function f is such that no two of its ordered pairs with different first elements have the same second element, then corresponding to f there is the inverse function f^{-1}.* It is obtained by interchanging first and second elements of each ordered pair in f.

If a function is designated by

$$f = \{(x, y) \mid y = f(x)\},$$

then the equation that defines the inverse relation is $x = f(y)$. If this equation is solvable for y, the solution is usually expressed in the form $y = f^{-1}(x)$, and the inverse relation is written as

$$f^{-1} = \{(x, y) \mid y = f^{-1}(x)\}$$

and is the inverse function, provided $y = f^{-1}(x)$ defines a function.

EXAMPLE 1

Find the inverse of

$$f = \{(x, y) \mid y = f(x) = 2x - 3\}$$

and sketch the graphs of the function and the inverse.

Solution: The defining equation in f is $y = 2x - 3$; hence, interchanging x and y, we see that the defining equation for the inverse f^{-1} is $x = 2y - 3$. If we solve the latter for y, we get $y = \frac{1}{2}(x + 3)$. Consequently, the inverse of f is

$$f^{-1} = \{(x, y) \mid y = f^{-1}(x) = \tfrac{1}{2}(x + 3)\}.$$

The graphs of f and f^{-1} are the lines in Figure 9.1. It should be noted that the

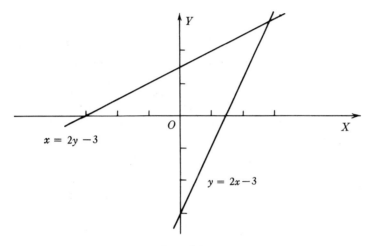

Figure 9.1

position of the graph of f^{-1} with respect to the Y-axis is the same as that of f with respect to the X-axis.

EXAMPLE 2

Find the inverse of

$$f = \{(x, y) \mid y = x^2 - 1\}. \tag{1}$$

Is the inverse a function or a relation?

Solution: The defining equation in (1) is

$$y = x^2 - 1.$$

Hence the defining equation of the inverse is

$$x = y^2 - 1,$$

and if we solve this equation for y we get

$$y = \pm\sqrt{x + 1}. \tag{2}$$

Consequently, for each value assigned to x, equation (2) determines two values of y which differ only in sign. Hence the inverse of (1) is

$$f^{-1} = \{(x, y) \mid y = \sqrt{x + 1}\} \cup \{(x, y) \mid y = -\sqrt{x + 1}\}.$$

This is a relation and not a function, since, for each value of $x > -1$, f^{-1} contains the pairs $(x, \sqrt{x + 1})$ and $(x, -\sqrt{x + 1})$ in which the first elements are the same and the second elements are different.

EXAMPLE 3

Restrict the domain of the function (1) so that the inverse relation f^{-1} is a function.

Solution: Since f^{-1} is obtained from the function f by interchanging the elements of each ordered pair, then f^{-1} is not a function unless the domain of f is restricted so that no two pairs of f have the same *second* element. If $x \geq 0$, $X > 0$, and $x \neq X$, then $x^2 - 1 \neq X^2 - 1$. Consequently, no two ordered pairs of f will have the same second element if the domain is restricted to the set $\{x \mid x \geq 0\}$. Therefore, the inverse of

$$f = \{(x, y) \mid y = x^2 - 1, x \geq 0\}$$

is a function. Then, since the range of the inverse of f is the domain of f, the range of f^{-1} is $\{y \mid y \geq 0\}$. Therefore in (2), we must use the positive sign before the radical, and thus obtain

$$f^{-1} = \{(x, y) \mid y = \sqrt{x + 1}\}.$$

9.2. The Inverse Trigonometric Functions

We shall now consider the sine function

$$\sin = \{(x, y) \mid y = \sin x\}$$

and the inverse relation

$$\sin^{-1} = \{(x, y) \mid x = \sin y\}.$$

The term *arcsine* is also used to indicate the inverse relation of sine. This relation is not a function since there is more than one value of y for some value of x, actually for each x. For example, if $x = .5$, then $.5 = \sin y$, and y may be $\pi/6$ or any angle whose sine is positive and whose related or reference angle is $\pi/6$. Thus, $y = \pi/6 + 2n\pi$ and $y = 5\pi/6 + 2n\pi$, where n is an integer. In set notation, $y = \{\pi/6 + 2n\pi\} \cup \{(5\pi/6 + 2n\pi\}$.

We can obtain a function from the relation

$$\sin^{-1} = \{(x, y) \mid x = \sin y\}$$

by so restricting the set of values that y may assume in such manner that there is only one value of y in the set for each x in the domain. This restriction could be accomplished in many ways, but the usual one is to choose the value of y so that $-\pi/2 \le y \le \pi/2$. We shall use the abbreviation Arcsine or Sin^{-1} to represent the arcsine function, and shall define it by

$$\text{Arcsine or Sin}^{-1} = \{(x, y) \mid x = \sin y, \ -\pi/2 \le y \le \pi/2\}. \quad (9.1)$$

The domain of the function is $\{x \mid -1 \le x \le 1\}$, and the range is

$$\{y \mid -\pi/2 \le y \le \pi/2\}.$$

Usually, instead of using $x = \sin y$, we use $y = \text{Arcsin } x$ or $y = \text{Sin}^{-1} x$ in the definition of the Arcsine.

Similar notations are used for the other inverse trigonometric functions. We shall now give the definition of each of them.

capital letter means you are on principle branch

$$\text{Arccosine} = \{(x, y) \mid y = \text{Cos}^{-1} x, \ 0 \le y \le \pi\} \quad (9.2)$$

$$\text{Arctangent} = \{(x, y) \mid y = \text{Tan}^{-1} x, \ -\pi/2 < y < \pi/2\} \quad (9.3)$$

$$\text{Arccotangent} = \{(x, y) \mid y = \text{Cot}^{-1} x, \ 0 < y < \pi\} \quad (9.4)$$

$$\text{Arcsecant} = \{(x, y) \mid y = \text{Sec}^{-1} x, \ 0 \le y \le \pi, \ y \ne \pi/2\} \quad (9.5)$$

$$\text{Arccosecant} = \{(x, y) \mid y = \text{Csc}^{-1} x, \ -\pi/2 \le y \le \pi/2, \ y \ne 0\} \quad (9.6)$$

Many older books, and some current ones, refer to the inverse trigonometric relations as "the inverse functions," and to the inverse functions as "the principal values of the inverse functions."

If we make use of (9.1), (9.3), and (9.6) as well as the function values of $\pi/6$, $\pi/4$ and their multiples, we find that

$$\text{Arcsin } \sqrt{3}/2 = \text{Sin}^{-1} \sqrt{3}/2 = \pi/3 \qquad \text{by (9.1)}$$

$$\text{Arctan } (-1) = \text{Tan}^{-1} (-1) = -\pi/4 \qquad \text{by (9.3)}$$

$$\text{Arccsc } 2 = \text{Csc}^{-1} 2 = \pi/6 \qquad \text{by (9.6)}$$

9.3. Graphs of the Inverse Trigonometric Functions

The two equations

$$y = \text{Arcsin } x \quad (1)$$

and

$$x = \sin y \quad (2)$$

"Arcsin" means "the angle whose sin is ..."

express the same relation between x and y for $-\pi/2 \le y \le \pi/2$ and consequently have the same graph. We shall sketch it by use of (2) and shall restrict y so that $-\pi/2 \le y \le \pi/2$, since the range of y in (1) is $-\pi/2$ to $\pi/2$. We shall make a table of corresponding values of y and x by assigning the multiples of $\pi/6$ and $\pi/4$ between $-\pi/2$ and $\pi/2$ to y and computing each corresponding value of $x = \sin y$. Thus we obtain

y	$-\pi/2$	$-\pi/3$	$-\pi/4$	$-\pi/6$	0	$\pi/6$	$\pi/4$	$\pi/3$	$\pi/2$
$x = \sin y$	-1	$-\sqrt{3}/2$	$-\sqrt{2}/2$	$-1/2$	0	$1/2$	$\sqrt{2}/2$	$\sqrt{3}/2$	1

If we locate the set of points determined by these ordered number pairs and draw a smooth curve through them, we get the solid part of the curve shown in Figure 9.2. The dotted portions can be obtained by assigning values less than $-\pi/2$ and greater than $\pi/2$ to y.

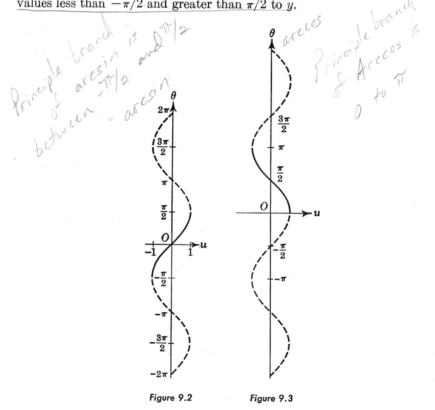

Figure 9.2   Figure 9.3

The graphs of the relations determined by $y = \arccos x$ and $y = \arctan x$ are shown in Figures 9.3 and 9.4. The graphs of the functions

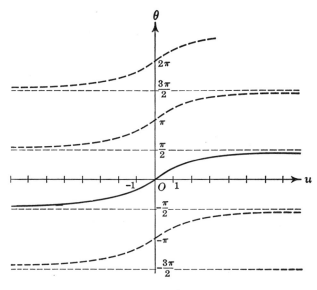

Figure 9.4

determined by $y = \text{Arccos } x$ and $y = \text{Arctan } x$ are shown as the solid portions of these figures.

9.4. Transformation of Inverse Trigonometric Functions

It is frequently desirable to express one inverse trigonometric function value in terms of one or more others. Another problem arising in connection with inverse trigonometric functions is that of finding the trigonometric function value of an inverse trigonometric function value. The methods to be applied to problems of this nature are illustrated in the following examples:

EXAMPLE 1

Find the value of $\sin (\text{Cos}^{-1} u)$.

— asking for sin of ∠ whose cos is u

Solution 1: If we let $\theta = \text{Cos}^{-1} u$, then $\cos \theta = u$, and we may construct the angle θ, $0 \leq \theta \leq \pi$, in standard position as in Figure 9.5. Obviously,

$$y = \sqrt{1 - u^2}.$$

Hence,

$$\sin \theta = \sqrt{1 - u^2}.$$

*sin (Arcsin u) is
sin of the angle
whose sin is
u
∴ sin(arcsin u)
= u.*

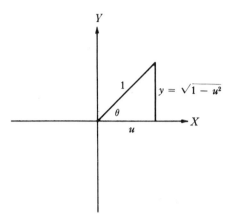

Figure 9.5

Note. In Figure 9.5, θ was constructed in Q_1. However, the same result would have been obtained if we had constructed θ in Q_2, since y is positive in both cases.

Solution 2: If we let $\theta = \text{Cos}^{-1} u$, then $\cos \theta = u$, and

$$\sin \theta = \sqrt{1 - \cos^2 \theta} = \sqrt{1 - u^2}.$$

The positive sign is used with the radical since $\text{Cos}^{-1} u$ is in the interval from 0 to π and the sine of any angle in this interval is positive or zero.

EXAMPLE 2

Arcsin

Find the value of $\tan 2[\text{Sin}^{-1}(-u)]$, $u > 0$.

Solution: If $\theta = \text{Sin}^{-1}(-u)$, $u > 0$, it follows that $\sin \theta = -u$ and that θ is in the interval from $-\pi/2$ to 0. Hence, if we construct the angle θ in standard

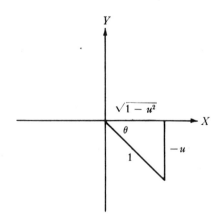

Figure 9.6

position as in Figure 9.6, we can readily determine any other trigonometric function value of θ. Thus, we have

$$\tan 2\theta = \frac{2\tan\theta}{1 - \tan^2\theta}$$

$$= \frac{2\left(\dfrac{-u}{\sqrt{1-u^2}}\right)}{1 - \dfrac{u^2}{1-u^2}}$$

$$= \frac{\dfrac{-2u}{\sqrt{1-u^2}}}{\dfrac{1-u^2-u^2}{1-u^2}}$$

$$= \frac{-2u(1-u^2)}{(1-2u^2)(\sqrt{1-u^2})}$$

$$= \frac{-2u\sqrt{1-u^2}}{1-2u^2}.$$

EXAMPLE 3

Express $\mathrm{Sin}^{-1} u - \mathrm{Cos}^{-1} u$, with $u > 0$, as an inverse tangent.

Solution: If $\alpha = \mathrm{Sin}^{-1} u$ and $\beta = \mathrm{Cos}^{-1} u$, $u > 0$, then

$$0 < \alpha \le \pi/2, \qquad 0 \le \beta < \pi/2,$$

$$\sin\alpha = u, \qquad \text{and} \qquad \cos\beta = u. \tag{1}$$

Furthermore,

$$\mathrm{Sin}^{-1} u - \mathrm{Cos}^{-1} u = \alpha - \beta. \tag{2}$$

We next take the tangent of each member of (2) and get

$$\tan(\mathrm{Sin}^{-1} u - \mathrm{Cos}^{-1} u) = \tan(\alpha - \beta) = \frac{\tan\alpha - \tan\beta}{1 + \tan\alpha\tan\beta}.$$

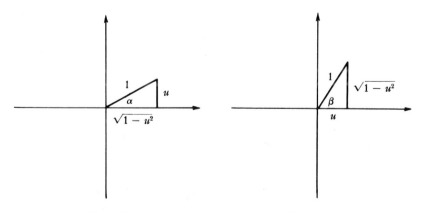

Figure 9.7   Figure 9.8

If the angles α and β are constructed in standard position by means of the information in (1), we obtain Figures 9.7 and 9.8. From them we can see that

$$\tan \alpha = \frac{u}{\sqrt{1 - u^2}} \quad \text{and} \quad \tan \beta = \frac{\sqrt{1 - u^2}}{u}.$$

Then,

$$\tan (\alpha - \beta) = \frac{\dfrac{u}{\sqrt{1 - u^2}} - \dfrac{\sqrt{1 - u^2}}{u}}{1 + \left(\dfrac{u}{\sqrt{1 - u^2}}\right)\left(\dfrac{\sqrt{1 - u^2}}{u}\right)}$$

$$= \frac{\dfrac{u^2 - 1 + u^2}{u\sqrt{1 - u^2}}}{1 + 1}$$

$$= \frac{2u^2 - 1}{2u\sqrt{1 - u^2}}.$$

Hence,

$$\alpha - \beta = \text{Sin}^{-1} u - \text{Cos}^{-1} u = \text{Arctan} \frac{2u^2 - 1}{2u\sqrt{1 - u^2}}.$$

The value of u is restricted by $-1 \le u \le 1$ since that is the domain of u in Arcsin u and Arccos u.

EXERCISE 9.1

Find the value of the inverse function in each of problems 1 to 12.

1. Arcsin $1/2$ 2. Arccos $\sqrt{3}/2$ 3. Arctan $\sqrt{3}$

4. Arccot $\sqrt{3}$ 5. Sec^{-1} (-2) 6. Csc^{-1} $(-2\sqrt{3}/3)$

7. Tan^{-1} $(-1/\sqrt{3})$ 8. Sec^{-1} $(-2\sqrt{3}/3)$ 9. Arccos $(-1/2)$

10. Arccot $(-1/\sqrt{3})$ 11. Arccot (-1) 12. Arcsin $(-1/\sqrt{2})$

Find the trigonometric function values in problems 13 to 24.

13. cos Arccos $.3$ 14. tan Arctan $.7$

15. cot Arccot (-2) 16. sin Arcsin $(-.6)$

17. sin Arccsc 3 18. cos Arcsec 5

19. tan Arccot 4 20. csc Arcsin $.4$

21. tan Arccos $(-1/2)$ 22. sin Arccos $(-3/5)$

23. sec Arcsin $(-5/13)$ 24. cot Arcsec $(-\sqrt{10}/3)$

If u and v are positive, find each function value called for in problems 25 to 52.

25. $\tan \operatorname{Arctan} u$	**26.** $\tan \operatorname{Arccot} u$
27. $\sin \operatorname{Arcsin} u$	**28.** $\cos \operatorname{Arccos} u$
29. $\cos \operatorname{Arccot} u$	**30.** $\sin \operatorname{Arcsec} u$
31. $\sin \operatorname{Arctan} u$	**32.** $\tan \operatorname{Arcsin} u$
33. $\sin \operatorname{Arccos}(-u)$	**34.** $\cos \operatorname{Arctan}(-u)$
35. $\tan \operatorname{Arccos}(-u)$	**36.** $\sin \operatorname{Arcsec}(-u)$
37. $\cos 2\operatorname{Arccos}(1/u)$	**38.** $\tan 2\operatorname{Arccot} u$
39. $\sin 2\operatorname{Arccos}(-u)$	**40.** $\sin 2\operatorname{Arcsin}(1/u)$
41. $\sin(\pi + \operatorname{Arcsin} u)$	**42.** $\cos(\pi - \operatorname{Arctan} u)$
43. $\cot(\pi + \operatorname{Arcsin} u)$	**44.** $\cos(\pi - \operatorname{Arctan} u)$
45. $\cos \frac{1}{2}(\operatorname{Arccos} u)$	**46.** $\sin \frac{1}{2}[\operatorname{Arccos}(-u)]$
47. $\tan \frac{1}{2}(\operatorname{Arctan} u)$	**48.** $\cos \frac{1}{2}[\operatorname{Arccos}(-u)]$
49. $\cos(\operatorname{Arccos} u + \operatorname{Arccos} v)$	**50.** $\sin(\operatorname{Arcsin} u + \operatorname{Arccos} v)$
51. $\tan(\operatorname{Arctan} u + \operatorname{Arccot} v)$	**52.** $\tan(\operatorname{Arcsec} u - \operatorname{Arctan} v)$

If θ is a positive acute angle, find the inverse function value in each of problems 53 to 60.

53. $\operatorname{Arcsin}(\sin \theta)$	**54.** $\operatorname{Arccos}(\sin \theta)$
55. $\operatorname{Arctan}(\cot \theta)$	**56.** $\operatorname{Arcsec}(\csc \theta)$
57. $\operatorname{Arccos}[\sin(-\theta)]$	**58.** $\operatorname{Arcsin}(-\sin \theta)$
59. $\operatorname{Arcsin}[-\sin(-\theta)]$	**60.** $\operatorname{Arccot}[\tan(-\theta)]$

Express the angle in each of problems 61 to 64 as an inverse sine value under the assumption that (x, y) is a point on the radius vector of length r.

61. $\operatorname{Arccos} x/r,\ x > 0$	**62.** $\operatorname{Arccos} x/r,\ x < 0$
63. $\operatorname{Arctan} y/x$	**64.** $\operatorname{Arccsc} r/y$

Express the sum or difference in each of problems 65 to 68 as a single angle.

65. $\operatorname{Arcsin} \frac{3}{5} + \operatorname{Arccos} \frac{3}{5}$	**66.** $\operatorname{Arcsin} \frac{5}{13} - \operatorname{Arcsin} \frac{3}{5}$
67. $\operatorname{Arccos} \frac{12}{13} - \operatorname{Arctan} \frac{3}{4}$	**68.** $\operatorname{Arctan} \frac{8}{15} + \operatorname{Arctan} \frac{3}{4}$

9.5. Inverse Trigonometric Identities

An equation that involves inverse trigonometric function values is called an *identity* if it is true for all admissible values of the variable. By the admissible values we mean the domain if only one function is involved and the intersection of the domains if more than one function is involved.

In order to prove that an equation that involves inverse trigonometric function values is an identity, we must show that, for every admissible value of the variable, an inverse trigonometric function value exists such that the two members of the equation are equal.

EXAMPLE 1

Prove that

$$\text{Arcsin } u = \frac{\pi}{2} - \text{Arccos } u \qquad -1 \le u \le 1. \tag{1}$$

Solution: If we let Arcsin $u = \theta$, then $u = \sin \theta$, and

$$\text{Arccos } u = \text{Arccos } (\sin \theta) = \frac{\pi}{2} - \theta.$$

Consequently, the right member of the given equation is

$$\frac{\pi}{2} - \left(\frac{\pi}{2} - \theta\right) = \theta.$$

Therefore, the given equation is an identity since the left member is also θ.

EXAMPLE 2

Prove that

$$2 \text{ Sin}^{-1} u = \text{Sin}^{-1} (2u\sqrt{1 - u^2}) \tag{2}$$

is an identity.

Solution: If we let

$$\text{Sin}^{-1} u = \theta,$$

then the left member of (2) becomes 2θ. Furthermore, for values of θ in the interval $-\pi/2 \le \theta \le \pi/2$,

$$u = \sin \theta, \quad \text{and} \quad \sqrt{1 - u^2} = \cos \theta,$$

and, for every value of u between -1 and 1, we have

$$\text{Sin}^{-1} (2u\sqrt{1 - u^2}) = \text{Sin}^{-1} (2 \sin \theta \cos \theta)$$
$$= \text{Sin}^{-1} (\sin 2\theta)$$
$$= 2\theta.$$

Hence, for every admissible value of u, the right and left members of (2) are equal to the same angle, and, therefore, it is an identity.

EXAMPLE 3

Prove that for $u \geq 0$,

$$\text{Arctan } u = \text{Arcsec } \sqrt{1 + u^2}. \tag{3}$$

Solution: The inverse function values in this equation have a meaning for every value of $u \geq 0$. Also, for each of these values, there is an angle θ in the interval from 0 to $\pi/2$ such that

$$\text{Arctan } u = \theta,$$

and

$$u = \tan \theta.$$

For these values of u and θ,

$$\text{Arcsec } \sqrt{1 + u^2} = \text{Arcsec } \sqrt{1 + \tan^2 \theta}$$
$$= \text{Arcsec } (\sec \theta)$$
$$= \theta.$$

Hence, (3) is an identity since, for every value of $u \geq 0$, both members are equal to the same angle.

Sometimes it is desirable to show that an expression containing inverse function values of constants is equal to another such expression. The method employed is similar to that used in identities, and we shall give one example of that type.

EXAMPLE 4

Prove that

$$\text{Arccos } \frac{2\sqrt{5}}{5} + \text{Arcsin } \left(\frac{-2\sqrt{13}}{13} \right) = \text{Arcsin } \frac{-\sqrt{65}}{65}. \tag{4}$$

Solution: If

$$\alpha = \text{Arccos } \frac{2\sqrt{5}}{5},$$

$$\beta = \text{Arcsin } \left(\frac{-2\sqrt{13}}{13} \right),$$

and

$$\gamma = \text{Arcsin } \frac{-\sqrt{65}}{65},$$

then,

$$\cos \alpha = \frac{2\sqrt{5}}{5},$$

$$\sin \beta = \frac{-2\sqrt{13}}{13},$$

and

$$\sin \gamma = \frac{-\sqrt{65}}{65}.$$

Furthermore,

$$0 < \alpha < \tfrac{1}{2}\pi, \tag{5}$$

and

$$-\tfrac{1}{2}\pi < \beta < 0. \tag{6}$$

Hence,

$$-\tfrac{1}{2}\pi < (\alpha + \beta) < \tfrac{1}{2}\pi,$$

as obtained by adding corresponding members of (5) and (6). Also,

$$-\tfrac{1}{2}\pi < \gamma < 0.$$

Hence, if $\sin (\alpha + \beta) = \sin \gamma$, then $\alpha + \beta = \gamma$. By (5), α is in the first quadrant, and

$$\sin \alpha = \sqrt{1 - \cos^2 \alpha} = \sqrt{1 - (\tfrac{2}{5}\sqrt{5})^2} = \sqrt{1 - \tfrac{4}{5}} = \sqrt{\tfrac{1}{5}} = \frac{\sqrt{5}}{5}.$$

Furthermore, by (6), β is in the fourth quadrant. Therefore,

$$\cos \beta = \sqrt{1 - \sin^2 \beta} = \sqrt{1 - \left(\frac{-2\sqrt{13}}{13}\right)^2} = \sqrt{1 - \frac{4}{13}} = \frac{3\sqrt{13}}{13}.$$

Therefore,

$$\sin (\alpha + \beta) = \sin \alpha \cos \beta + \cos \alpha \sin \beta$$

$$= \left(\frac{\sqrt{5}}{5}\right)\left(\frac{3\sqrt{13}}{13}\right) + \left(\frac{2\sqrt{5}}{5}\right)\left(\frac{-2\sqrt{13}}{13}\right)$$

$$= \frac{3\sqrt{65}}{65} - \frac{4\sqrt{65}}{65}$$

$$= \frac{-\sqrt{65}}{65}.$$

Hence, since $\sin \gamma = \frac{-\sqrt{65}}{65}$, statement (4) is true.

EXERCISE 9.2

Prove that the statement in each of problems 1 to 12 is true.

1. $\text{Arccos } (-3/\sqrt{13}) - \text{Arcsin } (3/\sqrt{13}) = \pi/2$
2. $\text{Arcsin } (4/5) - \text{Arcsin } (-3/5) = \pi/2$
3. $\text{Arctan } (5/3) - \text{Arctan } (1/4) = \pi/4$
4. $\text{Arctan } (3/7) + \text{Arctan } (2/5) = \pi/4$
5. $\text{Arctan } 2 - \text{Arctan } (1/3) = \pi/4$
6. $\text{Arcsin } (3/5) + \text{Arcsin } (\sqrt{2}/10) = \pi/4$
7. $\text{Arcsin } (1/\sqrt{10}) + \text{Arcsin } (3/\sqrt{10}) = \pi/2$
8. $\text{Arccos } (-2/\sqrt{13}) + \text{Arcsin } (3/\sqrt{13}) = \pi$
9. $\text{Arccos } (4/5) + \text{Arccos } (12/13) = \text{Arccos } (33/65)$
10. $\text{Arctan } (3/4) - \text{Arctan } (5/8) = \text{Arctan } (4/47)$
11. $\text{Arcsin } (1/2) + \text{Arcsin } (1/7) = \text{Arccos } (11/14)$
12. $\text{Arcsin } (1/3) + \text{Arccos } (7/11) = \text{Arcsin } (31/33)$

Prove that the following statements are true for every value of u and v or for all values in the specified domain.

13. $\text{Arccos } u = \text{Arcsec } (1/u), u \leq 1$
14. $\text{Arcsin } (1/u) = \text{Arccsc } u, -1 \leq u \leq 1$
15. $\text{Arctan } u = \text{Arccot } (1/u)$
16. $\text{Arcsin } u = \pi/2 - \text{Arccos } u, -1 \leq u \leq 1$
17. $\text{Arccos } (-u) + \text{Arccos } u = \pi, -1 \leq u \leq 1$
18. $\text{Arctan } u + \text{Arctan } (1/u) = \pi/2$
19. $\text{Arctan } u + \text{Arctan } (-u) = 0, -\infty < u < \infty$
20. $\text{Arcsin } u + \text{Arcsin } (-u) = 0, -1 \leq u \leq 1$
21. $2 \text{ Arcsin } \dfrac{1}{\sqrt{u^2 + 1}} = \text{Arccos } \dfrac{u^2 - 1}{u^2 + 1}$
22. $2 \text{ Arctan } (1/u) = \text{Arcsin } \dfrac{2u}{u^2 + 1}$
23. $2 \text{ Arccos } \dfrac{u}{\sqrt{u^2 + 1}} = \text{Arcsin } \dfrac{2u}{u^2 + 1}, 0 \leq u \leq 1$
24. $2 \text{ Arcsin } \dfrac{1}{\sqrt{u^2 + 1}} = \text{Arcsin } \dfrac{2u}{u^2 + 1}, u \geq 1$
25. $\dfrac{1}{2} \text{ Arccos } \dfrac{1 - u^2}{1 + u^2} = \text{Arctan } u$

26. $\dfrac{1}{2} \operatorname{Arctan} \dfrac{2u}{u^2 - 1} = \operatorname{Arctan}(1/u)$

27. $\dfrac{1}{2} \operatorname{Arcsin} \dfrac{-4u}{u^2 + 4} = \operatorname{Arcsin} \dfrac{-u}{\sqrt{u^2 + 4}},\ |u| \le 2$

28. $\dfrac{1}{2} \operatorname{Arctan} \dfrac{-4u}{4 - u^2} = \operatorname{Arctan}\left(\dfrac{-u}{2}\right),\ |u| \le 2$

29. $\operatorname{Arcsin} \dfrac{u}{\sqrt{u^2 + 1}} - \operatorname{Arcsin} \dfrac{1}{\sqrt{u^2 + 1}} = \operatorname{Arcsin} \dfrac{u^2 - 1}{u^2 + 1},\ u \ge 1$

30. $\operatorname{Arccos} \dfrac{1}{\sqrt{u^2 + 1}} - \operatorname{Arccos} \dfrac{u}{\sqrt{u^2 + 1}} = \operatorname{Arccos} \dfrac{2u}{u^2 + 1},\ u \ge 1$

31. $\operatorname{Arcsin} \sqrt{\dfrac{u}{1 + u}} - \operatorname{Arcsin} \dfrac{1}{\sqrt{1 + u}} = \operatorname{Arcsin} \dfrac{u - 1}{1 + u},\ u \ge 1$

32. $\operatorname{Arcsin} \dfrac{1}{u} + \operatorname{Arctan} \dfrac{1}{u} = \operatorname{Arcsin} \dfrac{u + \sqrt{u^2 - 1}}{u\sqrt{u^2 + 1}},\ u > 1$

33. $\operatorname{Arcsec} u - \operatorname{Arccos} \dfrac{\sqrt{u^2 - 1}}{u} = \operatorname{Arcsin} \dfrac{u^2 - 2}{u^2},\ u > \sqrt{2}$

34. $\operatorname{Arcsec} \sqrt{1 + u^2} + \operatorname{Arccos} \dfrac{2u}{1 + u^2} = \operatorname{Arcsin} \dfrac{1}{\sqrt{1 + u^2}},\ u > 1/\sqrt{3}$

35. $\operatorname{Arcsin} \dfrac{1}{\sqrt{1 + u^2}} - \operatorname{Arctan} u = \operatorname{Arccot} \dfrac{2u}{1 - u^2},\ u > 1$

36. $\operatorname{Arccos} \dfrac{1}{u} - \operatorname{Arcsec} \dfrac{u}{\sqrt{u^2 - 1}} = \operatorname{Arccsc} \dfrac{u^2}{u^2 - 2},\ u > \sqrt{2}$

37. $\operatorname{Arcsin} u - \operatorname{Arcsin} v = \operatorname{Arcsin}(u\sqrt{1 - v^2} - v\sqrt{1 - u^2})$

38. $\operatorname{Arctan}(1/u) - \operatorname{Arctan} v = \operatorname{Arctan} \dfrac{1 - uv}{u + v},\ u > 0,\ v > 0$

39. $\operatorname{Arcsin} \dfrac{u}{\sqrt{1 + u^2}} - \operatorname{Arccos} \dfrac{\sqrt{v^2 - 1}}{v} = \operatorname{Arccos} \dfrac{\sqrt{v^2 - 1} + u}{v\sqrt{u^2 + 1}},\ u > 1,\ v > \sqrt{2}$

40. $\operatorname{Arccos} \dfrac{1}{\sqrt{1 + u^2}} + \operatorname{Arcsin} \dfrac{v}{\sqrt{1 + v^2}} = \operatorname{Arcsin} \dfrac{u + v}{\sqrt{(1 + u^2)(1 + v^2)}}$

10

COMPLEX
NUMBERS

10.1. Introduction

If we use the quadratic formula to solve $x^2 - 6x + 13 = 0$, we find that $x = 3 + \sqrt{-4}$ and $x = 3 - \sqrt{-4}$. The square root of -4 is not a real number, since the square of a real number is positive or zero. Any negative number can be thought of and written as a positive number times -1. If this is done, we have

$$\sqrt{-P} = \sqrt{(-1)P}$$
$$= \sqrt{-1}\,\sqrt{P},$$

and this is written in the form $i\sqrt{P}$ since it is customary in mathematics to represent $\sqrt{-1}$ by i. If this symbol is used, the roots of the equation given above are $3 \pm 2i$ and are called *complex numbers*. In fact, *any number of the form* **a + bi** *with* **a** *and* **b** *real and* **i** $= \sqrt{-1}$ *is called* **a complex number.** The complex number $a + ib$ is sometimes written as the ordered number pair (a, b). If this is done, the first number in the pair represents the real part of the complex number and the second number represents the coefficient of the imaginary part. If $a = 0$, then the complex number $bi = (0, b)$ is called a *pure imaginary* number; if $b = 0$, then the complex number $a = (a, 0)$ is a real number. Consequently, the real numbers are a subset of the complex numbers.

The complex number $a + bi = (a, b)$ may be thought of as a vector with horizontal and vertical components a and b, respectively.

10.2. Equal Complex Numbers; Conjugate Complex Numbers

Two complex numbers are *equal* if, and only if, their real parts are equal and the coefficients of their imaginary parts are equal. For example, the two complex numbers $a + bi = (a, b)$ and $c + di = (c, d)$ are equal if, and only if, $a = c$ and $b = d$.

EXAMPLE

Determine x and y so that

$$x + 2 + 4i = 5 + (y - 3)i.$$

Solution: In order for two complex numbers to be equal, their real parts must be equal, and the coefficients of their imaginary parts must be equal. Hence, in this problem, we must have

$$x + 2 = 5 \qquad \text{from the real parts,}$$

and

$$4 = y - 3 \qquad \text{from the imaginary parts.}$$

Consequently, the given equation is satisfied if $x = 3$ and $y = 7$.

Two complex numbers are *conjugate* if, and only if, their real parts are equal and the coefficients of their imaginary parts differ only in algebraic sign. Thus, $a + bi$ and $a - bi$ are conjugate complex numbers, and the conjugate of $(2, 5)$ is $(2, -5)$.

10.3. Graphical Representation

We employ a pair of directed perpendicular lines in a plane, as indicated in Figure 10.1, as a frame of reference for representing complex numbers

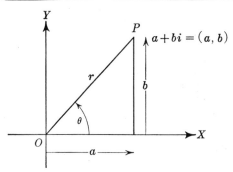

Figure 10.1

graphically. The horizontal line is called the *real axis;* the other, the *pure imaginary axis.* The real part and the coefficient of the imaginary part are used as the coordinates of a point in a rectangular coordinate system. For example, in Figure 10.1, the point P, whose coordinates are (a, b), is the graphical representation of the complex number $a + bi$. The length of the line from the origin to the point is called the *absolute value* or *modulus* of the complex number. Any angle from the positive real axis to the line joining the origin and the complex number is called its *amplitude* or *argument.*

The definitions in the above paragraph are illustrated in Figure 10.1. The absolute value of $a + bi = (a, b)$ is represented by

$$r = \sqrt{a^2 + b^2}, \tag{10.1}$$

as found by using the distance formula to find the distance between (a, b) and $(0, 0)$, and its argument is represented by

$$\theta = \arctan \frac{b}{a}. \tag{10.2}$$

The absolute value of $3 - 4i$ is $r = \sqrt{3^2 + (-4)^2} = 5$ and the argument is $\theta = \text{Arctan} (-4/3)$, but is not $\text{Arctan} [-(4/3)]$ and not $\text{Arctan} 4/(-3)$. It is the angle whose terminal side passes through $(3, -4)$.

In Figure 10.1, the line segment OP may be considered a vector, whose magnitude is $r = \sqrt{a^2 + b^2}$ and whose direction with respect to the axis of reals is determined by $\theta = \arctan b/a$. Then $a + bi$ is at the terminus of this vector. Hence, the position of the point in the plane that represents $a + bi$ is determined by the ordered number pair (a, b) or by the vector OP.

The following statements may be of use in determining the angle θ such that $0 \le \theta < 360°$, to be used in connection with (10.2):

If $a > 0$ and $b > 0$, then $0 < \theta < 90°$,
if $a < 0$ and $b > 0$, then $90° < \theta < 180°$,
if $a < 0$ and $b < 0$, then $180° < \theta < 270°$,
if $a > 0$ and $b < 0$, then $270° < \theta < 360°$.

Thus, if θ is determined by (10.2) for the number $3 - 4i$, it is between $270°$ and $360°$, since $a = 3 > 0$ and $b = -4 < 0$. The tangent of the related angle R is $4/3$; hence, $R = 53°10'$. Consequently, $\theta = 360° - 53°10' = 306°50'$. Therefore, we have found that $\theta = \text{Arctan} (-4)/(3) = 306°50'$.

EXERCISE 10.1

Plot the point that represents the given complex number in each of problems 1 to 16.

1. $2 + 3i$	2. $3 - 4i$	3. $5 + 2i$	4. $1 + 4i$
5. $(3, 4)$	6. $(2, 1)$	7. $(-1, -2)$	8. $(4, -3)$
9. $(5 - 2i)$	10. $(4 + i)$	11. $(3, 0)$	12. $(0, 2)$
13. $(4, -3)$	14. $(0, -3)$	15. $(-2 + 3i)$	16. $-5 - 4i$

Plot the point that represents the conjugate of the complex number in each of problems 17 to 24.

17. $5 - 6i$	18. $-2 + 3i$	19. $4 + 7i$	20. $-2 - 3i$
21. $(-2, 1)$	22. $(3, -4)$	23. $(-2, -1)$	24. $(5, 3)$

Find the absolute value and the argument of each complex number in problems 25 to 40.

25. $3 + 4i$	26. $2 - 5i$	27. $-4 + i$	28. $-2 - 3i$
29. $5 - 2i$	30. $4 + 5i$	31. $-3 - 7i$	32. $-3 + 5i$
33. $1 - i$	34. $-1 + \sqrt{3}i$	35. $-2 - 2i$	36. $\sqrt{3} - i$
37. $\sqrt{3} + i$	38. $-1 + i$	39. $-1 - \sqrt{3}i$	40. $3 - 3i$

Determine x and y in each of problems 41 to 52 so that the statement is true.

41. $x + 3i = 2 + yi$ 42. $2x - 6i = 4 + yi$
43. $5x - 8i = 10 + 2yi$ 44. $3x + 14i = 12 - 7yi$

45. $(x + y, -2) = (3, -2x)$ **46.** $(x, x + y) = 3 + i$

47. $x + (x - y)i = 3 + i$ **48.** $3x + 5i = 6 + (3x + y)i$

49. $2x + y + 3i = 4 + (x + y)i$ **50.** $2 + (2x + y)i = x - y + i$

51. $-1 + (x + 3y)i = x + y + i$ **52.** $x + 3y + 3i = 3 + (x - y)i$

10.4. Algebraic Addition and Subtraction

The real part of the sum of two complex numbers is the sum of their real parts, and the coefficient of the imaginary part is the sum of the coefficients of the imaginary parts. A similar statement is true for the difference of two complex numbers. In keeping with these statements, we have

$$(2 + 3i) + (4 + 5i) = (2 + 4) + (3 + 5)i$$
$$= 6 + 8i,$$

and $$(4, 5) - (-3, 4) = [4 - (-3), 5 - 4] = (7, 1)$$
$$= 7 + i.$$

10.5. Geometric Addition and Subtraction

The sum of two complex numbers may be found geometrically by plotting either of them and then using this number as the beginning point for laying off the other. The last point located is their sum. The length of the line from the origin to this final point is the absolute value of the sum. Figure 10.2 shows the sum of the two numbers $C = a + bi$ and $C' = c + di$.

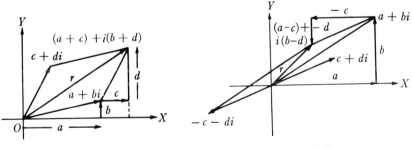

Figure 10.2   Figure 10.3

It should be noticed that, in Figure 10.2, the vector that represents the sum $(a + bi) + (c + di)$ is a diagonal of the parallelogram with the vectors that represent $a + bi$ and $c + di$ as sides.

The difference of two complex numbers may be obtained in the same manner, if only one recalls that $C - C' = C + (-C')$. The difference, $C - C'$, is shown in Figure 10.3.

We can see from Figure 10.3 that the point that represents $(a + bi) - (c + di)$ is determined by use of the parallelogram law for finding the vector that represents $(a + bi) + [-(c + di)] = (a - c, b + d)$.

10.6. Algebraic Multiplication and Division

The product of two complex numbers is obtained in the same manner as the product of any other two binomials. The product of $C = a + bi$ and $C' = c + di$ is $CC' = (a + bi)(c + di) = ac + adi + bci + bdi^2 = ac - bd + (ad + bc)i$, since $i^2 = -1$. Hence, we know that $(a, b)(c, d) = (ac - bd, ad + bc)$.

In order to express the quotient of two complex numbers in the form $a + bi$, we multiply the numerator and the denominator by the conjugate of the denominator and thus get a real number for the denominator. For example,

$$\frac{a + bi}{c + di} = \frac{a + bi}{c + di}\frac{c - di}{c - di} = \frac{ac + bd + (bc - ad)i}{c^2 + d^2}.$$

Consequently, we can say that

$$\frac{(a, b)}{(c, d)} = \left(\frac{ac + bd}{c^2 + d^2}, \frac{bc - ad}{c^2 + d^2}\right).$$

EXERCISE 10.2

Perform the indicated operation in each of problems 1 to 32 and leave the result in the form $a + bi$ or (a, b), according as the problem is in that form.

1. $(2 + 3i) + (4 + 5i)$
2. $(3 + 4i) + (5 - 2i)$
3. $(5 - 6i) + (-3 - 2i)$
4. $(-2 - i) + (3 + 4i)$
5. $(3, 1) + (5, 2)$
6. $(-2, -3) + (4, 1)$
7. $(4, 2) + (7, -3)$
8. $(7, 3) + (-4, -1)$
9. $(3 + 4i) - (2 + 6i)$
10. $(5 - 3i) - (2 + i)$
11. $(5 + 4i) - (3 - 2i)$
12. $(7 + 6i) - (-2 + 3i)$
13. $(4, 2) - (3, 0)$
14. $(5, -1) - (2, 3)$
15. $(-2, -5) - (-3, 4)$
16. $(-7, 3) - (-9, -1)$

17. $(3 + 2i)(5 + 3i)$ 18. $(2 - 4i)(3 + 5i)$

19. $(-6 + 3i)(3 - 4i)$ 20. $(-4 + 7i)(-2 - 3i)$

21. $(2, 3)(5, 1)$ 22. $(3, 0)(0, 2)$

23. $(4, -3)(-3, 4)$ 24. $(-2, -5)(2, -3)$

25. $\dfrac{3 + 4i}{2 + 3i}$ 26. $\dfrac{3 - 2i}{2 - 5i}$

27. $\dfrac{4 - i}{3 + 2i}$ 28. $\dfrac{4 + 3i}{3 - 4i}$

29. $(2, 3)/(3, 4)$ 30. $(2, -5)/(3, -2)$

31. $(3, 2)/(4, -1)$ 32. $(3, -4)/(4, 3)$

Find the value of x and of y for which the statement in each of problems 33 to 40 is true.

33. $(3 + yi)(2 + 3i) = x + 17i$ 34. $(x + 2i)(2 - 3i) = 12 - yi$

35. $(3 - i)(4 + 3i) = x + yi$ 36. $(2 + yi)(1 - 2i) = x + i$

37. $\dfrac{1 + yi}{x + 2i} = 3 + i$ 38. $\dfrac{5 - yi}{1 - 3i} = x + i$

39. $\dfrac{x + 7i}{3 + yi} = 4 - 3i$ 40. $\dfrac{x - 4i}{3 - 2i} = 5 + yi$

10.7. Trigonometric Representation

$$C = x + yi = r(\cos\theta + i\sin\theta)$$

There is another way to represent a complex number that has advantages over the forms $a + bi$ and (a, b). It is called the *trigonometric* or *polar form*, and makes use of the absolute value and the argument of the number. If the trigonometric form is used, the labor of multiplication, division and raising to a power is materially reduced.

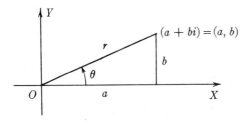

Figure 10.4

If we represent $a + bi = (a, b)$ graphically, we see from Figure 10.4 and the definition of $\sin \theta$ and $\cos \theta$ that

$$\cos \theta = \frac{a}{r} \qquad \text{and} \qquad \sin \theta = \frac{b}{r}.$$

Hence,

$$a = r \cos \theta, \qquad \text{and} \qquad b = r \sin \theta.$$

Therefore,

$$a + ib = r \cos \theta + ir \sin \theta$$
$$= r(\cos \theta + i \sin \theta).$$

The expression $r(\cos \theta + i \sin \theta)$ is the polar or trigonometric form. By use of (10.1) and (10.2) of article 10.3,

$$r = \sqrt{a^2 + b^2},$$

and

$$\theta = \arctan \frac{b}{a},$$

we may express the value of arctan (b/a) in degrees or radians, using a table of trigonometric function values. The polar form

$$r(\cos \theta + i \sin \theta)$$

is often abbreviated as r cis θ.

EXAMPLE

Express $1 - i\sqrt{3}$ in trigonometric form.

Solution: In this problem, $r = \sqrt{1^2 + (-\sqrt{3})^2} = 2$; furthermore, $\theta = $ arctan $b/a = $ arctan $(-\sqrt{3}/1)$. Therefore, θ is a fourth quadrant angle and is 300° or $-60°$ or any angle coterminal with them. We shall use $\theta = 300°$; hence,

$$1 - i\sqrt{3} = 2[\cos (300°) + i \sin (300°)].$$

10.8. The Product of Two Complex Numbers

The product of the two complex numbers

$$C = r(\cos \theta + i \sin \theta)$$

and

$$C' = r'(\cos \theta' + i \sin \theta')$$

is

$$CC' = r(\cos\theta + i\sin\theta)r'(\cos\theta' + i\sin\theta')$$
$$= rr'(\cos\theta\cos\theta' + i\cos\theta\sin\theta' + i\sin\theta\cos\theta' + i^2\sin\theta\sin\theta')$$
$$= rr'[(\cos\theta\cos\theta' - \sin\theta\sin\theta') + i(\sin\theta\cos\theta' + \cos\theta\sin\theta')],$$

which could have been obtained from the formula for the product as given in article 10.6. By equation (4.6), the expression in the first parentheses of the last line above is equal to $\cos(\theta + \theta')$ and, by (4.9), the expression in the second parentheses is equal to $\sin(\theta + \theta')$. Hence,

$$CC' = rr'[\cos(\theta + \theta') + i\sin(\theta + \theta')]. \qquad (10.3)$$

This formula for the product of two complex numbers should be remembered in words as:

> The **absolute value** of the product of two complex numbers is the product of their absolute values. An **argument** of the product is the sum of their arguments.

Since the product of two complex numbers is a complex number, we may obtain the product of any finite number of complex numbers by a repeated use of this formula. For example,

$$[2(\cos 30° + i\sin 30°)][\sqrt{2}(\cos 40° + i\sin 40°)]$$
$$[\sqrt{3}(\cos 50° + i\sin 50°)]$$
$$= [2\sqrt{2}(\cos 70° + i\sin 70°)][\sqrt{3}(\cos 50° + i\sin 50°)]$$
$$= 2\sqrt{6}(\cos 120° + i\sin 120°)$$
$$= 2\sqrt{6}\left(-\frac{1}{2} + \frac{i\sqrt{3}}{2}\right)$$
$$= \sqrt{6}(-1 + i\sqrt{3})$$
$$= -\sqrt{6} + 3i\sqrt{2}.$$

10.9. The Quotient of Two Complex Numbers

As was pointed out in article 10.6, the quotient of two complex numbers is obtained by multiplying both numerator and denominator by the conjugate of the denominator. Thus, we have

$$\frac{r_1(\cos\theta_1 + i\sin\theta_1)}{r_2(\cos\theta_2 + i\sin\theta_2)} = \frac{r_1(\cos\theta_1 + i\sin\theta_1)\,r_2(\cos\theta_2 - i\sin\theta_2)}{r_2(\cos\theta_2 + i\sin\theta_2)\,r_2(\cos\theta_2 - i\sin\theta_2)}$$

$$= \frac{r_1 r_2[\cos\theta_1\cos\theta_2 - i\cos\theta_1\sin\theta_2 + i\sin\theta_1\cos\theta_2 - i^2\sin\theta_1\sin\theta_2]}{r_2^2[\cos\theta_2\cos\theta_2 - i\cos\theta_2\sin\theta_2 + i\sin\theta_2\cos\theta_2 - i^2\sin\theta_2\sin\theta_2]}$$

$$= \frac{r_1[\cos\theta_1\cos\theta_2 + \sin\theta_1\sin\theta_2 + i(\sin\theta_1\cos\theta_2 - \cos\theta_1\sin\theta_2)]}{r_2(\cos^2\theta_2 + \sin^2\theta_2)}$$

$$= \frac{r_1}{r_2}\,[\cos(\theta_1 - \theta_2) + i\sin(\theta_1 - \theta_2)],$$

by (4.1), (4.12) and (3.6). Therefore,

$$\frac{r_1(\cos\theta_1 + i\sin\theta_1)}{r_2(\cos\theta_2 + i\sin\theta_2)} = \frac{r_1}{r_2}\,[\cos(\theta_1 - \theta_2) + i\sin(\theta_1 - \theta_2)]. \qquad (10.4)$$

This should be remembered in words as:

*The **absolute value** of the quotient of two complex numbers is the quotient of their absolute values. An **argument** of the quotient is the difference of their arguments.*

10.10. The Product and Quotient Graphically

The polar form of complex numbers can be used to obtain a graphical representation of the product or quotient of two complex numbers. We shall consider

$$Z = a + bi = (a,\, b) = r\ cis\ \theta$$

and

$$W = c + di = (c,\, d) = R\ cis\ \varphi$$

Each of the points, located in Figure 10.5, is connected to the origin. Consequently, $OZ = r$, $OW = R$, angles XOZ, and XOW are θ and φ, respectively. We now plot $(1, 0)$ and call the point T. Next, we join Z and T. Further, we draw a ray from W that makes an angle with OW that is equal angle OTZ, and a ray from O that makes an angle equal to θ with OW. Then, if Q is the intersection of these rays, it follows that triangles OTZ and OWQ are similar. Consequently,

$$\frac{OQ}{r} = \frac{OW}{1} = \frac{R}{1}$$

and $OQ = rR$. Furthermore, since angle XOQ is $\theta + \varphi$, it is clear that the point Q represents the product $(r\ cis\ \theta)(R\ cis\ \varphi) = rR\ cis\ (\theta + \varphi)$.

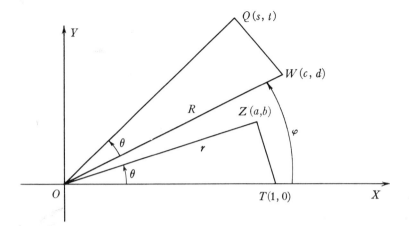

Figure 10.5

The construction for obtaining the point Q that represents the quotient

$$\frac{Z}{W} = \frac{(a, b)}{(c, d)} = \frac{r \; cis \; \theta}{R \; cis \; \varphi}$$

is shown in Figure 10.6. In it, angle OZQ = angle OWT by construction,

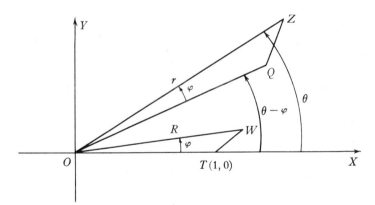

Figure 10.6

and angle TOW = angle QOZ = φ. Hence triangles TOW and QOZ are similar. Consequently,

$$\frac{OQ}{1} = \frac{OZ}{OW} = \frac{r}{R}$$

and, since angle $XOQ = \theta - \varphi$, the point Q represents the quotient $r \; cis \; \theta / R \; cis \; \varphi = (r/R) \; cis \; (\theta - \varphi)$.

EXERCISE 10.3

Express the number in each of problems 1 to 16 in polar form. Evaluate the angles in problems 1 to 8 exactly and the others to the nearest 10 minutes.

1. $1 + i$ **2.** $-1 + \sqrt{3}i$ **3.** $\sqrt{3} - i$ **4.** $-1 - i$

5. $(-\sqrt{3}, 1)$ **6.** $(1, -\sqrt{3})$ **7.** $(-1, -1)$ **8.** $(\sqrt{3}, 3)$

9. $(2, -1)$ **10.** $(-3, -2)$ **11.** $(2, 5)$ **12.** $(-3, 4)$

13. $-1 - 3i$ **14.** $2 + 3i$ **15.** $-3 + 5i$ **16.** $5 + 2i$

Perform the operations indicated in problems 17 to 40 and express each result in the form $a + bi$.

17. $(2 \ cis \ 27°)(5 \ cis \ 18°)$ **18.** $(3 \ cis \ 16°)(2 \ cis \ 14°)$

19. $(3 \ cis \ 32°)(4 \ cis \ 28°)$ **20.** $(7 \ cis \ 39°)(3 \ cis \ 6°)$

21. $(5 \ cis \ 76°)(6 \ cis \ 44°)$ **22.** $(2 \ cis \ 113°)(2 \ cis \ 22°)$

23. $(8 \ cis \ 148°)(3 \ cis \ 62°)$ **24.** $(4 \ cis \ 197°)(3 \ cis \ 133°)$

25. $\dfrac{18 \ cis \ 126°}{9 \ cis \ 81°}$ **26.** $\dfrac{15 \ cis \ 77°}{3 \ cis \ 17°}$

27. $\dfrac{8 \ cis \ 72°}{2 \ cis \ 42°}$ **28.** $\dfrac{6 \ cis \ 83°}{2 \ cis \ 38°}$

29. $\dfrac{28 \ cis \ 237°}{7 \ cis \ 12°}$ **30.** $\dfrac{14 \ cis \ 171°}{7 \ cis \ 51°}$

31. $\dfrac{30 \ cis \ 276°}{12 \ cis \ 66°}$ **32.** $\dfrac{108 \ cis \ 354°}{36 \ cis \ 24°}$

33. $\dfrac{(2 \ cis \ 41°)(3 \ cis \ 22°)}{6 \ cis \ 33°}$ **34.** $\dfrac{(5 \ cis \ 17°)(4 \ cis \ 89°)}{10 \ cis \ 16°}$

35. $\dfrac{(3 \ cis \ 28°)(5 \ cis \ 42°)}{15 \ cis \ 25°}$ **36.** $\dfrac{(12 \ cis \ 72°)(15 \ cis \ 81°)}{90 \ cis \ 93°}$

37. $\dfrac{36 \ cis \ 49°}{(9 \ cis \ 13°)(2 \ cis \ 6°)}$ **38.** $\dfrac{30 \ cis \ 105°}{(5 \ cis \ 20°)(3 \ cis \ 25°)}$

39. $\dfrac{66 \ cis \ 49°}{(6 \ cis \ 37°)(11 \ cis \ 42°)}$ **40.** $\dfrac{72 \ cis \ 27°}{(18 \ cis \ 81°)(2 \ cis \ 66°)}$

Convert the complex numbers in each of problems 41 to 48 to polar form, perform the indicated operation, and leave the result in the form $r \ cis \ \theta$.

41. $(1 + i)(\sqrt{3} - i)$ **42.** $(-1 + \sqrt{3}i)(\sqrt{3} + i)$

43. $(-1 - \sqrt{3}i)(-1 + i)$ **44.** $(1 - i)(-1 - \sqrt{3}i)$

45. $\dfrac{-1 - i}{\sqrt{3} + i}$ **46.** $\dfrac{\sqrt{3} - i}{1 + i}$

47. $\dfrac{-1 + \sqrt{3}i}{\sqrt{3} - i}$ 　　　　　　　　　　**48.** $\dfrac{1 + \sqrt{3}i}{-\sqrt{3} + i}$

10.11. DeMoivre's Theorem

If, in (10.3), we let $C' = C$, we have

$$CC' = C^2 = r^2(\cos 2\theta + i \sin 2\theta).$$

Furthermore,

$$
\begin{aligned}
C^3 &= C^2C \\
&= r^2(\cos 2\theta + i \sin 2\theta)\, r(\cos \theta + i \sin \theta) \\
&= r^3(\cos 3\theta + i \sin 3\theta).
\end{aligned}
$$

This process can be continued until we have raised C to any desired positive integral power. We then have

$$[r(\cos \theta + i \sin \theta)]^n = r^n(\cos n\theta + i \sin n\theta). \qquad (10.5)$$

This formula can be proved for integral values of n by mathematical induction, and is known as *DeMoivre's theorem*.

EXAMPLE 1

Raise $(1 + i)$ to the fourth power.

Solution:

$$
\begin{aligned}
(1 + i)^4 &= [\sqrt{2}(\cos \arctan 1 + i \sin \arctan 1)]^4 \\
&= [\sqrt{2}(\cos 45° + i \sin 45°)]^4 \\
&= (\sqrt{2})^4[\cos 4(45°) + i \sin 4(45°)] \\
&= 4(\cos 180° + i \sin 180°) \\
&= -4.
\end{aligned}
$$

10.12. Roots of Complex Numbers

In the field of real numbers, there is no square root of -4, no fourth root of -16, and, in general, no even root of a negative number. Moreover, there is only one cube root of 27 and only one fifth root of -32. If, however, we employ complex numbers, we can obtain n nth roots of any number by use of DeMoivre's theorem.

We shall illustrate the procedure to be followed by solving an example.

EXAMPLE 1

Find the cube roots of 27.

Solution: If we express $27 = 27 + 0i$ in trigonometric form, we have

$$27 = 27(\cos 0° + i \sin 0°)$$
$$= 27[\cos (0° + n\ 360°) + i \sin (0° + n\ 360°)] \tag{1}$$

for n any integer, since $\sin \theta$ and $\cos \theta$ are periodic with 360° as a period. Now, if we let

$$27^{1/3} = r(\cos \theta + i \sin \theta)$$

and equate the cubes of the members, we have

$$27 = [r(\cos \theta + i \sin \theta)]^3 \qquad \text{cubing}$$
$$= r^3(\cos 3\theta + i \sin 3\theta), \tag{2}$$

by applying DeMoivre's theorem. Equating the expressions for 27, as given by (1) and (2), we have

$$r^3(\cos 3\theta + i \sin 3\theta) = 27[\cos (0° + n\ 360°) + i \sin (0° + n\ 360°)].$$

Now $r^3 = 27$ since they are two expressions for the absolute value of the same complex number; hence, $r = 3$. Furthermore 3θ and $0° + n\ 360° = n\ 360°$ are two expressions for the argument of the complex number regardless of the integral value of n; hence,

$$3\theta = n\ 360°,$$
$$\theta = n\ 120°$$
$$= 0° \qquad \text{for } n = 0$$
$$= 120° \qquad \text{for } n = 1$$
$$= 240° \qquad \text{for } n = 2.$$

The cube roots of 27 which correspond to these values of θ are

$$3(\cos 0° + i \sin 0°) = 3(1 + i0) = 3,$$
$$3(\cos 120° + i \sin 120°) = 3\left(-\frac{1}{2} + i\frac{\sqrt{3}}{2}\right),$$
$$3(\cos 240° + i \sin 240°) = 3\left(-\frac{1}{2} - i\frac{\sqrt{3}}{2}\right).$$

There is no point in using additional values of n, since doing so would yield angles that differ from one of those already obtained by an integral multiple of 360°; consequently, each additional value of $\sqrt[3]{27}$ would be one already obtained.

EXAMPLE 2

Find the fifth roots of $1 + i$.

Solution:

$$1 + i = \sqrt{2}(\cos \arctan 1 + i \sin \arctan 1)$$
$$= \sqrt{2}(\cos 45° + i \sin 45°)$$
$$= \sqrt{2}[\cos (45° + n\,360°) + i \sin (45° + n\,360°)] \qquad (3)$$

for all integral values of n, since $\sin \theta$ and $\cos \theta$ have 360° as a period. If we represent $(1 + i)^{1/5}$ by $r(\cos \theta + i \sin \theta)$ and apply DeMoivre's theorem, we get

$$1 + i = r^5(\cos 5\theta + i \sin 5\theta). \qquad (4)$$

Now, equating the expressions for $1 + i$, as given by (3) and (4), we have

$$r^5(\cos 5\theta + i \sin 5\theta) = 2^{1/2}[\cos (45° + n\,360°) + i \sin (45° + n\,360°)]$$

for all integral values of n; furthermore, equating the absolute values of these two forms for $1 + i$ and also equating their arguments, gives

$$r^5 = \sqrt{2} \quad \text{and} \quad 5\theta = 45° + n\,360°.$$

Hence,

$$r = 2^{1/10} \quad \text{and} \quad \theta = 9° + n\,72°.$$

Therefore,

$$\theta = 9° \qquad \text{using } n = 0$$
$$= 81° \qquad \text{using } n = 1$$
$$= 153° \qquad \text{using } n = 2$$
$$= 225° \qquad \text{using } n = 3$$
$$= 297° \qquad \text{using } n = 4.$$

Therefore, the fifth roots of $1 + i$ are

$$2^{1/10}(\cos 9° + i \sin 9°),$$
$$2^{1/10}(\cos 81° + i \sin 81°),$$
$$2^{1/10}(\cos 153° + i \sin 153°),$$
$$2^{1/10}(\cos 225° + i \sin 225°),$$

and

$$2^{1/10}(\cos 297° + i \sin 297°).$$

The roots of a complex number, if known in trigonometric form, may be reduced to approximate algebraic form by obtaining the approximate trigonometric function value for each angle from a table of sines and cosines and by calculating the real root of r by use of logarithms.

It is interesting to note that the five fifth roots of $1 + i$ are equally spaced about the circumference of a circle of radius $r^{1/5} = (\sqrt{2})^{1/5}$, the argument of the first one being $45°/5$. This is a special case of the general statement:

> *The n nth roots of $a + bi$ are equally spaced about the circumference of a circle of radius $r^{1/n} = (\sqrt{a^2 + b^2})^{1/n}$ and center at the origin, the argument of the first being θ/n.*

EXERCISE 10.4

Use DeMoivre's theorem to raise the number in each of problems 1 to 16 to the indicated power. Use Table III to get the angle in problems 13 to 16 to the nearest $10'$. Leave each answer in polar form.

1. $(1 + i)^3$ 2. $(1 - i)^4$ 3. $(-1 - i)^5$

4. $(-1 + i)^4$ 5. $(1 - \sqrt{3}i)^3$ 6. $(-1 + \sqrt{3}i)^2$

7. $(1 + \sqrt{3}i)^5$ 8. $(-1 - \sqrt{3}i)^6$ 9. $(\sqrt{3} + i)^3$

10. $(-\sqrt{3} - i)^4$ 11. $(-\sqrt{3} + i)^5$ 12. $(\sqrt{3} - i)^4$

13. $(4 - 3i)^3$ 14. $(5 + 12i)^2$

15. $(-15 - 8i)^3$ 16. $(-24 + 7i)^2$

Find the designated roots in problems 17 to 32.

17. Cube roots of $\sqrt{3} - i$ 18. Cube roots of $-1 + i$

19. Cube roots of $-1 - \sqrt{3}i$ 20. Cube roots of $\sqrt{3} + i$

21. Fourth roots of $1 + \sqrt{3}i$ 22. Fourth roots of $1 - \sqrt{3}i$

23. Fourth roots of 16   24. Fourth roots of $81i$

25. Fifth roots of $1 + i$ 26. Fifth roots of $\sqrt{3} - i$

27. Sixth roots of $-\sqrt{3} - i$ 28. Sixth roots of $-64i$

29. Eighth roots of $-1 + \sqrt{3}i$ 30. Eighth roots of $-1 - \sqrt{3}i$

31. Ninth roots of $-1 - i$ 32. Tenth roots of i

Find the solution set of the following equations by use of DeMoivre's theorem.

33. $x^2 - 9 = 0$ 34. $x^3 - 27 = 0$ 35. $x^4 + 16 = 0$

36. $x^5 + 32 = 0$ 37. $x^6 - 64i = 0$ 38. $x^3 - 64i = 0$

39. $x^4 - i = 0$ 40. $x^5 + 32i = 0$

11

POLAR
COORDINATES

11.1. Some Definitions and Concepts

We have encountered the rectangular coordinate system on many occasions. We are able to locate points and sketch the graph of a curve by use of the ordered pairs of numbers used in a rectangular coordinate system. There are, however, other types of coordinate systems that are often more desirable under some circumstances. We shall now present one of them that is known as the *polar coordinate system*, or merely *polar coordinates*. In this system, the frame of reference is a pair of perpendicular directed rays extending from the same point. The horizontal ray is called the *polar axis*, the vertical is known as the *normal axis*, and their intersection is the *pole*. A point is located by giving its distance r from the pole and the angle θ from the polar axis to the ray from the pole to the point, as indicated in Figure 11.1. The distance r is called the *radius vector*, and the angle θ is called the

207

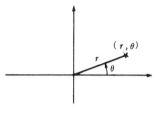

Figure 11.1

amplitude or *vectorial angle*. The vectorial angle may be positive, zero, or negative. Distances measured along the radius vector from the pole are positive, and a negative sign before the value of the radius vector indicates that the distance is to be measured along the extension of the radius vector through the pole.

11.2. Pairs of Coordinates for a Point

In the rectangular coordinate system, there is only one point for an ordered pair of numbers, and only one ordered pair of numbers for a point. Thus, there is a one-to-one correspondence between points and ordered pairs of numbers in connection with the rectangular coordinate system. This is not the situation if the polar coordinate system is used. There is only one point for an ordered pair of numbers, but a point may have an unlimited number of pairs of coordinates. If k is an integer, then $(r, \theta + k360°)$ represents the same point as (r, θ), since r is the radius vector in each case and the angles θ and $\theta + k360°$ are coterminal. If we bear in mind that, when r is preceded by a negative sign, then the distance is measured along the extension of the radius vector through the origin, we see that $(-r, \theta + 180°)$ represents the same point as (r, θ); furthermore, $(-r, \theta + 180° + k360°)$ also represents the point for k an integer.

EXAMPLE

Locate the point $(3, 30°)$, and give three other pairs of coordinates for it.

Solution: The point is shown in Figure 11.2, as are three other pairs of coordinates for it. The point $(3, 390°)$ is the same one as $(3, 30°)$, since $r = 3$ for each and 390° differs from 30° by an integral multiple of 360°; furthermore, $(-3, 210°)$ and $(-3, -150°)$ also represent the same point as $(3, 30°)$, since $r = -3$ for each of them and the terminal side of each angle is the extension through the origin of the terminal side of 30°.

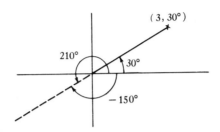

Figure 11.2

11.3. Symmetry and Intercepts

This article will be devoted to a concept which, if understood and used, makes the sketching of a curve simpler than it otherwise would be.

> *We say that two points P_1 and P_2 are **located symmetrically with respect to a third point P** if P is the midpoint of the segment that joins them.*

The points P_1 and P_2 in Figure 11.3 are located symmetrically with respect to P, since P is the midpoint of the segment between P_1 and P_2.

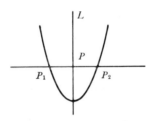

Figure 11.3

> *Two points P_1 and P_2 are **located symmetrically with respect to a line** if the line is the perpendicular bisector of the segment that joins them.*

The points P_1 and P_2 in Figure 11.3 are symmetrically located with respect to the line L, since L is the perpendicular bisector of the segment between P_1 and P_2.

> *A curve is symmetrical with respect to a point **P** if, for each point P_1 on the curve, there is a point P_2 on it such that P_1 and P_2 are symmetrically located with respect to the point **P**. The point is called a **center of symmetry**.*

A circle is symmetrical with respect to its center.

A curve is **symmetrical with respect to a line** *if, for each point P_1 on the curve, there is a point P_2 on it such that P_1 and P_2 are symmetrically located with respect to the line. The line is called an* **axis of symmetry.**

The curve in Figure 11.3 is symmetrical with respect to the line L, since corresponding to each point on the curve there is a second point on it such that L is the perpendicular bisector of the segment between the two points.

We shall be interested primarily in determining whether a curve is symmetrical with respect to the origin or a coordinate axis and shall give three theorems that furnish appropriate tests.

If a polar equation is unchanged or multiplied by -1

(a) *by replacing θ by $180° + \theta$, the curve is symmetrical with respect to the pole;*

(b) *by replacing θ by $-\theta$, the curve is symmetrical with respect to the polar axis, or polar axis produced through the pole;*

(c) *by replacing θ by $180° - \theta$, the curve is symmetrical with respect to the normal axis, or normal axis produced through the pole.*

If we look at Figure 11.4, we see that O is the midpoint of P_2P, since OP and OP_2 are each r in length; hence, test (a) is valid. Furthermore,

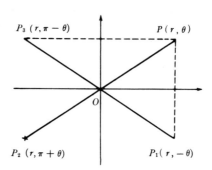

Figure 11.4

by use of the congruent triangles, we can prove that the polar axis is the perpendicular bisector of the segment PP_1; hence, test (b) is valid. Similarly, the normal axis is the perpendicular bisector of PP_3; therefore, we have test (c).

As pointed out in article 11.2, a point has an infinite number of pairs of coordinates in the polar coordinate system; hence, it should not come as a surprise that there are other tests for symmetry besides those given above.

For example, a curve is symmetrical with respect to the normal axis if its polar equation is unchanged by replacing r by $-r$ and θ by $2\pi - \theta$ simultaneously, even though symmetry with respect to the normal axis may not be revealed when the test (c) is applied. The curious and capable student can devise other tests for symmetry if he makes use of the various pairs of coordinates that P_1, P_2 and P_3 of Figure 11.4 may have.

EXAMPLE 1

Test $f(r, \theta) = r - 3 \cos 2\theta = 0$ for symmetry.

Solution: In testing for symmetry with respect to the pole, we shall see if the equation is changed by replacing θ by $180° + \theta$, as suggested in (a). Thus, we have

$$f(r, 180° + \theta) = r - 3 \cos 2(180° + \theta)$$
$$= r - 3 \cos (360° + 2\theta)$$
$$= r - 3 \cos 2\theta, \qquad \text{since } 360° + 2\theta \text{ and } 2\theta \text{ are coterminal,}$$
$$= f(r, \theta).$$

Consequently, the curve is symmetrical with respect to the pole.

As stated in (b), we can test for symmetry with respect to the polar axis by replacing θ by $-\theta$. Thus, we get

$$f(r, -\theta) = r - 3 \cos 2(-\theta)$$
$$= r - 3 \cos 2\theta, \qquad \text{since } \cos (-2\theta) = \cos 2\theta,$$
$$= f(r, \theta).$$

Therefore, the curve is symmetrical with respect to the polar axis.

To test for symmetry with respect to the normal axis, we shall use the test (c) and have

$$f(r, 180° - \theta) = r - 3 \cos 2(180° - \theta)$$
$$= r - 3 \cos (360° - 2\theta)$$
$$= r - 3 \cos (-2\theta), \qquad \text{since } 360° - 2\theta \text{ and } -2\theta \text{ are coterminal,}$$
$$= r - 3 \cos 2\theta, \qquad \text{since } \cos (-2\theta) = \cos 2\theta.$$

Consequently, the curve is symmetrical with respect to the normal axis.

The r-coordinate of each intersection of $f(r, \theta) = 0$ and the axes or their extensions through the pole is called an *intercept* of $f(r, \theta) = 0$. Consequently, we obtain intercepts by solving the equations in r that are obtained by replacing θ successively by $0°$ and the positive integral multiples of $90°$ in $f(r, \theta) = 0$; furthermore, $r = 0$ is an intercept if $f(0, \theta) = 0$ has a solution. The points $(r, n\pi/2)$ are called *intercept points*.

EXAMPLE 2

Find the intercept points of $f(r, \theta) = r - 3 \cos 2\theta = 0$.

Solution:

$f(r, 0°) = r - 3 \cos 0° = r - 3 = 0$ for $r = 3$; hence, $(3, 0°)$ is an intercept point.

$f(r, 90°) = r - 3 \cos 180° = r + 3 = 0$ for $r = -3$; hence, $(-3, 90°)$ is an intercept point.

$f(r, 180°) = r - 3 \cos 360° = r - 3 = 0$ for $r = 3$; hence, $(3, 180°)$ is an intercept point.

$f(r, 270°) = r - 3 \cos 540° = r + 3 = 0$ for $r = -3$; hence, $(-3, 270°)$ is an intercept point.

Furthermore,

$r = 0$ is an intercept, since $3 \cos 2\theta = 0$ has solutions.

11.4. Construction of the Graph

If we have an equation $f(r, \theta) = 0$, we can sketch its graph by assigning successively the elements of a set of values to θ, evaluating each corresponding r, plotting the points (r, θ), and drawing a smooth curve through them. The amount of labor is often reduced if we determine and use the intercepts and symmetry.

The problem of determining the set of values to assign to θ is often a troublesome one. It is ordinarily sufficient to begin with $\theta = 0°$ and assign the multiples of 30° and 45° to θ until we have enough points to determine the graph or the desired part of it. At times, it may be advisable to assign values between the multiples of 30° and 45°. This is likely to be the situation if a multiple of θ occurs in the given equation.

EXAMPLE 1

Sketch the graph of $f(r, \theta) = r - 3 \cos 2\theta = 0$.

Solution. We found in the examples in article 11.3 that the curve is symmetrical with respect to both coordinate axes and the origin, and that the intercept points are $(3, 0°)$, $(-3, 90°)$, $(3, 180°)$, $(-3, 270°)$, and the origin. We shall now continue collecting data on the curve by making a table of values of r and θ that has 2θ and $\cos 2\theta$ as intermediate steps between the values of θ and the corresponding values of r.

θ	0°	15°	22½°	30°	45°	60°	67½°	75°	90°
2θ	0°	30°	45°	60°	90°	120°	135°	150°	180°
$\cos 2\theta$	1	$\sqrt{3}/2$	$\sqrt{2}/2$	1/2	0	−1/2	$-\sqrt{2}/2$	$-\sqrt{3}/2$	−1
r	3	$3\sqrt{3}/2$	$3\sqrt{2}/2$	1.5	0	−1.5	$-3\sqrt{3}/2$	$-3\sqrt{3}/2$	−3

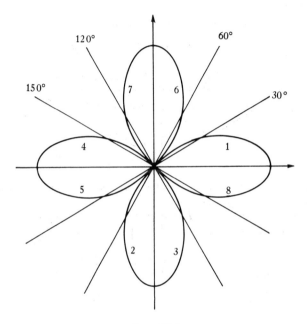

Figure 11.5

The half loops numbered 1 and 2 in Figure 11.5 are obtained by locating the points (r, θ) indicated in the table. Those numbered 8 and 7 can be obtained from half loops 1 and 2 by use of the symmetry with respect to the polar axis, and the others from these four by use of symmetry with respect to the normal axis. We could have begun with half loops 1 and 2 and have obtained 5 and 6 by use of symmetry with respect to the pole: then the others could have been obtained by use of symmetry with respect to either axis. The half loops are numbered in the order in which they would have been obtained by using increasing values of θ. The curve is called a *four-leaf rose*.

EXAMPLE 2

Test $f(r, \theta) = r - 2 \cos \theta - 3 = 0$ for symmetry, find its intercept points and sketch the curve.

Solution: Since

$$f(r, 180° + \theta) = r - 2 \cos (180° + \theta) - 3$$
$$= r + 2 \cos \theta - 3 \neq f(r, \theta),$$

we can only say that, if symmetry exists with respect to the pole, it is not revealed by use of test (a).

We shall now test for symmetry with respect to the polar axis by replacing θ by $-\theta$. Thus, since

$$f(r, -\theta) = r - 2 \cos (-\theta) - 3$$
$$= r - 2 \cos \theta - 3 = f(r, \theta),$$

the curve is symmetrical with respect to the polar axis. Consequently, if we get the graph for values of θ from 0° to 180°, the part for θ from 180° to 360° can be obtained by use of symmetry. Finally, replacing θ by $180° - \theta$, we have

$$f(r, 180° - \theta) = r - 2 \cos (180° - \theta) - 3$$
$$= r + 2 \cos \theta - 3 \neq f(r, \theta).$$

Consequently, we know that any symmetry that may exist relative to the normal axis is not revealed by use of test (c).

We find the intercept by assigning an integral multiple of 90° to θ and solving the resulting equation for r. Thus,

$f(r, 0°) = r - 2 \cos 0° - 3 = r - 5 = 0$ for $r = 5$; hence, $(5, 0°)$ is an intercept point.

$f(r, 90°) = r - 2 \cos 90° - 3 = r - 3 = 0$ for $r = 3$; hence, $(3, 90°)$ is an intercept point.

$f(r, 180°) = r - 2 \cos 180° - 3 = r - 1 = 0$ for $r = 1$; hence, $(1, 180°)$ is an intercept point.

$f(r, 270°) = r - 2 \cos 270° - 3 = r - 3 = 0$ for $r = 3$; hence, $(3, 270°)$ is an intercept point.

We shall now make a table of values of θ and r that gives $\cos \theta$ as an intermediate value, plot the points determined by the ordered pairs thus obtained, draw a smooth curve through them, and thus get the upper half of the curve in Figure 11.6. The lower half was obtained by use of symmetry.

θ	0	30°	45°	60°	90°	120°	135°	150°	180°
$\cos \theta$	1	$\sqrt{3}/2$	$\sqrt{2}/2$	1/2	0	$-1/2$	$-\sqrt{2}/2$	$-\sqrt{3}/2$	-1
r	5	$\sqrt{3} + 3$	$\sqrt{2} + 3$	4	3	2	$-\sqrt{2} + 3$	$-\sqrt{3} + 3$	1

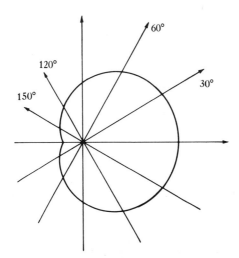

Figure 11.6

EXERCISE 11.1

Locate the points given in problems 1 to 8 and write three other pairs of coordinates for each.

1. $(3, 45°)$, $(3, -60°)$, $(-2, 75°)$, $(-4, -45°)$
2. $(5, 110°)$, $(-5, 70°)$, $(6, -120°)$, $(-5, -30°)$
3. $(-2, -50°)$, $(-7, 160°)$, $(5, -140°)$, $(2, 70°)$
4. $(4, -210°)$, $(-5, 300°)$, $(5, 330°)$, $(-3, -190°)$
5. $(-7, 315°)$, $(7, -150°)$, $(-4, -120°)$, $(6, 270°)$
6. $(9, -100°)$, $(-2, -20°)$, $(5, 310°)$, $(-3, 135°)$
7. $(-3, -225°)$, $(4, 40°)$, $(-2, 230°)$, $(5, -170°)$
8. $(5, 250°)$, $(-4, 340°)$, $(3, -260°)$, $(-2, -200°)$

Find the intercept points, test for symmetry with respect to the pole and coordinate axes by use of tests (a), (b), and (c), and sketch the graph of the equation given in each of the following problems.

9. $\theta = 45°$ 　　　　　 10. $\theta = 150°$ 　　　　　 11. $\theta = -120°$

12. $\theta = 390°$ 　　　　 13. $r = 3$ 　　　　　　　 14. $r = 5$

15. $r = -2$ 　　　　　　 16. $r = -4$ 　　　　　　 17. $r = 2 \sin \theta$

18. $r = 4 \sin \theta$ 　　　 19. $r = 6 \cos \theta$ 　　　 20. $r = 5 \cos \theta$

21. $r = 2 \cos 2\theta$ 　　 22. $r = 3 \cos 3\theta$ 　　 23. $r = 5 \cos 4\theta$

24. $r = 6 \cos 5\theta$ 　　 25. $r = 4 \sin 5\theta$ 　　 26. $r = 3 \sin 4\theta$

27. $r = 2 \sin 3\theta$ 　　 28. $r = 5 \sin 2\theta$ 　　 29. $r = 2(\sin \theta - 1)$

30. $r = 3(\sin \theta + 1)$ **31.** $r = 4(\cos \theta + 1)$ **32.** $r = 2(\cos \theta - 1)$

33. $r = 2 \cos \theta - 3$ **34.** $r = 3 \cos \theta + 2$ **35.** $r = 4 \sin \theta - 3$

36. $r = 2 \sin \theta + 3$ **37.** $r = \dfrac{3}{2 - \cos \theta}$ **38.** $r = \dfrac{2}{3 + \cos \theta}$

39. $r = \dfrac{4}{3 - \sin \theta}$ **40.** $r = \dfrac{3}{2 + \sin \theta}$ **41.** $r = \dfrac{2}{1 + \sin \theta}$

42. $r = \dfrac{5}{1 - \sin \theta}$ **43.** $r = \dfrac{4}{1 + \cos \theta}$ **44.** $r = \dfrac{3}{1 - \cos \theta}$

45. $r = \dfrac{2}{1 + 2 \sin \theta}$ **46.** $r = \dfrac{5}{1 + 3 \sin \theta}$ **47.** $r = \dfrac{4}{2 - 3 \cos \theta}$

48. $r = \dfrac{6}{3 + 4 \cos \theta}$ **49.** $r = \tan \theta$ **50.** $r = \sec \theta$

51. $r = \csc \theta$ **52.** $r = \cot \theta$ **53.** $r = \cos^2 \theta$

54. $r = \sin^2 \theta$ **55.** $r = \sin^2 2\theta$ **56.** $r = \cos^2 3\theta$

57. $r^2 = \sin \theta$ **58.** $r^2 = \cos \theta$

59. $r^2 = 9 \sin \theta$ **60.** $r^2 = 16 \cos \theta$

61. $r = 3 \sin \theta \tan \theta$ **62.** $r = 2\theta$, θ in radians

63. $r^2 = 3^2 \sin 2\theta$ **64.** $r^2\theta = \pi$, θ in radians

11.5. Relations between Rectangular and Polar Coordinates

We shall now use Figure 11.7 in deriving four equations which can be used for changing an equation from rectangular or polar coordinates to the other

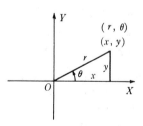

Figure 11.7

form. The figure was drawn by superimposing the polar frame of reference on the rectangular with the pole on the origin and the polar axis along the positive X-axis. If we use the definition of the cosine and sine ratios, we have $\cos \theta = x/r$ and $\sin \theta = y/r$. Now, solving for x and y, we have

$$x = r \cos \theta \quad \text{and} \quad y = r \sin \theta. \tag{11.1}$$

Equations (11.1) can be used for changing an equation from the rectangular to the polar coordinate system. If we use the Pythagorean theorem and the definition of the tangent of an angle, we have $r^2 = x^2 + y^2$ and $\tan \theta = y/x$. Solving for r and θ, we have

$$r = \sqrt{x^2 + y^2} \quad \text{and} \quad \theta = \arctan \frac{y}{x}. \qquad (11.2)$$

Equations (11.2) can be used for changing from polar to rectangular form. Quite often the work of changing from one type of coordinate system to the other is simplified if a combination of equations (11.1) and (11.2) is used rather than either pair alone.

EXAMPLE 1

Express $x^2 - 2y + y^2 = 0$ in terms of polar coordinates.

Solution: If we replace y by $r \sin \theta$ as given in equations (11.1) and $x^2 + y^2$ by r^2 as given in equations (11.2), the equation

$$x^2 - 2y + y^2 = 0$$

becomes

$$r^2 - 2r \sin \theta = 0.$$

Now, dividing by r, we find that the polar form of the given equation is

$$r - 2 \sin \theta = 0.$$

No part of the curve was lost in dividing through by r, since the graph of $r - 2 \sin \theta = 0$ passes through the pole.

EXAMPLE 2

Express

$$r = \frac{2}{1 + \sin \theta}$$

in rectangular form.

Solution: We shall multiply both members by $1 + \sin \theta$ and get

$$r + r \sin \theta = 2.$$

We can now replace r by $\sqrt{x^2 + y^2}$ and $r \sin \theta$ by y and have

$$\sqrt{x^2 + y^2} + y = 2;$$

hence,

$$\sqrt{x^2 + y^2} = 2 - y.$$

Now, squaring each member, we have

$$x^2 + y^2 = 4 - 4y + y^2,$$

and, collecting terms, we find that

$$x^2 = 4 - 4y$$

is the rectangular form of the given equation.

EXERCISE 11.2

Express each of the following equations in terms of a different type of coordinate system.

1. $x = 2$ **2.** $y = -3$

3. $y = 5$ **4.** $x = -4$

5. $x^2 + y^2 = 16$ **6.** $x^2 + y^2 = 9$

7. $(x - 1)^2 + (y + 2)^2 = 5$ **8.** $(x + 3)^2 + (y - 4)^2 = 25$

9. $x + 2y = 1$ **10.** $2x - 3y = 5$

11. $y^2 = 4x$ **12.** $x^2 = -6y$

13. $4x^2 + 9y^2 = 36$ **14.** $x^2 + 4y^2 = 4$

15. $9x^2 - y^2 = 9$ **16.** $4y^2 - 25x^2 = 100$

17. $y^2 = ax^3$ **18.** $(x - a)^2 + (y - b)^2 = a^2 + b^2$

19. $ax + by = c$ **20.** $x^4 - y^4 = 2axy$

21. $(x^2 + y^2)^2 = 2a^2xy$ **22.** $x(x^2 + y^2) = y^2$

23. $x(x^2 + y^2) = a(3x^2 - y^2)$ **24.** $(x^2 + y^2)^{3/2} = x^2 - y^2 - 2xy$

25. $r \cos \theta = 2$ **26.** $r = -3 \csc \theta$

27. $r \sin \theta = 5$ **28.** $r = -4 \sec \theta$

29. $r = 4$ **30.** $r = 3$

31. $r = 2 \cos \theta - 4 \sin \theta$ **32.** $r = 8 \sin \theta - 6 \cos \theta$

33. $r = 1/(\cos \theta + 2 \sin \theta)$ **34.** $r = 5/(2 \cos \theta - 3 \sin \theta)$

35. $r \sin^2 \theta = 4 \cos \theta$ **36.** $r = -6 \sec \theta \tan \theta$

37. $4r^2 + 5r^2 \sin^2 \theta = 36$ **38.** $r^2(1 + 3 \sin^2 \theta) = 4$

39. $r^2 = 9/(9 - 10 \sin^2 \theta)$ **40.** $r^2 = -100/(25 \cos^2 \theta - 4 \sin^2 \theta)$

41. $\sin^2 \theta = ar \cos^3 \theta$ **42.** $r = 2(a \cos \theta + b \sin \theta)$

43. $r = c/(a \cos \theta + b \sin \theta)$ **44.** $r^2 = a \tan 2\theta$

45. $r^2 = a^2 \sin 2\theta$ **46.** $r = \sin \theta \tan \theta$

47. $r \cos \theta = a(3 \cos^2 \theta - \sin^2 \theta)$ **48.** $r = \cos 2\theta - \sin 2\theta$

MOLLWEIDE'S EQUATIONS

If we multiply each of the first two members of (7.1), by $(\sin A)/b$, we obtain

$$\frac{a}{b} = \frac{\sin A}{\sin B}.$$

Subtracting 1 from each member of this equation and reducing each member to a common denominator, we have

$$\frac{a - b}{b} = \frac{\sin A - \sin B}{\sin B}.$$

Hence,

$$\frac{a - b}{\sin A - \sin B} = \frac{b}{\sin B},$$

$$= \frac{c}{\sin C} \qquad \text{by (7.1)}.$$

Consequently,

$$\frac{a-b}{c} = \frac{\sin A - \sin B}{\sin C}$$

$$= \frac{2 \cos \dfrac{A+B}{2} \sin \dfrac{A-B}{2}}{\sin 2 \dfrac{C}{2}} \qquad \text{by (5.22)}$$

$$= \frac{2 \cos \dfrac{A+B}{2} \sin \dfrac{A-B}{2}}{2 \sin \dfrac{C}{2} \cos \dfrac{C}{2}} \qquad \text{by (5.10)}$$

$$= \frac{\cos \dfrac{A+B}{2} \sin \dfrac{A-B}{2}}{\cos \dfrac{A+B}{2} \cos \dfrac{C}{2}},$$

since $C/2$ and $(A+B)/2$ are complementary angles, and sine and cosine are co-functions. Therefore,

$$\frac{a-b}{c} = \frac{\sin \dfrac{A-B}{2}}{\cos \dfrac{C}{2}}. \qquad \textbf{(M}_1\textbf{)}$$

The equation

$$\frac{a+b}{c} = \frac{\cos \dfrac{A-B}{2}}{\sin \dfrac{C}{2}} \qquad \textbf{(M}_2\textbf{)}$$

may be derived in a similar manner.

The equations (M₁) and (M₂) are known as *Mollweide's equations*, and either of them may be used in checking the solution of a triangle. Either furnishes an adequate check, since all six parts are involved.

LOGARITHMIC
AND
TRIGONOMETRIC
TABLES

TABLE I. LOGARITHMS OF NUMBERS

0-509

Do not interpolate on this page

N.	L. 0	1	2	3	4	5	6	7	8	9
0		00 000	30 103	47 712	60 206	69 897	77 815	84 510	90 309	95 424
1	00 000	04 139	07 918	11 394	14 613	17 609	20 412	23 045	25 527	27 875
2	30 103	32 222	34 242	36 173	38 021	39 794	41 497	43 136	44 716	46 240
3	47 712	49 136	50 515	51 851	53 148	54 407	55 630	56 820	57 978	59 106
4	60 206	61 278	62 325	63 347	64 345	65 321	66 276	67 210	68 124	69 020
5	69 897	70 757	71 600	72 428	73 239	74 036	74 819	75 587	76 343	77 085
6	77 815	78 533	79 239	79 934	80 618	81 291	81 954	82 607	83 251	83 885
7	84 510	85 126	85 733	86 332	86 923	87 506	88 081	88 649	89 209	89 763
8	90 309·	90 849	91 381	91 908	92 428	92 942	93 450	93 952	94 448	94 939
9	95 424	95 904	96 379	96 848	97 313	97 772	98 227	98 677	99 123	99 564
10	00 000	00 432	00 860	01 284	01 703	02 119	02 531	02 938	03 342	03 743
11	04 139	04 532	04 922	05 308	05 690	06 070	06 446	06 819	07 188	07 555
12	07 918	08 279	08 636	08 991	09 342	09 691	10 037	10 380	10 721	11 059
13	11 394	11 727	12 057	12 385	12 710	13 033	13 354	13 672	13 988	14 301
14	14 613	14 922	15 229	15 534	15 836	16 137	16 435	16 732	17 026	17 319
15	17 609	17 898	18 184	18 469	18 752	19 033	19 312	19 590	19 866	20 140
16	20 412	20 683	20 952	21 219	21 484	21 748	22 011	22 272	22 531	22 789
17	23 045	23 300	23 553	23 805	24 055	24 304	24 551	24 797	25 042	25 285
18	25 527	25 768	26 007	26 245	26 482	26 717	26 951	27 184	27 416	27 646
19	27 875	28 103	28 330	28 556	28 780	29 003	29 226	29 447	29 667	29 885
20	30 103	30 320	30 535	30 750	30 963	31 175	31 387	31 597	31 806	32 015
21	32 222	32 428	32 634	32 838	33 041	33 244	33 445	33 646	33 846	34 044
22	34 242	34 439	34 635	34 830	35 025	35 218	35 411	35 603	35 793	35 984
23	36 173	36 361	36 549	36 736	36 922	37 107	37 291	37 475	37 658	37 840
24	38 021	38 202	38 382	38 561	38 739	38 917	39 094	39 270	39 445	39 620
25	39 794	39 967	40 140	40 312	40 483	40 654	40 824	40 993	41 162	41 330
26	41 497	41 664	41 830	41 996	42 160	42 325	42 488	42 651	42 813	42 975
27	43 136	43 297	43 457	43 616	43 775	43 933	44 091	44 248	44 404	44 560
28	44 716	44 871	45 025	45 179	45 332	45 484	45 637	45 788	45 939	46 090
29	46 240	46 389	46 538	46 687	46 835	46 982	47 129	47 276	47 422	47 567
30	47 712	47 857	48 001	48 144	48 287	48 430	48 572	48 714	48 855	48 996
31	49 136	49 276	49 415	49 554	49 693	49 831	49 969	50 106	50 243	50 379
32	50 515	50 651	50 786	50 920	51 055	51 188	51 322	51 455	51 587	51 720
33	51 851	51 983	52 114	52 244	52 375	52 504	52 634	52 763	52 892	53 020
34	53 148	53 275	53 403	53 529	53 656	53 782	53 908	54 033	54 158	54 283
35	54 407	54 531	54 654	54 777	54 900	55 023	55 145	55 267	55 388	55 509
36	55 630	55 751	55 871	55 991	56 110	56 229	56 348	56 467	56 585	56 703
37	56 820	56 937	57 054	57 171	57 287	57 403	57 519	57 634	57 749	57 864
38	57 978	58 092	58 206	58 320	58 433	58 546	58 659	58 771	58 883	58 995
39	59 106	59 218	59 329	59 439	59 550	59 660	59 770	59 879	59 988	60 097
40	60 206	60 314	60 423	60 531	60 638	60 746	60 853	60 959	61 066	61 172
41	61 278	61 384	61 490	61 595	61 700	61 805	61 909	62 014	62 118	62 221
42	62 325	62 428	62 531	62 634	62 737	62 839	62 941	63 043	63 144	63 246
43	63 347	63 448	63 548	63 649	63 749	63 849	63 949	64 048	64 147	64 246
44	64 345	64 444	64 542	64 640	64 738	64 836	64 933	65 031	65 128	65 225
45	65 321	65 418	65 514	65 610	65 706	65 801	65 896	65 992	66 087	66 181
46	66 276	66 370	66 464	66 558	66 652	66 745	66 839	66 932	67 025	67 117
47	67 210	67 302	67 394	67 486	67 578	67 669	67 761	67 852	67 943	68 034
48	68 124	68 215	68 305	68 395	68 485	68 574	68 664	68 753	68 842	68 931
49	69 020	69 108	69 197	69 285	69 373	69 461	69 548	69 636	69 723	69 810
50	69 897	69 984	70 070	70 157	70 243	70 329	70 415	70 501	70 586	70 672
N.	L. 0	1	2	3	4	5	6	7	8	9

Do not interpolate on this page

N.	L. 0	1	2	3	4	5	6	7	8	9
50	69 897	69 984	70 070	70 157	70 243	70 329	70 415	70 501	70 586	70 672
51	70 757	70 842	70 927	71 012	71 096	71 181	71 265	71 349	71 433	71 517
52	71 600	71 684	71 767	71 850	71 933	72 016	72 099	72 181	72 263	72 346
53	72 428	72 509	72 591	72 673	72 754	72 835	72 916	72 997	73 078	73 159
54	73 239	73 320	73 400	73 480	73 560	73 640	73 719	73 799	73 878	73 957
55	74 036	74 115	74 194	74 273	74 351	74 429	74 507	74 586	74 663	74 741
56	74 819	74 896	74 974	75 051	75 128	75 205	75 282	75 358	75 435	75 511
57	75 587	75 664	75 740	75 815	75 891	75 967	76 042	76 118	76 193	76 268
58	76 343	76 418	76 492	76 567	76 641	76 716	76 790	76 864	76 938	77 012
59	77 085	77 159	77 232	77 305	77 379	77 452	77 525	77 597	77 670	77 743
60	77 815	77 887	77 960	78 032	78 104	78 176	78 247	78 319	78 390	78 462
61	78 533	78 604	78 675	78 746	78 817	78 888	78 958	79 029	79 099	79 169
62	79 239	79 309	79 379	79 449	79 518	79 588	79 657	79 727	79 796	79 865
63	79 934	80 003	80 072	80 140	80 209	80 277	80 346	80 414	80 482	80 550
64	80 618	80 686	80 754	80 821	80 889	80 956	81 023	81 090	81 158	81 224
65	81 291	81 358	81 425	81 491	81 558	81 624	81 690	81 757	81 823	81 889
66	81 954	82 020	82 086	82 151	82 217	82 282	82 347	82 413	82 478	82 543
67	82 607	82 672	82 737	82 802	82 866	82 930	82 995	83 059	83 123	83 187
68	83 251	83 315	83 378	83 442	83 506	83 569	83 632	83 696	83 759	83 822
69	83 885	83 948	84 011	84 073	84 136	84 198	84 261	84 323	84 386	84 448
70	84 510	84 572	84 634	84 696	84 757	84 819	84 880	84 942	85 003	85 065
71	85 126	85 187	85 248	85 309	85 370	85 431	85 491	85 552	85 612	85 673
72	85 733	85 794	85 854	85 914	85 974	86 034	86 094	86 153	86 213	86 273
73	86 332	86 392	86 451	86 510	86 570	86 629	86 688	86 747	86 806	86 864
74	86 923	86 982	87 040	87 099	87 157	87 216	87 274	87 332	87 390	87 448
75	87 506	87 564	87 622	87 679	87 737	87 795	87 852	87 910	87 967	88 024
76	88 081	88 138	88 195	88 252	88 309	88 366	88 423	88 480	88 536	88 593
77	88 649	88 705	88 762	88 818	88 874	88 930	88 986	89 042	89 098	89 154
78	89 209	89 265	89 321	89 376	89 432	89 487	89 542	89 597	89 653	89 708
79	89 763	89 818	89 873	89 927	89 982	90 037	90 091	90 146	90 200	90 255
80	90 309	90 363	90 417	90 472	90 526	90 580	90 634	90 687	90 741	90 795
81	90 849	90 902	90 956	91 009	91 062	91 116	91 169	91 222	91 275	91 328
82	91 381	91 434	91 487	91 540	91 593	91 645	91 698	91 751	91 803	91 855
83	91 908	91 960	92 012	92 065	92 117	92 169	92 221	92 273	92 324	92 376
84	92 428	92 480	92 531	92 583	92 634	92 686	92 737	92 788	92 840	92 891
85	92 942	92 993	93 044	93 095	93 146	93 197	93 247	93 298	93 349	93 399
86	93 450	93 500	93 551	93 601	93 651	93 702	93 752	93 802	93 852	93 902
87	93 952	94 002	94 052	94 101	94 151	94 201	94 250	94 300	94 349	94 399
88	94 448	94 498	94 547	94 596	94 645	94 694	94 743	94 792	94 841	94 890
89	94 939	94 988	95 036	95 085	95 134	95 182	95 231	95 279	95 328	95 376
90	95 424	95 472	95 521	95 569	95 617	95 665	95 713	95 761	95 809	95 856
91	95 904	95 952	95 999	96 047	96 095	96 142	96 190	96 237	96 284	96 332
92	96 379	96 426	96 473	96 520	96 567	96 614	96 661	96 708	96 755	96 802
93	96 848	96 895	96 942	96 988	97 035	97 081	97 128	97 174	97 220	97 267
94	97 313	97 359	97 405	97 451	97 497	97 543	97 589	97 635	97 681	97 727
95	97 772	97 818	97 864	97 909	97 955	98 000	98 046	98 091	98 137	98 182
96	98 227	98 272	98 318	98 363	98 408	98 453	98 498	98 543	98 588	98 632
97	98 677	98 722	98 767	98 811	98 856	98 900	98 945	98 989	99 034	99 078
98	99 123	99 167	99 211	99 255	99 300	99 344	99 388	99 432	99 476	99 520
99	99 564	99 607	99 651	99 695	99 739	99 782	99 826	99 870	99 913	99 957
100	00 000	00 043	00 087	00 130	00 173	00 217	00 260	00 303	00 346	00 389
N.	L. 0	1	2	3	4	5	6	7	8	9

N.	L. 0	1	2	3	4	5	6	7	8	9	P. P.
100	00 000	043	087	130	173	217	260	303	346	389	
101	432	475	518	561	604	647	689	732	775	817	
102	860	903	945	988	*030	*072	*115	*157	*199	*242	
103	01 284	326	368	410	452	494	536	578	620	662	
104	703	745	787	828	870	912	953	995	*036	*078	
105	02 119	160	202	243	284	325	366	407	449	490	
106	531	572	612	653	694	735	776	816	857	898	
107	938	979	*019	*060	*100	*141	*181	*222	*262	*302	
108	03 342	383	423	463	503	543	583	623	663	703	
109	743	782	822	862	902	941	981	*021	*060	*100	
110	04 139	179	218	258	297	336	376	415	454	493	
111	532	571	610	650	689	727	766	805	844	883	
112	922	961	999	*038	*077	*115	*154	*192	*231	*269	
113	05 308	346	385	423	461	500	538	576	614	652	
114	690	729	767	805	843	881	918	956	994	*032	
115	06 070	108	145	183	221	258	296	333	371	408	
116	446	483	521	558	595	633	670	707	744	781	
117	819	856	893	930	967	*004	*041	*078	*115	*151	
118	07 188	225	262	298	335	372	408	445	482	518	
119	555	591	628	664	700	737	773	809	846	882	
120	918	954	990	*027	*063	*099	*135	*171	*207	*243	
121	08 279	314	350	386	422	458	493	529	565	600	
122	636	672	707	743	778	814	849	884	920	955	
123	991	*026	*061	*096	*132	*167	*202	*237	*272	*307	
124	09 342	377	412	447	482	517	552	587	621	656	
125	691	726	760	795	830	864	899	934	968	*003	
126	10 037	072	106	140	175	209	243	278	312	346	
127	380	415	449	483	517	551	585	619	653	687	
128	721	755	789	823	857	890	924	958	992	*025	
129	11 059	093	126	160	193	227	261	294	327	361	
130	394	428	461	494	528	561	594	628	661	694	
131	727	760	793	826	860	893	926	959	992	*024	
132	12 057	090	123	156	189	222	254	287	320	352	
133	385	418	450	483	516	548	581	613	646	678	
134	710	743	775	808	840	872	905	937	969	*001	
135	13 033	066	098	130	162	194	226	258	290	322	
136	354	386	418	450	481	513	545	577	609	640	
137	672	704	735	767	799	830	862	893	925	956	
138	988	*019	*051	*082	*114	*145	*176	*208	*239	*270	
139	14 301	333	364	395	426	457	489	520	551	582	
140	613	644	675	706	737	768	799	829	860	891	
141	922	953	983	*014	*045	*076	*106	*137	*168	*198	
142	15 229	259	290	320	351	381	412	442	473	503	
143	534	564	594	625	655	685	715	746	776	806	
144	836	866	897	927	957	987	*017	*047	*077	*107	
145	16 137	167	197	227	256	286	316	346	376	406	
146	435	465	495	524	554	584	613	643	673	702	
147	732	761	791	820	850	879	909	938	967	997	
148	17 026	056	085	114	143	173	202	231	260	289	
149	319	348	377	406	435	464	493	522	551	580	
150	609	638	667	696	725	754	782	811	840	869	
N.	L. 0	1	2	3	4	5	6	7	8	9	P. P.

P. P.

	44	43	42
1	4.4	4.3	4.2
2	8.8	8.6	8.4
3	13.2	12.9	12.6
4	17.6	17.2	16.8
5	22.0	21.5	21.0
6	26.4	25.8	25.2
7	30.8	30.1	29.4
8	35.2	34.4	33.6
9	39.6	38.7	37.8

	41	40	39
1	4.1	4.0	3.9
2	8.2	8.0	7.8
3	12.3	12.0	11.7
4	16.4	16.0	15.6
5	20.5	20.0	19.5
6	24.6	24.0	23.4
7	28.7	28.0	27.3
8	32.8	32.0	31.2
9	36.9	36.0	35.1

	38	37	36
1	3.8	3.7	3.6
2	7.6	7.4	7.2
3	11.4	11.1	10.8
4	15.2	14.8	14.4
5	19.0	18.5	18.0
6	22.8	22.2	21.6
7	26.6	25.9	25.2
8	30.4	29.6	28.8
9	34.2	33.3	32.4

	35	34	33
1	3.5	3.4	3.3
2	7.0	6.8	6.6
3	10.5	10.2	9.9
4	14.0	13.6	13.2
5	17.5	17.0	16.5
6	21.0	20.4	19.8
7	24.5	23.8	23.1
8	28.0	27.2	26.4
9	31.5	30.6	29.7

	32	31	30
1	3.2	3.1	3.0
2	6.4	6.2	6.0
3	9.6	9.3	9.0
4	12.8	12.4	12.0
5	16.0	15.5	15.0
6	19.2	18.6	18.0
7	22.4	21.7	21.0
8	25.6	24.8	24.0
9	28.8	27.9	27.0

N.	L. 0	1	2	3	4	5	6	7	8	9
150	17 609	638	667	696	725	754	782	811	840	869
151	898	926	955	984	*013	*041	*070	*099	*127	*156
152	18 184	213	241	270	298	327	355	384	412	441
153	469	498	526	554	583	611	639	667	696	724
154	752	780	808	837	865	893	921	949	977	*005
155	19 033	061	089	117	145	173	201	229	257	285
156	312	340	368	396	424	451	479	507	535	562
157	590	618	645	673	700	728	756	783	811	838
158	866	893	921	948	976	*003	*030	*058	*085	*112
159	20 140	167	194	222	249	276	303	330	358	385
160	412	439	466	493	520	548	575	602	629	656
161	683	710	737	763	790	817	844	871	898	925
162	952	978	*005	*032	*059	*085	*112	*139	*165	*192
163	21 219	245	272	299	325	352	378	405	431	458
164	484	511	537	564	590	617	643	669	696	722
165	748	775	801	827	854	880	906	932	958	985
166	22 011	037	063	089	115	141	167	194	220	246
167	272	298	324	350	376	401	427	453	479	505
168	531	557	583	608	634	660	686	712	737	763
169	789	814	840	866	891	917	943	968	994	*019
170	23 045	070	096	121	147	172	198	223	249	274
171	300	325	350	376	401	426	452	477	502	528
172	553	578	603	629	654	679	704	729	754	779
173	805	830	855	880	905	930	955	980	*005	*030
174	24 055	080	105	130	155	180	204	229	254	279
175	304	329	353	378	403	428	452	477	502	527
176	551	576	601	625	650	674	699	724	748	773
177	797	822	846	871	895	920	944	969	993	*018
178	25 042	066	091	115	139	164	188	212	237	261
179	285	310	334	358	382	406	431	455	479	503
180	527	551	575	600	624	648	672	696	720	744
181	768	792	816	840	864	888	912	935	959	983
182	26 007	031	055	079	102	126	150	174	198	221
183	245	269	293	316	340	364	387	411	435	458
184	482	505	529	553	576	600	623	647	670	694
185	717	741	764	788	811	834	858	881	905	928
186	951	975	998	*021	*045	*068	*091	*114	*138	*161
187	27 184	207	231	254	277	300	323	346	370	393
188	416	439	462	485	508	531	554	577	600	623
189	646	669	692	715	738	761	784	807	830	852
190	875	898	921	944	967	989	*012	*035	*058	*081
191	28 103	126	149	171	194	217	240	262	285	307
192	330	353	375	398	421	443	466	488	511	533
193	556	578	601	623	646	668	691	713	735	758
194	780	803	825	847	870	892	914	937	959	981
195	29 003	026	048	070	092	115	137	159	181	203
196	226	248	270	292	314	336	358	380	403	425
197	447	469	491	513	535	557	579	601	623	645
198	667	688	710	732	754	776	798	820	842	863
199	885	907	929	951	973	994	*016	*038	*060	*081
200	30 103	125	146	168	190	211	233	255	276	298
N.	L. 0	1	2	3	4	5	6	7	8	9

P. P.

	29	28
1	2.9	2.8
2	5.8	5.6
3	8.7	8.4
4	11.6	11.2
5	14.5	14.0
6	17.4	16.8
7	20.3	19.6
8	23.2	22.4
9	26.1	25.2

	27	26
1	2.7	2.6
2	5.4	5.2
3	8.1	7.8
4	10.8	10.4
5	13.5	13.0
6	16.2	15.6
7	18.9	18.2
8	21.6	20.8
9	24.3	23.4

	25
1	2.5
2	5.0
3	7.5
4	10.0
5	12.5
6	15.0
7	17.5
8	20.0
9	22.5

	24	23
1	2.4	2.3
2	4.8	4.6
3	7.2	6.9
4	9.6	9.2
5	12.0	11.5
6	14.4	13.8
7	16.8	16.1
8	19.2	18.4
9	21.6	20.7

	22	21
1	2.2	2.1
2	4.4	4.2
3	6.6	6.3
4	8.8	8.4
5	11.0	10.5
6	13.2	12.6
7	15.4	14.7
8	17.6	16.8
9	19.8	18.9

N.	L. 0	1	2	3	4	5	6	7	8	9
200	30 103	125	146	168	190	211	233	255	276	298
201	320	341	363	384	406	428	449	471	492	514
202	535	557	578	600	621	643	664	685	707	728
203	750	771	792	814	835	856	878	899	920	942
204	963	984	*006	*027	*048	*069	*091	*112	*133	*154
205	31 175	197	218	239	260	281	302	323	345	366
206	387	408	429	450	471	492	513	534	555	576
207	597	618	639	660	681	702	723	744	765	785
208	806	827	848	869	890	911	931	952	973	994
209	32 015	035	056	077	098	118	139	160	181	201
210	222	243	263	284	305	325	346	366	387	408
211	428	449	469	490	510	531	552	572	593	613
212	634	654	675	695	715	736	756	777	797	818
213	838	858	879	899	919	940	960	980	*001	*021
214	33 041	062	082	102	122	143	163	183	203	224
215	244	264	284	304	325	345	365	385	405	425
216	445	465	486	506	526	546	566	586	606	626
217	646	666	686	706	726	746	766	786	806	826
218	846	866	885	905	925	945	965	985	*005	*025
219	34 044	064	084	104	124	143	163	183	203	223
220	242	262	282	301	321	341	361	380	400	420
221	439	459	479	498	518	537	557	577	596	616
222	635	655	674	694	713	733	753	772	792	811
223	830	850	869	889	908	928	947	967	986	*005
224	35 025	044	064	083	102	122	141	160	180	199
225	218	238	257	276	295	315	334	353	372	392
226	411	430	449	468	488	507	526	545	564	583
227	603	622	641	660	679	698	717	736	755	774
228	793	813	832	851	870	889	908	927	946	965
229	984	*003	*021	*040	*059	*078	*097	*116	*135	*154
230	36 173	192	211	229	248	267	286	305	324	342
231	361	380	399	418	436	455	474	493	511	530
232	549	568	586	605	624	642	661	680	698	717
233	736	754	773	791	810	829	847	866	884	903
234	922	940	959	977	996	*014	*033	*051	*070	*088
235	37 107	125	144	162	181	199	218	236	254	273
236	291	310	328	346	365	383	401	420	438	457
237	475	493	511	530	548	566	585	603	621	639
238	658	676	694	712	731	749	767	785	803	822
239	840	858	876	894	912	931	949	967	985	*003
240	38 021	039	057	075	093	112	130	148	166	184
241	202	220	238	256	274	292	310	328	346	364
242	382	399	417	435	453	471	489	507	525	543
243	561	578	596	614	632	650	668	686	703	721
244	739	757	775	792	810	828	846	863	881	899
245	917	934	952	970	987	*005	*023	*041	*058	*076
246	39 094	111	129	146	164	182	199	217	235	252
247	270	287	305	322	340	358	375	393	410	428
248	445	463	480	498	515	533	550	568	585	602
249	620	637	655	672	690	707	724	742	759	777
250	794	811	829	846	863	881	898	915	933	950

N.	L. 0	1	2	3	4	5	6	7	8	9

P. P.

	22	21
1	2.2	2.1
2	4.4	4.2
3	6.6	6.3
4	8.8	8.4
5	11.0	10.5
6	13.2	12.6
7	15.4	14.7
8	17.6	16.8
9	19.8	18.9

	20
1	2.0
2	4.0
3	6.0
4	8.0
5	10.0
6	12.0
7	14.0
8	16.0
9	18.0

	19
1	1.9
2	3.8
3	5.7
4	7.6
5	9.5
6	11.4
7	13.3
8	15.2
9	17.1

	18
1	1.8
2	3.6
3	5.4
4	7.2
5	9.0
6	10.8
7	12.6
8	14.4
9	16.2

	17
1	1.7
2	3.4
3	5.1
4	6.8
5	8.5
6	10.2
7	11.9
8	13.6
9	15.3

N.	L. 0	1	2	3	4	5	6	7	8	9
250	39 794	811	829	846	863	881	898	915	933	950
251	967	985	*002	*019	*037	*054	*071	*088	*106	*123
252	40 140	157	175	192	209	226	243	261	278	295
253	312	329	346	364	381	398	415	432	449	466
254	483	500	518	535	552	569	586	603	620	637
255	654	671	688	705	722	739	756	773	790	807
256	824	841	858	875	892	909	926	943	960	976·
257	993	*010	*027	*044	*061	*078	*095	*111	*128	*145
258	41 162	179	196	212	229	246	263	280	296	313
259	330	347	363	380	397	414	430	447	464	481
260	497	514	531	547	564	581	597	614	631	647
261	664	681	697	714	731	747	764	780	797	814
262	830	847	863	880	896	913	929	946	963	979
263	996	*012	*029	*045	*062	*078	*095	*111	*127	*144
264	42 160	177	193	210	226	243	259	275	292	308
265	325	341	357	374	390	406	423	439	455	472
266	488	504	521	537	553	570	586	602	619	635
267	651	667	684	700	716	732	749	765	781	797
268	813	830	846	862	878	894	911	927	943	959
269	975	991	*008	*024	*040	*056	*072	*088	*104	*120
270	43 136	152	169	185	201	217	233	249	265	281
271	297	313	329	345	361	377	393	409	425	441
272	457	473	489	505	521	537	553	569	584	600
273	616	632	648	664	680	696	712	727	743	759
274	775	791	807	823	838	854	870	886	902	917
275	933	949	965	981	996	*012	*028	*044	*059	*075
276	44 091	107	122	138	154	170	185	201	217	232
277	248	264	279	295	311	326	342	358	373	389
278	404	420	436	451	467	483	498	514	529	545
279	560	576	592	607	623	638	654	669	685	700
280	716	731	747	762	778	793	809	824	840	855
281	871	886	902	917	932	948	963	979	994	*010
282	45 025	040	056	071	086	102	117	133	148	163
283	179	194	209	225	240	255	271	286	301	317
284	332	347	362	378	393	408	423	439	454	469
285	484	500	515	530	545	561	576	591	606	621
286	637	652	667	682	697	712	728	743	758	773
287	788	803	818	834	849	864	879	894	909	924
288	939	954	969	984	*000	*015	*030	*045	*060	*075
289	46 090	105	120	135	150	165	180	195	210	225
290	240	255	270	285	300	315	330	345	359	374
291	389	404	419	434	449	464	479	494	509	523
292	538	553	568	583	598	613	627	642	657	672
293	687	702	716	731	746	761	776	790	805	820
294	835	850	864	879	894	909	923	938	953	967
295	982	997	*012	*026	*041	*056	*070	*085	*100	*114
296	47 129	144	159	173	188	202	217	232	246	261
297	276	290	305	319	334	349	363	378	392	407
298	422	436	451	465	480	494	509	524	538	553
299	567	582	596	611	625	640	654	669	683	698
300	712	727	741	756	770	784	799	813	828	842

N.	L. 0	1	2	3	4	5	6	7	8	9

P. P.

	18	17	16	15	14
1	1.8	1.7	1.6	1.5	1.4
2	3.6	3.4	3.2	3.0	2.8
3	5.4	5.1	4.8	4.5	4.2
4	7.2	6.8	6.4	6.0	5.6
5	9.0	8.5	8.0	7.5	7.0
6	10.8	10.2	9.6	9.0	8.4
7	12.6	11.9	11.2	10.5	9.8
8	14.4	13.6	12.8	12.0	11.2
9	16.2	15.3	14.4	13.5	12.6

Log e = .43429

N.	L. 0	1	2	3	4	5	6	7	8	9	P. P.
300	47 712	727	741	756	770	784	799	813	828	842	
301	857	871	885	900	914	929	943	958	972	986	
302	48 001	015	029	044	058	073	087	101	116	130	
303	144	159	173	187	202	216	230	244	259	273	
304	287	302	316	330	344	359	373	387	401	416	
305	430	444	458	473	487	501	515	530	544	558	
306	572	586	601	615	629	643	657	671	686	700	
307	714	728	742	756	770	785	799	813	827	841	
308	855	869	883	897	911	926	940	954	968	982	
309	996	*010	*024	*038	*052	*066	*080	*094	*108	*122	
310	49 136	150	164	178	192	206	220	234	248	262	
311	276	290	304	318	332	346	360	374	388	402	
312	415	429	443	457	471	485	499	513	527	541	
313	554	568	582	596	610	624	638	651	665	679	
314	693	707	721	734	748	762	776	790	803	817	
315	831	845	859	872	886	900	914	927	941	955	
316	969	982	996	*010	*024	*037	*051	*065	*079	*092	
317	50 106	120	133	147	161	174	188	202	215	229	
318	243	256	270	284	297	311	325	338	352	365	
319	379	393	406	420	433	447	461	474	488	501	
320	515	529	542	556	569	583	596	610	623	637	
321	651	664	678	691	705	718	732	745	759	772	
322	786	799	813	826	840	853	866	880	893	907	
323	920	934	947	961	974	987	*001	*014	*028	*041	
324	51 055	068	081	095	108	121	135	148	162	175	
325	188	202	215	228	242	255	268	282	295	308	
326	322	335	348	362	375	388	402	415	428	441	
327	455	468	481	495	508	521	534	548	561	574	
328	587	601	614	627	640	654	667	680	693	706	
329	720	733	746	759	772	786	799	812	825	838	
330	851	865	878	891	904	917	930	943	957	970	
331	983	996	*009	*022	*035	*048	*061	*075	*088	*101	
332	52 114	127	140	153	166	179	192	205	218	231	
333	244	257	270	284	297	310	323	336	349	362	
334	375	388	401	414	427	440	453	466	479	492	
335	504	517	530	543	556	569	582	595	608	621	
336	634	647	660	673	686	699	711	724	737	750	
337	763	776	789	802	815	827	840	853	866	879	
338	892	905	917	930	943	956	969	982	994	*007	
339	53 020	033	046	058	071	084	097	110	122	135	
340	148	161	173	186	199	212	224	237	250	263	
341	275	288	301	314	326	339	352	364	377	390	
342	403	415	428	441	453	466	479	491	504	517	
343	529	542	555	567	580	593	605	618	631	643	
344	656	668	681	694	706	719	732	744	757	769	
345	782	794	807	820	832	845	857	870	882	895	
346	908	920	933	945	958	970	983	995	*008	*020	
347	54 033	045	058	070	083	095	108	120	133	145	
348	158	170	183	195	208	220	233	245	258	270	
349	283	295	307	320	332	345	357	370	382	394	
350	407	419	432	444	456	469	481	494	506	518	
N.	L. 0	1	2	3	4	5	6	7	8	9	P. P.

P. P. columns:

15
1	1.5
2	3.0
3	4.5
4	6.0
5	7.5
6	9.0
7	10.5
8	12.0
9	13.5

14
1	1.4
2	2.8
3	4.2
4	5.6
5	7.0
6	8.4
7	9.8
8	11.2
9	12.6

13
1	1.3
2	2.6
3	3.9
4	5.2
5	6.5
6	7.8
7	9.1
8	10.4
9	11.7

12
1	1.2
2	2.4
3	3.6
4	4.8
5	6.0
6	7.2
7	8.4
8	9.6
9	10.8

Log π = .49715

N.	L. 0	1	2	3	4	5	6	7	8	9	P. P.
350	54 407	419	432	444	456	469	481	494	506	518	
351	531	543	555	568	580	593	605	617	630	642	
352	654	667	679	691	704	716	728	741	753	765	
353	777	790	802	814	827	839	851	864	876	888	
354	900	913	925	937	949	962	974	986	998	*011	
355	55 023	035	047	060	072	084	096	108	121	133	
356	145	157	169	182	194	206	218	230	242	255	
357	267	279	291	303	315	328	340	352	364	376	
358	388	400	413	425	437	449	461	473	485	497	
359	509	522	534	546	558	570	582	594	606	618	
360	630	642	654	666	678	691	703	715	727	739	
361	751	763	775	787	799	811	823	835	847	859	
362	871	883	895	907	919	931	943	955	967	979	
363	991	*003	*015	*027	*038	*050	*062	*074	*086	*098	
364	56 110	122	134	146	158	170	182	194	205	217	
365	229	241	253	265	277	289	301	312	324	336	
366	348	360	372	384	396	407	419	431	443	455	
367	467	478	490	502	514	526	538	549	561	573	
368	585	597	608	620	632	644	656	667	679	691	
369	703	714	726	738	750	761	773	785	797	808	
370	820	832	844	855	867	879	891	902	914	926	
371	937	949	961	972	984	996	*008	*019	*031	*043	
372	57 054	066	078	089	101	113	124	136	148	159	
373	171	183	194	206	217	229	241	252	264	276	
374	287	299	310	322	334	345	357	368	380	392	
375	403	415	426	438	449	461	473	484	496	507	
376	519	530	542	553	565	576	588	600	611	623	
377	634	646	657	669	680	692	703	715	726	738	
378	749	761	772	784	795	807	818	830	841	852	
379	864	875	887	898	910	921	933	944	955	967	
380	978	990	*001	*013	*024	*035	*047	*058	*070	*081	
381	58 092	104	115	127	138	149	161	172	184	195	
382	206	218	229	240	252	263	274	286	297	309	
383	320	331	343	354	365	377	388	399	410	422	
384	433	444	456	467	478	490	501	512	524	535	
385	546	557	569	580	591	602	614	625	636	647	
386	659	670	681	692	704	715	726	737	749	760	
387	771	782	794	805	816	827	838	850	861	872	
388	883	894	906	917	928	939	950	961	973	984	
389	995	*006	*017	*028	*040	*051	*062	*073	*084	*095	
390	59 106	118	129	140	151	162	173	184	195	207	
391	218	229	240	251	262	273	284	295	306	318	
392	329	340	351	362	373	384	395	406	417	428	
393	439	450	461	472	483	494	506	517	528	539	
394	550	561	572	583	594	605	616	627	638	649	
395	660	671	682	693	704	715	726	737	748	759	
396	770	780	791	802	813	824	835	846	857	868	
397	879	890	901	912	923	934	945	956	966	977	
398	988	999	*010	*021	*032	*043	*054	*065	*076	*086	
399	60 097	108	119	130	141	152	163	173	184	195	
400	206	217	228	239	249	260	271	282	293	304	
N.	L. 0	1	2	3	4	5	6	7	8	9	P. P.

P. P.

	13
1	1.3
2	2.6
3	3.9
4	5.2
5	6.5
6	7.8
7	9.1
8	10.4
9	11.7

	12
1	1.2
2	2.4
3	3.6
4	4.8
5	6.0
6	7.2
7	8.4
8	9.6
9	10.8

	11
1	1.1
2	2.2
3	3.3
4	4.4
5	5.5
6	6.6
7	7.7
8	8.8
9	9.9

	10
1	1.0
2	2.0
3	3.0
4	4.0
5	5.0
6	6.0
7	7.0
8	8.0
9	9.0

N.	L. 0	1	2	3	4	5	6	7	8	9	P. P.
400	60 206	217	228	239	249	260	271	282	293	304	
401	314	325	336	347	358	369	379	390	401	412	
402	423	433	444	455	466	477	487	498	509	520	
403	531	541	552	563	574	584	595	606	617	627	
404	638	649	660	670	681	692	703	713	724	735	
405	746	756	767	778	788	799	810	821	831	842	
406	853	863	874	885	895	906	917	927	938	949	
407	959	970	981	991	*002	*013	*023	*034	*045	*055	
408	61 066	077	087	098	109	119	130	140	151	162	
409	172	183	194	204	215	225	236	247	257	268	
410	278	289	300	310	321	331	342	352	363	374	
411	384	395	405	416	426	437	448	458	469	479	
412	490	500	511	521	532	542	553	563	574	584	
413	595	606	616	627	637	648	658	669	679	690	
414	700	711	721	731	742	752	763	773	784	794	
415	805	815	826	836	847	857	868	878	888	899	
416	909	920	930	941	951	962	972	982	993	*003	
417	62 014	024	034	045	055	066	076	086	097	107	
418	118	128	138	149	159	170	180	190	201	211	
419	221	232	242	252	263	273	284	294	304	315	
420	325	335	346	356	366	377	387	397	408	418	
421	428	439	449	459	469	480	490	500	511	521	
422	531	542	552	562	572	583	593	603	613	624	
423	634	644	655	665	675	685	696	706	716	726	
424	737	747	757	767	778	788	798	808	818	829	
425	839	849	859	870	880	890	900	910	921	931	
426	941	951	961	972	982	992	*002	*012	*022	*033	
427	63 043	053	063	073	083	094	104	114	124	134	
428	144	155	165	175	185	195	205	215	225	236	
429	246	256	266	276	286	296	306	317	327	337	
430	347	357	367	377	387	397	407	417	428	438	
431	448	458	468	478	488	498	508	518	528	538	
432	548	558	568	579	589	599	609	619	629	639	
433	649	659	669	679	689	699	709	719	729	739	
434	749	759	769	779	789	799	809	819	829	839	
435	849	859	869	879	889	899	909	919	929	939	
436	949	959	969	979	988	998	*008	*018	*028	*038	
437	64 048	058	068	078	088	098	108	118	128	137	
438	147	157	167	177	187	197	207	217	227	237	
439	246	256	266	276	286	296	306	316	326	335	
440	345	355	365	375	385	395	404	414	424	434	
441	444	454	464	473	483	493	503	513	523	532	
442	542	552	562	572	582	591	601	611	621	631	
443	640	650	660	670	680	689	699	709	719	729	
444	738	748	758	768	777	787	797	807	816	826	
445	836	846	856	865	875	885	895	904	914	924	
446	933	943	953	963	972	982	992	*002	*011	*021	
447	65 031	040	050	060	070	079	089	099	108	118	
448	128	137	147	157	167	176	186	196	205	ʼ215	
449	225	234	244	254	263	273	283	292	302	312	
450	321	331	341	350	360	369	379	389	398	408	
N.	L. 0	1	2	3	4	5	6	7	8	9	P. P.

P. P.

	11
1	1.1
2	2.2
3	3.3
4	4.4
5	5.5
6	6.6
7	7.7
8	8.8
9	9.9

	10
1	1.0
2	2.0
3	3.0
4	4.0
5	5.0
6	6.0
7	7.0
8	8.0
9	9.0

	9
1	0.9
2	1.8
3	2.7
4	3.6
5	4.5
6	5.4
7	6.3
8	7.2
9	8.1

N.	L. 0	1	2	3	4	5	6	7	8	9
450	65 321	331	341	350	360	369	379	389	398	408
451	418	427	437	447	456	466	475	485	495	504
452	514	523	533	543	552	562	571	581	591	600
453	610	619	629	639	648	658	667	677	686	696
454	706	715	725	734	744	753	763	772	782	792
455	801	811	820	830	839	849	858	868	877	887
456	896	906	916	925	935	944	954	963	973	982
457	992	*001	*011	*020	*030	*039	*049	*058	*068	*077
458	66 087	096	106	115	124	134	143	153	162	172
459	181	191	200	210	219	229	238	247	257	266
460	276	285	295	304	314	323	332	342	351	361
461	370	380	389	398	408	417	427	436	445	455
462	464	474	483	492	502	511	521	530	539	549
463	558	567	577	586	596	605	614	624	633	642
464	652	661	671	680	689	699	708	717	727	736
465	745	755	764	773	783	792	801	811	820	829
466	839	848	857	867	876	885	894	904	913	922
467	932	941	950	960	969	978	987	997	*006	*015
468	67 025	034	043	052	062	071	080	089	099	108
469	117	127	136	145	154	164	173	182	191	201
470	210	219	228	237	247	256	265	274	284	293
471	302	311	321	330	339	348	357	367	376	385
472	394	403	413	422	431	440	449	459	468	477
473	486	495	504	514	523	532	541	550	560	569
474	578	587	596	605	614	624	633	642	651	660
475	669	679	688	697	706	715	724	733	742	752
476	761	770	779	788	797	806	815	825	834	843
477	852	861	870	879	888	897	906	916	925	934
478	943	952	961	970	979	988	997	*006	*015	*024
479	68 034	043	052	061	070	079	088	097	106	115
480	124	133	142	151	160	169	178	187	196	205
481	215	224	233	242	251	260	269	278	287	296
482	305	314	323	332	341	350	359	368	377	386
483	395	404	413	422	431	440	449	458	467	476
484	485	494	502	511	520	529	538	547	556	565
485	574	583	592	601	610	619	628	637	646	655
486	664	673	681	690	699	708	717	726	735	744
487	753	762	771	780	789	797	806	815	824	833
488	842	851	860	869	878	886	895	904	913	922
489	931	940	949	958	966	975	984	993	*002	*011
490	69 020	028	037	046	055	064	073	082	090	099
491	108	117	126	135	144	152	161	170	179	188
492	197	205	214	223	232	241	249	258	267	276
493	285	294	302	311	320	329	338	346	355	364
494	373	381	390	399	408	417	425	434	443	452
495	461	469	478	487	496	504	513	522	531	539
496	548	557	566	574	583	592	601	609	618	627
497	636	644	653	662	671	679	688	697	705	714
498	723	732	740	749	758	767	775	784	793	801
499	810	819	827	836	845	854	862	871	880	888
500	897	906	914	923	932	940	949	958	966	975

P. P

10
1	1.0
2	2.0
3	3.0
4	4.0
5	5.0
6	6.0
7	7.0
8	8.0
9	9.0

9
1	0.9
2	1.8
3	2.7
4	3.6
5	4.5
6	5.4
7	6.3
8	7.2
9	8.1

8
1	0.8
2	1.6
3	2.4
4	3.2
5	4.0
6	4.8
7	5.6
8	6.4
9	7.2

N.	L. 0	1	2	3	4	5	6	7	8	9	P. P.

N.	L. 0.	1	2	3	4	5	6	7	8	9	P. P.
500	69 897	906	914	923	932	940	949	958	966	975	
501	984	992	*001	*010	*018	*027	*036	*044	*053	*062	
502	70 070	079	088	096	105	114	122	131	140	148	
503	157	165	174	183	191	200	209	217	226	234	
504	243	252	260	269	278	286	295	303	312	321	
505	329	338	346	355	364	372	381	389	398	406	
506	415	424	432	441	449	458	467	475	484	492	
507	501	509	518	526	535	544	552	561	569	578	
508	586	595	603	612	621	629	638	646	655	663	
509	672	680	689	697	706	714	723	731	740	749	
510	757	766	774	783	791	800	808	817	825	834	
511	842	851	859	868	876	885	893	902	910	919	
512	927	935	944	952	961	969	978	986	995	*003	
513	71 012	020	029	037	046	054	063	071	079	088	
514	096	105	113	122	130	139	147	155	164	172	
515	181	189	198	206	214	223	231	240	248	257	
516	265	273	282	290	299	307	315	324	332	341	
517	349	357	366	374	383	391	399	408	416	425	
518	433	441	450	458	466	475	483	492	500	508	
519	517	525	533	542	550	559	567	575	584	592	
520	600	609	617	625	634	642	650	659	667	675	
521	684	692	700	709	717	725	734	742	750	759	
522	767	775	784	792	800	809	817	825	834	842	
523	850	858	867	875	883	892	900	908	917	925	
524	933	941	950	958	966	975	983	991	999	*008	
525	72 016	024	032	041	049	057	066	074	082	090	
526	099	107	115	123	132	140	148	156	165	173	
527	181	189	198	206	214	222	230	239	247	255	
528	263	272	280	288	296	304	313	321	329	337	
529	346	354	362	370	378	387	395	403	411	419	
530	428	436	444	452	460	469	477	485	493	501	
531	509	518	526	534	542	550	558	567	575	583	
532	591	599	607	616	624	632	640	648	656	665	
533	673	681	689	697	705	713	722	730	738	746	
534	754	762	770	779	787	795	803	811	819	827	
535	835	843	852	860	868	876	884	892	900	908	
536	916	925	933	941	949	957	965	973	981	989	
537	997	*006	*014	*022	*030	*038	*046	*054	*062	*070	
538	73 078	086	094	102	111	119	127	135	143	151	
539	159	167	175	183	191	199	207	215	223	231	
540	239	247	255	263	272	280	288	296	304	312	
541	320	328	336	344	352	360	368	376	384	392	
542	400	408	416	424	432	440	448	456	464	472	
543	480	488	496	504	512	520	528	536	544	552	
544	560	568	576	584	592	600	608	616	624	632	
545	640	648	656	664	672	679	687	695	703	711	
546	719	727	735	743	751	759	767	775	783	791	
547	799	807	815	823	830	838	846	854	862	870	
548	878	886	894	902	910	918	926	933	941	949	
549	957	965	973	981	989	997	*005	*013	*020	*028	
550	74 036	044	052	060	068	076	084	092	099	107	
N.	L. 0	1	2	3	4	5	6	7	8	9	P. P.

P. P.

9
1 | 0.9
2 | 1.8
3 | 2.7
4 | 3.6
5 | 4.5
6 | 5.4
7 | 6.3
8 | 7.2
9 | 8.1

8
1 | 0.8
2 | 1.6
3 | 2.4
4 | 3.2
5 | 4.0
6 | 4.8
7 | 5.6
8 | 6.4
9 | 7.2

7
1 | 0.7
2 | 1.4
3 | 2.1
4 | 2.8
5 | 3.5
6 | 4.2
7 | 4.9
8 | 5.6
9 | 6.3

N.	L. 0	1	2	3	4	5	6	7	8	9	P. P.
550	74 036	044	052	060	068	076	084	092	099	107	
551	115	123	131	139	147	155	162	170	178	186	
552	194	202	210	218	225	233	241	249	257	265	
553	273	280	288	296	304	312	320	327	335	343	
554	351	359	367	374	382	390	398	406	414	421	
555	429	437	445	453	461	468	476	484	492	500	
556	507	515	523	531	539	547	554	562	570	578	
557	586	593	601	609	617	624	632	640	648	656	
558	663	671	679	687	695	702	710	718	726	733	
559	741	749	757	764	772	780	788	796	803	811	
560	819	827	834	842	850	858	865	873	881	889	
561	896	904	912	920	927	935	943	950	958	966	
562	974	981	989	997	*005	*012	*020	*028	*035	*043	
563	75 051	059	066	074	082	089	097	105	113	120	
564	128	136	143	151	159	166	174	182	189	197	
565	205	213	220	228	236	243	251	259	266	274	
566	282	289	297	305	312	320	328	335	343	351	
567	358	366	374	381	389	397	404	412	420	427	
568	435	442	450	458	465	473	481	488	496	504	
569	511	519	526	534	542	549	557	565	572	580	
570	587	595	603	610	618	626	633	641	648	656	
571	664	671	679	686	694	702	709	717	724	732	
572	740	747	755	762	770	778	785	793	800	808	
573	815	823	831	838	846	853	861	868	876	884	
574	891	899	906	914	921	929	937	944	952	959	
575	967	974	982	989	997	*005	*012	*020	*027	*035	
576	76 042	050	057	065	072	080	087	095	103	110	
577	118	125	133	140	148	155	163	170	178	185	
578	193	200	208	215	223	230	238	245	253	260	
579	268	275	283	290	298	305	313	320	328	335	
580	343	350	358	365	373	380	388	395	403	410	
581	418	425	433	440	448	455	462	470	477	485	
582	492	500	507	515	522	530	537	545	552	559	
583	567	574	582	589	597	604	612	619	626	634	
584	641	649	656	664	671	678	686	693	701	708	
585	716	723	730	738	745	753	760	768	775	782	
586	790	797	805	812	819	827	834	842	849	856	
587	864	871	879	886	893	901	908	916	923	930	
588	938	945	953	960	967	975	982	989	997	*004	
589	77 012	019	026	034	041	048	056	063	070	078	
590	085	093	100	107	115	122	129	137	144	151	
591	159	166	173	181	188	195	203	210	217	225	
592	232	240	247	254	262	269	276	283	291	298	
593	305	313	320	327	335	342	349	357	364	371	
594	379	386	393	401	408	415	422	430	437	444	
595	452	459	466	474	481	488	495	503	510	517	
596	525	532	539	546	554	561	568	576	583	590	
597	597	605	612	619	627	634	641	648	656	663	
598	670	677	685	692	699	706	714	721	728	735	
599	743	750	757	764	772	779	786	793	801	808	
600	815	822	830	837	844	851	859	866	873	880	
N.	L. 0	1	2	3	4	5	6	7	8	9	P. P.

P. P. column:

8
1	0.8
2	1.6
3	2.4
4	3.2
5	4.0
6	4.8
7	5.6
8	6.4
9	7.2

7
1	0.7
2	1.4
3	2.1
4	2.8
5	3.5
6	4.2
7	4.9
8	5.6
9	6.3

N.	L. 0	1	2	3	4	5	6	7	8	9	P. P.
600	77 815′	822	830	837	844	851	859	866	873	880	
601	887	895	902	909	916	924	931	938	945	952	
602	960	967	974	981	988	996	*003	*010	*017	*025	
603	78 032	039	046	053	061	068	075	082	089	097	
604	104	111	118	125	132	140	147	154	161	168	
605	176	183	190	197	204	211	219	226	233	240	
606	247	254	262	269	276	283	290	297	305	312	
607	319	326	333	340	347	355	362	369	376	383	
608	390	398	405	412	419	426	433	440	447	455	
609	462	469	476	483	490	497	504	512	519	526	
610	533	540	547	554	561	569	576	583	590	597	
611	604	611	618	625	633	640	647	654	661	668	
612	675	682	689	696	704	711	718	725	732	739	
613	746	753	760	767	774	781	789	796	803	810	
614	817	824	831	838	845	852	859	866	873	880	
615	888	895	902	909	916	923	930	937	944	951	
616	958	965	972	979	986	993	*000	*007	*014	*021	
617	79 029	036	043	050	057	064	071	078	085	092	
618	099	106	113	120	127	134	141	148	155	162	
619	169	176	183	190	197	204	211	218	225	232	
620	239	246	253	260	267	274	281	288	295	302	
621	309	316	323	330	337	344	351	358	365	372	
622	379	386	393	400	407	414	421	428	435	442	
623	449	456	463	470	477	484	491	498	505	511	
624	518	525	532	539	546	553	560	567	574	581	
625	588	595	602	609	616	623	630	637	644	650	
626	657	664	671	678	685	692	699	706	713	720	
627	727	734	741	748	754	761	768	775	782	789	
628	796	803	810	817	824	831	837	844	851	858	
629	865	872	879	886	893	900	906	913	920	927	
630	934	941	948	955	962	969	975	982	989	996	
631	80 003	010	017	024	030	037	044	051	058	065	
632	072	079	085	092	099	106	113	120	127	134	
633	140	147	154	161	168	175	182	188	195	202	
634	209	216	223	229	236	243	250	257	264	271	
635	277	284	291	298	305	312	318	325	332	339	
636	346	353	359	366	373	380	387	393	400	407	
637	414	421	428	434	441	448	455	462	468	475	
638	482	489	496	502	509	516	523	530	536	543	
639	550	557	564	570	577	584	591	598	604	611	
640	618	625	632	638	645	652	659	665	672	679	
641	686	693	699	706	713	720	726	733	740	747	
642	754	760	767	774	781	787	794	801	808	814	
643	821	828	835	841	848	855	862	868	875	882	
644	889	895	902	909	916	922	929	936	943	949	
645	956	963	969	976	983	990	996	*003	*010	*017	
646	81 023	030	037	043	050	057	064	070	077	084	
647	090	097	104	111	117	124	131	137	144	151	
648	158	164	171	178	184	191	198	204	211	218	
649	224	231	238	245	251	258	265	271	278	285	
650	291	298	305	311	318	325	331	338	345	351	
N.	L. 0	1	2	3	4	5	6	7	8	9	P. P.

P. P.

8
1	0.8
2	1.6
3	2.4
4	3.2
5	4.0
6	4.8
7	5.6
8	6.4
9	7.2

7
1	0.7
2	1.4
3	2.1
4	2.8
5	3.5
6	4.2
7	4.9
8	5.6
9	6.3

6
1	0.6
2	1.2
3	1.8
4	2.4
5	3.0
6	3.6
7	4.2
8	4.8
9	5.4

N.	L. 0	1	2	3	4	5	6	7	8	9	P. P.
650	81 291	298	305	311	318	325	331	338	345	351	
651	358	365	371	378	385	391	398	405	411	418	
652	425	431	438	445	451	458	465	471	478	485	
653	491	498	505	511	518	525	531	538	544	551	
654	558	564	571	578	584	591	598	604	611	617	
655	624	631	637	644	651	657	664	671	677	684	
656	690	697	704	710	717	723	730	737	743	750	
657	757	763	770	776	783	790	796	803	809	816	
658	823	829	836	842	849	856	862	869	875	882	
659	889	895	902	908	915	921	928	935	941	948	
660	954	961	968	974	981	987	994	*000	*007	*014	
661	82 020	027	033	040	046	053	060	066	073	079	**7**
662	086	092	099	105	112	119	125	132	138	145	1 0.7
663	151	158	164	171	178	184	191	197	204	210	2 1.4
664	217	223	230	236	243	249	256	263	269	276	3 2.1
665	282	289	295	302	308	315	321	328	334	341	4 2.8 5 3.5
666	347	354	360	367	373	380	387	393	400	406	6 4.2 7 4.9
667	413	419	426	432	439	445	452	458	465	471	8 5.6 9 6.3
668	478	484	491	497	504	510	517	523	530	536	
669	543	549	556	562	569	575	582	588	595	601	
670	607	614	620	627	633	640	646	653	659	666	
671	672	679	685	692	698	705	711	718	724	730	
672	737	743	750	756	763	769	776	782	789	795	
673	802	808	814	821	827	834	840	847	853	860	
674	866	872	879	885	892	898	905	911	918	924	
675	930	937	943	950	956	963	969	975	982	988	
676	995	*001	*008	*014	*020	*027	*033	*040	*046	*052	
677	83 059	065	072	078	085	091	097	104	110	117	
678	123	129	136	142	149	155	161	168	174	181	
679	187	193	200	206	213	219	225	232	238	245	
680	251	257	264	270	276	283	289	296	302	308	**6**
681	315	321	327	334	340	347	353	359	366	372	1 0.6
682	378	385	391	398	404	410	417	423	429	436	2 1.2 3 1.8
683	442	448	455	461	467	474	480	487	493	499	4 2.4
684	506	512	518	525	531	537	544	550	556	563	5 3.0 6 3.6
685	569	575	582	588	594	601	607	613	620	626	7 4.2
686	632	639	645	651	658	664	670	677	683	689	8 4.8 9 5.4
687	696	702	708	715	721	727	734	740	746	753	
688	759	765	771	778	784	790	797	803	809	816	
689	822	828	835	841	847	853	860	866	872	879	
690	885	891	897	904	910	916	923	929	935	942	
691	948	954	960	967	973	979	985	992	998	*004	
692	84 011	017	023	029	036	042	048	055	061	067	
693	073	080	086	092	098	105	111	117	123	130	
694	136	142	148	155	161	167	173	180	186	192	
695	198	205	211	217	223	230	236	242	248	255	
696	261	267	273	280	286	292	298	305	311	317	
697	323	330	336	342	348	354	361	367	373	379	
698	386	392	398	404	410	417	423	429	435	442	
699	448	454	460	466	473	479	485	491	497	504	
700	510	516	522	528	535	541	547	553	559	566	
N.	L. 0	1	2	3	4	5	6	7	8	9	P. P.

N.	L. 0	1	2	3	4	5	6	7	8	9	P. P.
700	84 510	516	522	528	535	541	547	553	559	566	
701	572	578	584	590	597	603	609	615	621	628	
702	634	640	646	652	658	665	671	677	683	689	
703	696	702	708	714	720	726	733	739	745	751	
704	757	763	770	776	782	788	794	800	807	813	
705	819	825	831	837	844	850	856	862	868	874	
706	880	887	893	899	905	911	917	924	930	936	
707	942	948	954	960	967	973	979	985	991	997	
708	85 003	009	016	022	028	034	040	046	052	058	
709	065	071	077	083	089	095	101	107	114	120	
710	126	132	138	144	150	156	163	169	175	181	
711	187	193	199	205	211	217	224	230	236	242	
712	248	254	260	266	272	278	285	291	297	303	
713	309	315	321	327	333	339	345	352	358	364	
714	370	376	382	388	394	400	406	412	418	425	
715	431	437	443	449	455	461	467	473	479	485	
716	491	497	503	509	516	522	528	534	540	546	
717	552	558	564	570	576	582	588	594	600	606	
718	612	618	625	631	637	643	649	655	661	667	
719	673	679	685	691	697	703	709	715	721	727	
720	733	739	745	751	757	763	769	775	781	788	
721	794	800	806	812	818	824	830	836	842	848	
722	854	860	866	872	878	884	890	896	902	908	
723	914	920	926	932	938	944	950	956	962	968	
724	974	980	986	992	998	*004	*010	*016	*022	*028	
725	86 034	040	046	052	058	064	070	076	082	088	
726	094	100	106	112	118	124	130	136	141	147	
727	153	159	165	171	177	183	189	195	201	207	
728	213	219	225	231	237	243	249	255	261	267	
729	273	279	285	291	297	303	308	314	320	326	
730	332	338	344	350	356	362	368	374	380	386	
731	392	398	404	410	415	421	427	433	439	445	
732	451	457	463	469	475	481	487	493	499	504	
733	510	516	522	528	534	540	546	552	558	564	
734	570	576	581	587	593	599	605	611	617	623	
735	629	635	641	646	652	658	664	670	676	682	
736	688	694	700	705	711	717	723	729	735	741	
737	747	753	759	764	770	776	782	788	794	800	
738	806	812	817	823	829	835	841	847	853	859	
739	864	870	876	882	888	894	900	906	911	917	
740	923	929	935	941	947	953	958	964	970	976	
741	982	988	994	999	*005	*011	*017	*023	*029	*035	
742	87 040	046	052	058	064	070	075	081	087	093	
743	099	105	111	116	122	128	134	140	146	151	
744	157	163	169	175	181	186	192	198	204	210	
745	216	221	227	233	239	245	251	256	262	268	
746	274	280	286	291	297	303	309	315	320	326	
747	332	338	344	349	355	361	367	373	379	384	
748	390	396	402	408	413	419	425	431	437	442	
749	448	454	460	466	471	477	483	489	495	500	
750	506	512	518	523	529	535	541	547	552	558	
N.	L. 0	1	2	3	4	5	6	7	8	9	P. P.

P. P.

7
1 | 0.7
2 | 1.4
3 | 2.1
4 | 2.8
5 | 3.5
6 | 4.2
7 | 4.9
8 | 5.6
9 | 6.3

6
1 | 0.6
2 | 1.2
3 | 1.8
4 | 2.4
5 | 3.0
6 | 3.6
7 | 4.2
8 | 4.8
9 | 5.4

5
1 | 0.5
2 | 1.0
3 | 1.5
4 | 2.0
5 | 2.5
6 | 3.0
7 | 3.5
8 | 4.0
9 | 4.5

N.	L. 0	1	2	3	4	5	6	7	8	9	P. P.
750	87 506	512	518	523	529	535	541	547	552	558	
751	564	570	576	581	587	593	599	604	610	616	
752	622	628	633	639	645	651	656	662	668	674	
753	679	685	691	697	703	708	714	720	726	731	
754	737	743	749	754	760	766	772	777	783	789	
755	795	800	806	812	818	823	829	835	841	846	
756	852	858	864	869	875	881	887	892	898	904	
757	910	915	921	927	933	938	944	950	955	961	
758	967	973	978	984	990	996	*001	*007	*013	*018	
759	88 024	030	036	041	047	053	058	064	070	076	
760	081	087	093	098	104	110	116	121	127	133	
761	138	144	150	156	161	167	173	178	184	190	
762	195	201	207	213	218	224	230	235	241	247	
763	252	258	264	270	275	281	287	292	298	304	
764	309	315	321	326	332	338	343	349	355	360	
765	366	372	377	383	389	395	400	406	412	417	
766	423	429	434	440	446	451	457	463	468	474	
767	480	485	491	497	502	508	513	519	525	530	
768	536	542	547	553	559	564	570	576	581	587	
769	593	598	604	610	615	621	627	632	638	643	
770	649	655	660	666	672	677	683	689	694	700	
771	705	711	717	722	728	734	739	745	750	756	
772	762	767	773	779	784	790	795	801	807	812	
773	818	824	829	835	840	846	852	857	863	868	
774	874	880	885	891	897	902	908	913	919	925	
775	930	936	941	947	953	958	964	969	975	981	
776	986	992	997	*003	*009	*014	*020	*025	*031	*037	
777	89 042	048	053	059	064	070	076	081	087	092	
778	098	104	109	115	120	126	131	137	143	148	
779	154	159	165	170	176	182	187	193	198	204	
780	209	215	221	226	232	237	243	248	254	260	
781	265	271	276	282	287	293	298	304	310	315	
782	321	326	332	337	343	348	354	360	365	371	
783	376	382	387	393	398	404	409	415	421	426	
784	432	437	443	448	454	459	465	470	476	481	
785	487	492	498	504	509	515	520	526	531	537	
786	542	548	553	559	564	570	575	581	586	592	
787	597	603	609	614	620	625	631	636	642	647	
788	653	658	664	669	675	680	686	691	697	702	
789	708	713	719	724	730	735	741	746	752	757	
790	763	768	774	779	785	790	796	801	807	812	
791	818	823	829	834	840	845	851	856	862	867	
792	873	878	883	889	894	900	905	911	916	922	
793	927	933	938	944	949	955	960	966	971	977	
794	982	988	993	998	*004	*009	*015	*020	*026	*031	
795	90 037	042	048	053	059	064	069	075	080	086	
796	091	097	102	108	113	119	124	129	135	140	
797	146	151	157	162	168	173	179	184	189	195	
798	200	206	211	217	222	227	233	238	244	249	
799	255	260	266	271	276	282	287	293	298	304	
800	309	314	320	325	331	336	342	347	352	358	
N.	L. 0	1	2	3	4	5	6	7	8	9	P. P.

P. P.

6
1 | 0.6
2 | 1.2
3 | 1.8
4 | 2.4
5 | 3.0
6 | 3.6
7 | 4.2
8 | 4.8
9 | 5.4

5
1 | 0.5
2 | 1.0
3 | 1.5
4 | 2.0
5 | 2.5
6 | 3.0
7 | 3.5
8 | 4.0
9 | 4.5

N.	L. 0	1	2	3	4	5	6	7	8	9	P. P.
800	90 309	314	320	325	331	336	342	347	352	358	
801	363	369	374	380	385	390	396	401	407	412	
802	417	423	428	434	439	445	450	455	461	466	
803	472	477	482	488	493	499	504	509	515	520	
804	526	531	536	542	547	553	558	563	569	574	
805	580	585	590	596	601	607	612	617	623	628	
806	634	639	644	650	655	660	666	671	677	682	
807	687	693	698	703	709	714	720	725	730	736	
808	741	747	752	757	763	768	773	779	784	789	
809	795	800	806	811	816	822	827	832	838	843	
810	849	854	859	865	870	875	881	886	891	897	
811	902	907	913	918	924	929	934	940	945	950	
812	956	961	966	972	977	982	988	993	998	*004	
813	91 009	014	020	025	030	036	041	046	052	057	
814	062	068	073	078	084	089	094	100	105	110	
815	116	121	126	132	137	142	148	153	158	164	
816	169	174	180	185	190	196	201	206	212	217	
817	222	228	233	238	243	249	254	259	265	270	
818	275	281	286	291	297	302	307	312	318	323	
819	328	334	339	344	350	355	360	365	371	376	
820	381	387	392	397	403	408	413	418	424	429	
821	434	440	445	450	455	461	466	471	477	482	
822	487	492	498	503	508	514	519	524	529	535	
823	540	545	551	556	561	566	572	577	582	587	
824	593	598	603	609	614	619	624	630	635	640	
825	645	651	656	661	666	672	677	682	687	693	
826	698	703	709	714	719	724	730	735	740	745	
827	751	756	761	766	772	777	782	787	793	798	
828	803	808	814	819	824	829	834	840	845	850	
829	855	861	866	871	876	882	887	892	897	903	
830	908	913	918	924	929	934	939	944	950	955	
831	960	965	971	976	981	986	991	997	*002	*007	
832	92 012	018	023	028	033	038	044	049	054	059	
833	065	070	075	080	085	091	096	101	106	111	
834	117	122	127	132	137	143	148	153	158	163	
835	169	174	179	184	189	195	200	205	210	215	
836	221	226	231	236	241	247	252	257	262	267	
837	273	278	283	288	293	298	304	309	314	319	
838	324	330	335	340	345	350	355	361	366	371	
839	376	381	387	392	397	402	407	412	418	423	
840	428	433	438	443	449	454	459	464	469	474	
841	480	485	490	495	500	505	511	516	521	526	
842	531	536	542	547	552	557	562	567	572	578	
843	583	588	593	598	603	609	614	619	624	629	
844	634	639	645	650	655	660	665	670	675	681	
845	686	691	696	701	706	711	716	722	727	732	
846	737	742	747	752	758	763	768	773	778	783	
847	788	793	799	804	809	814	819	824	829	834	
848	840	845	850	855	860	865	870	875	881	886	
849	891	896	901	906	911	916	921	927	932	937	
850	942	947	952	957	962	967	973	978	983	988	
N.	L. 0	1	2	3	4	5	6	7	8	9	P. P.

6

1	0.6
2	1.2
3	1.8
4	2.4
5	3.0
6	3.6
7	4.2
8	4.8
9	5.4

5

1	0.5
2	1.0
3	1.5
4	2.0
5	2.5
6	3.0
7	3.5
8	4.0
9	4.5

N.	L. 0	1	2	3	4	5	6	7	8	9	P. P.
850	92 942	947	952	957	962	967	973	978	983	988	
851	993	998	*003	*008	*013	*018	*024	*029	*034	*039	
852	93 044	049	054	059	064	069	075	080	085	090	
853	095	100	105	110	115	120	125	131	136	141	
854	146	151	156	161	166	171	176	181	186	192	
855	197	202	207	212	217	222	227	232	237	242	
856	247	252	258	263	268	273	278	283	288	293	
857	298	303	308	313	318	323	328	334	339	344	
858	349	354	359	364	369	374	379	384	389	394	
859	399	404	409	414	420	425	430	435	440	445	
860	450	455	460	465	470	475	480	485	490	495	
861	500	505	510	515	520	526	531	536	541	546	
862	551	556	561	566	571	576	581	586	591	596	
863	601	606	611	616	621	626	631	636	641	646	
864	651	656	661	666	671	676	682	687	692	697	
865	702	707	712	717	722	727	732	737	742	747	
866	752	757	762	767	772	777	782	787	792	797	
867	802	807	812	817	822	827	832	837	842	847	
868	852	857	862	867	872	877	882	887	892	897	
869	902	907	912	917	922	927	932	937	942	947	
870	952	957	962	967	972	977	982	987	992	997	
871	94 002	007	012	017	022	027	032	037	042	047	
872	052	057	062	067	072	077	082	086	091	096	
873	101	106	111	116	121	126	131	136	141	146	
874	151	156	161	166	171	176	181	186	191	196	
875	201	206	211	216	221	226	231	236	240	245	
876	250	255	260	265	270	275	280	285	290	295	
877	300	305	310	315	320	325	330	335	340	345	
878	349	354	359	364	369	374	379	384	389	394	
879	399	404	409	414	419	424	429	433	438	443	
880	448	453	458	463	468	473	478	483	488	493	
881	498	503	507	512	517	522	527	532	537	542	
882	547	552	557	562	567	571	576	581	586	591	
883	596	601	606	611	616	621	626	630	635	640	
884	645	650	655	660	665	670	675	680	685	689	
885	694	699	704	709	714	719	724	729	734	738	
886	743	748	753	758	763	768	773	778	783	787	
887	792	797	802	807	812	817	822	827	832	836	
888	841	846	851	856	861	866	871	876	880	885	
889	890	895	900	905	910	915	919	924	929	934	
890	939	944	949	954	959	963	968	973	978	983	
891	988	993	998	*002	*007	*012	*017	*022	*027	*032	
892	95 036	041	046	051	056	061	066	071	075	080	
893	085	090	095	100	105	109	114	119	124	129	
894	134	139	143	148	153	158	163	168	173	177	
895	182	187	192	197	202	207	211	216	221	226	
896	231	236	240	245	250	255	260	265	270	274	
897	279	284	289	294	299	303	308	313	318	323	
898	328	332	337	342	347	352	357	361	366	371	
899	376	381	386	390	395	400	405	410	415	419	
900	424	429	434	439	444	448	453	458	463	468	
N.	L. 0	1	2	3	4	5	6	7	8	9	P. P.

P. P.

6
1 | 0.6
2 | 1.2
3 | 1.8
4 | 2.4
5 | 3.0
6 | 3.6
7 | 4.2
8 | 4.8
9 | 5.4

5
1 | 0.5
2 | 1.0
3 | 1.5
4 | 2.0
5 | 2.5
6 | 3.0
7 | 3.5
8 | 4.0
9 | 4.5

4
1 | 0.4
2 | 0.8
3 | 1.2
4 | 1.6
5 | 2.0
6 | 2.4
7 | 2.8
8 | 3.2
9 | 3.6

N.	L. 0	1	2	3	4	5	6	7	8	9	P. P.
900	95 424	429	434	439	444	448	453	458	463	468	
901	472	477	482	487	492	497	501	506	511	516	
902	521	525	530	535	540	545	550	554	559	564	
903	569	574	578	583	588	593	598	602	607	612	
904	617	622	626	631	636	641	646	650	655	660	
905	665	670	674	679	684	689	694	698	703	708	
906	713	718	722	727	732	737	742	746	751	756	
907	761	766	770	775	780	785	789	794	799	804	
908	809	813	818	823	828	832	837	842	847	852	
909	856	861	866	871	875	880	885	890	895	899	
910	904	909	914	918	923	928	933	938	942	947	
911	952	957	961	966	971	976	980	985	990	995	
912	999	*004	*009	*014	*019	*023	*028	*033	*038	*042	
913	96 047	052	057	061	066	071	076	080	085	090	
914	095	099	104	109	114	118	123	128	133	137	
915	142	147	152	156	161	166	171	175	180	185	
916	190	194	199	204	209	213	218	223	227	232	
917	237	242	246	251	256	261	265	270	275	280	
918	284	289	294	298	303	308	313	317	322	327	
919	332	336	341	346	350	355	360	365	369	374	
920	379	384	388	393	398	402	407	412	417	421	
921	426	431	435	440	445	450	454	459	464	468	
922	473	478	483	487	492	497	501	506	511	515	
923	520	525	530	534	539	544	548	553	558	562	
924	567	572	577	581	586	591	595	600	605	609	
925	614	619	624	628	633	638	642	647	652	656	
926	661	666	670	675	680	685	689	694	699	703	
927	708	713	717	722	727	731	736	741	745	750	
928	755	759	764	769	774	778	783	788	792	797	
929	802	806	811	816	820	825	830	834	839	844	
930	848	853	858	862	867	872	876	881	886	890	
931	895	900	904	909	914	918	923	928	932	937	
932	942	946	951	956	960	965	970	974	979	984	
933	988	993	997	*002	*007	*011	*016	*021	*025	*030	
934	97 035	039	044	049	053	058	063	067	072	077	
935	081	086	090	095	100	104	109	114	118	123	
936	128	132	137	142	146	151	155	160	165	169	
937	174	179	183	188	192	197	202	206	211	216	
938	220	225	230	234	239	243	248	253	257	262	
939	267	271	276	280	285	290	294	299	304	308	
940	313	317	322	327	331	336	340	345	350	354	
941	359	364	368	373	377	382	387	391	396	400	
942	405	410	414	419	424	428	433	437	442	447	
943	451	456	460	465	470	474	479	483	488	493	
944	497	502	506	511	516	520	525	529	534	539	
945	543	548	552	557	562	566	571	575	580	585	
946	589	594	598	603	607	612	617	621	626	630	
947	635	640	644	649	653	658	663	667	672	676	
948	681	685	690	695	699	704	708	713	717	722	
949	727	731	736	740	745	749	754	759	763	768	
950	772	777	782	786	791	795	800	804	809	813	
N.	L. 0	1	2	3	4	5	6	7	8	9	P. P.

P. P.

	5
1	0.5
2	1.0
3	1.5
4	2.0
5	2.5
6	3.0
7	3.5
8	4.0
9	4.5

	4
1	0.4
2	0.8
3	1.2
4	1.6
5	2.0
6	2.4
7	2.8
8	3.2
9	3.6

N.	L. 0	1	2	3	4	5	6	7	8	9	P. P.
950	97 772	777	782	786	791	795	800	804	809	813	
951	818	823	827	832	836	841	845	850	855	859	
952	864	868	873	877	882	886	891	896	900	905	
953	909	914	918	923	928	932	937	941	946	950	
954	955	959	964	968	973	978	982	987	991	996	
955	98 000	005	009	014	019	023	028	032	037	041	
956	046	050	055	059	064	068	073	078	082	087	
957	091	096	100	105	109	114	118	123	127	132	
958	137	141	146	150	155	159	164	168	173	177	
959	182	186	191	195	200	204	209	214	218	223	
960	227	232	236	241	245	250	254	259	263	268	
961	272	277	281	286	290	295	299	304	308	313	
962	318	322	327	331	336	340	345	349	354	358	**5**
963	363	367	372	376	381	385	390	394	399	403	1 0.5
											2 1.0
964	408	412	417	421	426	430	435	439	444	448	3 1.5
965	453	457	462	466	471	475	480	484	489	493	4 2.0
966	498	502	507	511	516	520	525	529	534	538	5 2.5
											6 3.0
967	543	547	552	556	561	565	570	574	579	583	7 3.5
968	588	592	597	601	605	610	614	619	623	628	8 4.0
969	632	637	641	646	650	655	659	664	668	673	9 4.5
970	677	682	686	691	695	700	704	709	713	717	
971	722	726	731	735	740	744	749	753	758	762	
972	767	771	776	780	784	789	793	798	802	807	
973	811	816	820	825	829	834	838	843	847	851	
974	856	860	865	869	874	878	883	887	892	896	
975	900	905	909	914	918	923	927	932	936	941	
976	945	949	954	958	963	967	972	976	981	985	
977	989	994	998	*003	*007	*012	*016	*021	*025	*029	
978	99 034	038	043	047	052	056	061	065	069	074	
979	078	083	087	092	096	100	105	109	114	118	
980	123	127	131	136	140	145	149	154	158	162	
981	167	171	176	180	185	189	193	198	202	207	**4**
982	211	216	220	224	229	233	238	242	247	251	1 0.4
983	255	260	264	269	273	277	282	286	291	295	2 0.8
											3 1.2
984	300	304	308	313	317	322	326	330	335	339	4 1.6
985	344	348	352	357	361	366	370	374	379	383	5 2.0
986	388	392	396	401	405	410	414	419	423	427	6 2.4
											7 2.8
987	432	436	441	445	449	454	458	463	467	471	8 3.2
988	476	480	484	489	493	498	502	506	511	515	9 3.6
989	520	524	528	533	537	542	546	550	555	559	
990	564	568	572	577	581	585	590	594	599	603	
991	607	612	616	621	625	629	634	638	642	647	
992	651	656	660	664	669	673	677	682	686	691	
993	695	699	704	708	712	717	721	726	730	734	
994	739	743	747	752	756	760	765	769	774	778	
995	782	787	791	795	800	804	808	813	817	822	
996	826	830	835	839	843	848	852	856	861	865	
997	870	874	878	883	887	891	896	900	904	909	
998	913	917	922	926	930	935	939	944	948	952	
999	957	961	965	970	974	978	983	987	991	996	
1000	00 000	004	009	013	017	022	026	030	035	039	
N.	L. 0	1	2	3	4	5	6	7	8	9	P. P.

N.	L. 0	1	2	3	4	5	6	7	8	9	d.
1000	000 0000	0434	0869	1303	1737	2171	2605	3039	3473	3907	434
1001	4341	4775	5208	5642	6076	6510	6943	7377	7810	8244	434
1002	8677	9111	9544	9977	*0411	*0844	*1277	*1710	*2143	*2576	433
1003	001 3009	3442	3875	4308	4741	5174	5607	6039	6472	6905	433
1004	7337	7770	8202	8635	9067	9499	9932	*0364	*0796	*1228	432
1005	002 1661	2093	2525	2957	3389	3821	4253	4685	5116	5548	432
1006	5980	6411	6843	7275	7706	8138	8569	9001	9432	9863	431
1007	003 0295	0726	1157	1588	2019	2451	2882	3313	3744	4174	431
1008	4605	5036	5467	5898	6328	6759	7190	7620	8051	8481	431
1009	8912	9342	9772	*0203	*0633	*1063	*1493	*1924	*2354	*2784	430
1010	004 3214	3644	4074	4504	4933	5363	5793	6223	6652	7082	430
1011	7512	7941	8371	8800	9229	9659	*0088	*0517	*0947	*1376	429
1012	005 1805	2234	2663	3092	3521	3950	4379	4808	5237	5666	429
1013	6094	6523	6952	7380	7809	8238	8666	9094	9523	9951	429
1014	006 0380	0808	1236	1664	2092	2521	2949	3377	3805	4233	428
1015	4660	5088	5516	5944	6372	6799	7227	7655	8082	8510	428
1016	8937	9365	9792	*0219	*0647	*1074	*1501	*1928	*2355	*2782	427
1017	007 3210	3637	4064	4490	4917	5344	5771	6198	6624	7051	427
1018	7478	7904	8331	8757	9184	9610	*0037	*0463	*0889	*1316	426
1019	008 1742	2168	2594	3020	3446	3872	4298	4724	5150	5576	426
1020	6002	6427	6853	7279	7704	8130	8556	8981	9407	9832	426
1021	009 0257	0683	1108	1533	1959	2384	2809	3234	3659	4084	425
1022	4509	4934	5359	5784	6208	6633	7058	7483	7907	8332	425
1023	8756	9181	9605	*0030	*0454	*0878	*1303	*1727	*2151	*2575	424
1024	010 3000	3424	3848	4272	4696	5120	5544	5967	6391	6815	424
1025	7239	7662	8086	8510	8933	9357	9780	*0204	*0627	*1050	424
1026	011 1474	1897	2320	2743	3166	3590	4013	4436	4859	5282	423
1027	5704	6127	6550	6973	7396	7818	8241	8664	9086	9509	423
1028	9931	*0354	*0776	*1198	*1621	*2043	*2465	*2887	*3310	*3732	422
1029	012 4154	4576	4998	5420	5842	6264	6685	7107	7529	7951	422
1030	8372	8794	9215	9637	*0059	*0480	*0901	*1323	*1744	*2165	422
1031	013 2587	3008	3429	3850	4271	4692	5113	5534	5955	6376	421
1032	6797	7218	7639	8059	8480	8901	9321	9742	*0162	*0583	421
1033	014 1003	1424	1844	2264	2685	3105	3525	3945	4365	4785	420
1034	5205	5625	6045	6465	6885	7305	7725	8144	8564	8984	420
1035	9403	9823	*0243	*0662	*1082	*1501	*1920	*2340	*2759	*3178	420
1036	015 3598	4017	4436	4855	5274	5693	6112	6531	6950	7369	419
1037	7788	8206	8625	9044	9462	9881	*0300	*0718	*1137	*1555	419
1038	016 1974	2392	2810	3229	3647	4065	4483	4901	5319	5737	418
1039	6155	6573	6991	7409	7827	8245	8663	9080	9498	9916	418
1040	017 0333	0751	1168	1586	2003	2421	2838	3256	3673	4090	417
1041	4507	4924	5342	5759	6176	6593	7010	7427	7844	8260	417
1042	8677	9094	9511	9927	*0344	*0761	*1177	*1594	*2010	*2427	417
1043	018 2843	3259	3676	4092	4508	4925	5341	5757	6173	6589	416
1044	7005	7421	7837	8253	8669	9084	9500	9916	*0332	*0747	416
1045	019 1163	1578	1994	2410	2825	3240	3656	4071	4486	4902	415
1046	5317	5732	6147	6562	6977	7392	7807	8222	8637	9052	415
1047	9467	9882	*0296	*0711	*1126	*1540	*1955	*2369	*2784	*3198	415
1048	020 3613	4027	4442	4856	5270	5684	6099	6513	6927	7341	414
1049	7755	8169	8583	8997	9411	9824	*0238	*0652	*1066	*1479	414
1050	021 1893	2307	2720	3134	3547	3961	4374	4787	5201	5614	413
N.	L. 0	1	2	3	4	5	6	7	8	9	d.

N.	L. 0	1	2	3	4	5	6	7	8	9	d.
1050	021 1893	2307	2720	3134	3547	3961	4374	4787	5201	5614	413
1051	6027	6440	6854	7267	7680	8093	8506	8919	9332	9745	413
1052	022 0157	0570	0983	1396	1808	2221	2634	3046	3459	3871	413
1053	4284	4696	5109	5521	5933	6345	6758	7170	7582	7994	412
1054	8406	8818	9230	9642	*0054	*0466	*0878	*1289	*1701	*2113	412
1055	023 2525	2936	3348	3759	4171	4582	4994	5405	5817	6228	411
1056	6639	7050	7462	7873	8284	8695	9106	9517	9928	*0339	411
1057	024 0750	1161	1572	1982	2393	2804	3214	3625	4036	4446	411
1058	4857	5267	5678	6088	6498	6909	7319	7729	8139	8549	410
1059	8960	9370	9780	*0190	*0600	*1010	*1419	*1829	*2239	*2649	410
1060	025 3059	3468	3878	4288	4697	5107	5516	5926	6335	6744	410
1061	7154	7563	7972	8382	8791	9200	9609	*0018	*0427	*0836	409
1062	026 1245	1654	2063	2472	2881	3289	3698	4107	4515	4924	409
1063	5333	5741	6150	6558	6967	7375	7783	8192	8600	9008	408
1064	9416	9824	*0233	*0641	*1049	*1457	*1865	*2273	*2680	*3088	408
1065	027 3496	3904	4312	4719	5127	5535	5942	6350	6757	7165	408
1066	7572	7979	8387	8794	9201	9609	*0016	*0423	*0830	*1237	407
1067	028 1644	2051	2458	2865	3272	3679	4086	4492	4899	5306	407
1068	5713	6119	6526	6932	7339	7745	8152	8558	8964	9371	406
1069	9777	*0183	*0590	*0996	*1402	*1808	*2214	*2620	*3026	*3432	406
1070	029 3838	4244	4649	5055	5461	5867	6272	6678	7084	7489	406
1071	7895	8300	8706	9111	9516	9922	*0327	*0732	*1138	*1543	405
1072	030 1948	2353	2758	3163	3568	3973	4378	4783	5188	5592	405
1073	5997	6402	6807	7211	7616	8020	8425	8830	9234	9638	405
1074	031 0043	0447	0851	1256	1660	2064	2468	2872	3277	3681	404
1075	4085	4489	4893	5296	5700	6104	6508	6912	7315	7719	404
1076	8123	8526	8930	9333	9737	*0140	*0544	*0947	*1350	*1754	403
1077	032 2157	2560	2963	3367	3770	4173	4576	4979	5382	5785	403
1078	6188	6590	6993	7396	7799	8201	8604	9007	9409	9812	403
1079	033 0214	0617	1019	1422	1824	2226	2629	3031	3433	3835	402
1080	4238	4640	5042	5444	5846	6248	6650	7052	7453	7855	402
1081	8257	8659	9060	9462	9864	*0265	*0667	*1068	*1470	*1871	402
1082	034 2273	2674	3075	3477	3878	4279	4680	5081	5482	5884	401
1083	6285	6686	7087	7487	7888	8289	8690	9091	9491	9892	401
1084	035 0293	0693	1094	1495	1895	2296	2696	3096	3497	3897	400
1085	4297	4698	5098	5498	5898	6298	6698	7098	7498	7898	400
1086	8298	8698	9098	9498	9898	*0297	*0697	*1097	*1496	*1896	400
1087	036 2295	2695	3094	3494	3893	4293	4692	5091	5491	5890	399
1088	6289	6688	7087	7486	7885	8284	8683	9082	9481	9880	399
1089	037 0279	0678	1076	1475	1874	2272	2671	3070	3468	3867	399
1090	4265	4663	5062	5460	5858	6257	6655	7053	7451	7849	398
1091	8248	8646	9044	9442	9839	*0237	*0635	*1033	*1431	*1829	398
1092	038 2226	2624	3022	3419	3817	4214	4612	5009	5407	5804	398
1093	6202	6599	6996	7393	7791	8188	8585	8982	9379	9776	397
1094	039 0173	0570	0967	1364	1761	2158	2554	2951	3348	3745	397
1095	4141	4538	4934	5331	5727	6124	6520	6917	7313	7709	397
1096	8106	8502	8898	9294	9690	*0086	*0482	*0878	*1274	*1670	396
1097	040 2066	2462	2858	3254	3650	4045	4441	4837	5232	5628	396
1098	6023	6419	6814	7210	7605	8001	8396	8791	9187	9582	395
1099	9977	*0372	*0767	*1162	*1557	*1952	*2347	*2742	*3137	*3532	395
1100	041 3927	4322	4716	5111	5506	5900	6295	6690	7084	7479	395
N.	L. 0	1	2	3	4	5	6	7	8	9	d.

TABLE IIa. AUXILIARY TABLE OF S AND T FOR A IN MINUTES

A' = The number of minutes in A.
A_1' = The number of minutes in $90° - A$.

For angles near $0°$: $\log \sin A = S + \log A'$, and $\log \tan A = T + \log A'$.
For angles near $90°$: $\log \cos A = S_1 + \log A_1'$, and $\log \cot A = T_1 + \log A_1'$. S_1 and T_1 correspond to A_1'.

A'	$S + 10$	A'	$T + 10$	A'	$T + 10$
0'– 13'	6.46 373	0'– 26'	6.46 373	131'–133'	6.46 394
14'– 42'	372	27'– 39'	374	134'–136'	395
43'– 58'	371	40'– 48'	375	137'–139'	396
59'– 71'	6.46 370	49'– 56'	6.46 376	140'–142'	6.46 397
72'– 81'	369	57'– 63'	377	143'–145'	398
82'– 91'	368	64'– 69'	378	146'–148'	399
92'– 99'	6.46 367	70'– 74'	6.46 379	149'–150'	6.46 400
100'–107'	366	75'– 80'	380	151'–153'	401
108'–115'	365	81'– 85'	381	154'–156'	402
116'–121'	6.46 364	86'– 89'	6.46 382	157'–158'	6.46 403
122'–128'	363	90'– 94'	383	159'–161'	404
129'–134'	362	95'– 98'	384	162'–163'	405
135'–140'	6.46 361	99'–102'	6.46 385	164'–166'	6.46 406
141'–146'	360	103'–106'	386	167'–168'	407
147'–151'	359	107'–110'	387	169'–171'	408
152'–157'	6.46 358	111'–113'	6.46 388	172'–173'	6.46 409
158'–162'	357	114'–117'	389	174'–175'	410
163'–167'	356	118'–120'	390	176'–178'	411
168'–171'	6.46 355	121'–124'	6.46 391	179'–180'	6.46 412
172'–176'	354	125'–127'	392	181'–182'	413
177'–181'	353	128'–130'	393	183'–184'	414

TABLE II.   FIVE-PLACE LOGARITHMS OF THE TRIGONOMETRIC FUNCTIONS

Subtract 10 *from the characteristic of each logarithm.*

Do not interpolate on this page. Use Table IIa.

	L. Sin.	L. Tan.	L. Cot.	L. Cos.	
0	—	—	—	10.00 000	60
1	6.46 373	6.46 373	13.53 627	10.00 000	59
2	6.76 476	6.76 476	13.23 524	10.00 000	58
3	6.94 085	6.94 085	13.05 915	10.00 000	57
4	7.06 579	7.06 579	12.93 421	10.00 000	56
5	7.16 270	7.16 270	12.83 730	10.00 000	55
6	7.24 188	7.24 188	12.75 812	10.00 000	54
7	7.30 882	7.30 882	12.69 118	10.00 000	53
8	7.36 682	7.36 682	12.63 318	10.00 000	52
9	7.41 797	7.41 797	12.58 203	10.00 000	51
10	7.46 373	7.46 373	12.53 627	10.00 000	50
11	7.50 512	7.50 512	12.49 488	10.00 000	49
12	7.54 291	7.54 291	12.45 709	10.00 000	48
13	7.57 767	7.57 767	12.42 233	10.00 000	47
14	7.60 985	7.60 986	12.39 014	10.00 000	46
15	7.63 982	7.63 982	12.36 018	10.00 000	45
16	7.66 784	7.66 785	12.33 215	10.00 000	44
17	7.69 417	7.69 418	12.30 582	9.99 999	43
18	7.71 900	7.71 900	12.28 100	9.99 999	42
19	7.74 248	7.74 248	12.25 752	9.99 999	41
20	7.76 475	7.76 476	12.23 524	9.99 999	40
21	7.78 594	7.78 595	12.21 405	9.99 999	39
22	7.80 615	7.80 615	12.19 385	9.99 999	38
23	7.82 545	7.82 546	12.17 454	9.99 999	37
24	7.84 393	7.84 394	12.15 606	9.99 999	36
25	7.86 166	7.86 167	12.13 833	9.99 999	35
26	7.87 870	7.87 871	12.12 129	9.99 999	34
27	7.89 509	7.89 510	12.10 490	9.99 999	33
28	7.91 088	7.91 089	12.08 911	9.99 999	32
29	7.92 612	7.92 613	12.07 387	9.99 998	31
30	7.94 084	7.94 086	12.05 914	9.99 998	30
31	7.95 508	7.95 510	12.04 490	9.99 998	29
32	7.96 887	7.96 889	12.03 111	9.99 998	28
33	7.98 223	7.98 225	12.01 775	9.99 998	27
34	7.99 520	7.99 522	12.00 478	9.99 998	26
35	8.00 779	8.00 781	11.99 219	9.99 998	25
36	8.02 002	8.02 004	11.97 996	9.99 998	24
37	8.03 192	8.03 194	11.96 806	9.99 997	23
38	8.04 350	8.04 353	11.95 647	9.99 997	22
39	8.05 478	8.05 481	11.94 519	9.99 997	21
40	8.06 578	8.06 581	11.93 419	9.99 997	20
41	8.07 650	8.07 653	11.92 347	9.99 997	19
42	8.08 696	8.08 700	11.91 300	9.99 997	18
43	8.09 718	8.09 722	11.90 278	9.99 997	17
44	8.10 717	8.10 720	11.89 280	9.99 996	16
45	8.11 693	8.11 696	11.88 304	9.99 996	15
46	8.12 647	8.12 651	11.87 349	9.99 996	14
47	8.13 581	8.13 585	11.86 415	9.99 996	13
48	8.14 495	8.14 500	11.85 500	9.99 996	12
49	8.15 391	8.15 395	11.84 605	9.99 996	11
50	8.16 268	8.16 273	11.83 727	9.99 995	10
51	8.17 128	8.17 133	11.82 867	9.99 995	9
52	8.17 971	8.17 976	11.82 024	9.99 995	8
53	8.18 798	8.18 804	11.81 196	9.99 995	7
54	8.19 610	8.19 616	11.80 384	9.99 995	6
55	8.20 407	8.20 413	11.79 587	9.99 994	5
56	8.21 189	8.21 195	11.78 805	9.99 994	4
57	8.21 958	8.21 964	11.78 036	9.99 994	3
58	8.22 713	8.22 720	11.77 280	9.99 994	2
59	8.23 456	8.23 462	11.76 538	9.99 994	1
60	8.24 186	8.24 192	11.75 808	9.99 993	0
	L. Cos.	L. Cot.	L. Tan.	L. Sin.	′

Do not interpolate on this page.   Use Table IIa.

′	L. Sin.	L. Tan.	L. Cot.	L. Cos.	
0	8.24 186	8.24 192	11.75 808	9.99 993	**60**
1	8.24 903	8.24 910	11.75 090	9.99 993	59
2	8.25 609	8.25 616	11.74 384	9.99 993	58
3	8.26 304	8.26 312	11.73 688	9.99 993	57
4	8.26 988	8.26 996	11.73 004	9.99 992	56
5	8.27 661	8.27 669	11.72 331	9.99 992	55
6	8.28 324	8.28 332	11.71 668	9.99 992	54
7	8.28 977	8.28 986	11.71 014	9.99 992	53
8	8.29 621	8.29 629	11.70 371	9.99 992	52
9	8.30 255	8.30 263	11.69 737	9.99 991	51
10	8.30 879	8.30 888	11.69 112	9.99 991	**50**
11	8.31 495	8.31 505	11.68 495	9.99 991	49
12	8.32 103	8.32 112	11.67 888	9.99 990	48
13	8.32 702	8.32 711	11.67 289	9.99 990	47
14	8.33 292	8.33 302	11.66 698	9.99 990	46
15	8.33 875	8.33 886	11.66 114	9.99 990	45
16	8.34 450	8.34 461	11.65 539	9.99 989	44
17	8.35 018	8.35 029	11.64 971	9.99 989	43
18	8.35 578	8.35 590	11.64 410	9.99 989	42
19	8.36 131	8.36 143	11.63 857	9.99 989	41
20	8.36 678	8.36 689	11.63 311	9.99 988	**40**
21	8.37 217	8.37 229	11.62 771	9.99 988	39
22	8.37 750	8.37 762	11.62 238	9.99 988	38
23	8.38 276	8.38 289	11.61 711	9.99 987	37
24	8.38 796	8.38 809	11.61 191	9.99 987	36
25	8.39 310	8.39 323	11.60 677	9.99 987	35
26	8.39 818	8.39 832	11.60 168	9.99 986	34
27	8.40 320	8.40 334	11.59 666	9.99 986	33
28	8.40 816	8.40 830	11.59 170	9.99 986	32
29	8.41 307	8.41 321	11.58 679	9.99 985	31
30	8.41 792	8.41 807	11.58 193	9.99 985	**30**
31	8.42 272	8.42 287	11.57 713	9.99 985	29
32	8.42 746	8.42 762	11.57 238	9.99 984	28
33	8.43 216	8.43 232	11.56 768	9.99 984	27
34	8.43 680	8.43 696	11.56 304	9.99 984	26
35	8.44 139	8.44 156	11.55 844	9.99 983	25
36	8.44 594	8.44 611	11.55 389	9.99 983	24
37	8.45 044	8.45 061	11.54 939	9.99 983	23
38	8.45 489	8.45 507	11.54 493	9.99 982	22
39	8.45 930	8.45 948	11.54 052	9.99 982	21
40	8.46 366	8.46 385	11.53 615	9.99 982	**20**
41	8.46 799	8.46 817	11.53 183	9.99 981	19
42	8.47 226	8.47 245	11.52 755	9.99 981	18
43	8.47 650	8.47 669	11.52 331	9.99 981	17
44	8.48 069	8.48 089	11.51 911	9.99 980	16
45	8.48 485	8.48 505	11.51 495	9.99 980	15
46	8.48 896	8.48 917	11.51 083	9.99 979	14
47	8.49 304	8.49 325	11.50 675	9.99 979	13
48	8.49 708	8.49 729	11.50 271	9.99 979	12
49	8.50 108	8.50 130	11.49 870	9.99 978	11
50	8.50 504	8.50 527	11.49 473	9.99 978	**10**
51	8.50 897	8.50 920	11.49 080	9.99 977	9
52	8.51 287	8.51 310	11.48 690	9.99 977	8
53	8.51 673	8.51 696	11.48 304	9.99 977	7
54	8.52 055	8.52 079	11.47 921	9.99 976	6
55	8.52 434	8.52 459	11.47 541	9.99 976	5
56	8.52 810	8.52 835	11.47 165	9.99 975	4
57	8.53 183	8.53 208	11.46 792	9.99 975	3
58	8.53 552	8.53 578	11.46 422	9.99 974	2
59	8.53 919	8.53 945	11.46 055	9.99 974	1
60	8.54 282	8.54 308	11.45 692	9.99 974	**0**
	L. Cos.	L. Cot.	L. Tan.	L. Sin.	′

88°

2°

· Do not interpolate on this page. Use Table IIa.

′	L. Sin.	L. Tan.	L. Cot.	L. Cos.	
0	8.54 282	8.54 308	11.45 692	9.99 974	60
1	8.54 642	8.54 669	11.45 331	9.99 973	59
2	8.54 999	8.55 027	11.44 973	9.99 973	58
3	8.55 354	8.55 382	11.44 618	9.99 972	57
4	8.55 705	8.55 734	11.44 266	9.99 972	56
5	8.56 054	8.56 083	11.43 917	9.99 971	55
6	8.56 400	8.56 429	11.43 571	9.99 971	54
7	8.56 743	8.56 773	11.43 227	9.99 970	53
8	8.57 084	8.57 114	11.42 886	9.99 970	52
9	8.57 421	8.57 452	11.42 548	9.99 969	51
10	8.57 757	8.57 788	11.42 212	9.99 969	50
11	8.58 089	8.58 121	11.41 879	9.99 968	49
12	8.58 419	8.58 451	11.41 549	9.99 968	48
13	8.58 747	8.58 779	11.41 221	9.99 967	47
14	8.59 072	8.59 105	11.40 895	9.99 967	46
15	8.59 395	8.59 428	11.40 572	9.99 967	45
16	8.59 715	8.59 749	11.40 251	9.99 966	44
17	8.60 033	8.60 068	11.39 932	9.99 966	43
18	8.60 349	8.60 384	11.39 616	9.99 965	42
19	8.60 662	8.60 698	11.39 302	9.99 964	41
20	8.60 973	8.61 009	11.38 991	9.99 964	40
21	8.61 282	8.61 319	11.38 681	9.99 963	39
22	8.61 589	8.61 626	11.38 374	9.99 963	38
23	8.61 894	8.61 931	11.38 069	9.99 962	37
24	8.62 196	8.62 234	11.37 766	9.99 962	36
25	8.62 497	8.62 535	11.37 465	9.99 961	35
26	8.62 795	8.62 834	11.37 166	9.99 961	34
27	8.63 091	8.63 131	11.36 869	9.99 960	33
28	8.63 385	8.63 426	11.36 574	9.99 960	32
29	8.63 678	8.63 718	11.36 282	9.99 959	31
30	8.63 968	8.64 009	11.35 991	9.99 959	30
31	8.64 256	8.64 298	11.35 702	9.99 958	29
32	8.64 543	8.64 585	11.35 415	9.99 958	28
33	8.64 827	8.64 870	11.35 130	9.99 957	27
34	8.65 110	8.65 154	11.34 846	9.99 956	26
35	8.65 391	8.65 435	11.34 565	9.99 956	25
36	8.65 670	8.65 715	11.34 285	9.99 955	24
37	8.65 947	8.65 993	11.34 007	9.99 955	23
38	8.66 223	8.66 269	11.33 731	9.99 954	22
39	8.66 497	8.66 543	11.33 457	9.99 954	21
40	8.66 769	8.66 816	11.33 184	9.99 953	20
41	8.67 039	8.67 087	11.32 913	9.99 952	19
42	8.67 308	8.67 356	11.32 644	9.99 952	18
43	8.67 575	8.67 624	11.32 376	9.99 951	17
44	8.67 841	8.67 890	11.32 110	9.99 951	16
45	8.68 104	8.68 154	11.31 846	9.99 950	15
46	8.68 367	8.68 417	11.31 583	9.99 949	14
47	8.68 627	8.68 678	11.31 322	9.99 949	13
48	8.68 886	8.68 938	11.31 062	9.99 948	12
49	8.69 144	8.69 196	11.30 804	9.99 948	11
50	8.69 400	8.69 453	11.30 547	9.99 947	10
51	8.69 654	8.69 708	11.30 292	9.99 946	9
52	8.69 907	8.69 962	11.30 038	9.99 946	8
53	8.70 159	8.70 214	11.29 786	9.99 945	7
54	8.70 409	8.70 465	11.29 535	9.99 944	6
55	8.70 658	8.70 714	11.29 286	9.99 944	5
56	8.70 905	8.70 962	11.29 038	9.99 943	4
57	8.71 151	8.71 208	11.28 792	9.99 942	3
58	8.71 395	8.71 453	11.28 547	9.99 942	2
59	8.71 638	8.71 697	11.28 303	9.99 941	1
60	8.71 880	8.71 940	11.28 060	9.99 940	0
	L. Cos.	L. Cot.	L. Tan.	L. Sin.	′

′	L. Sin.	d.	L. Tan.	c. d.	L. Cot.	L. Cos.	′
0	8.71 880		8.71 940		11.28 060	9.99 940	60
1	8.72 120	240	8.72 181	241	11.27 819	9.99 940	59
2	8.72 359	239	8.72 420	239	11.27 580	9.99 939	58
3	8.72 597	238	8.72 659	239	11.27 341	9.99 938	57
4	8.72 834	237	8.72 896	237	11.27 104	9.99 938	56
5	8.73 069	235	8.73 132	236	11.26 868	9.99 937	55
6	8.73 303	234	8.73 366	234	11.26 634	9.99 936	54
7	8.73 535	232	8.73 600	234	11.26 400	9.99 936	53
8	8.73 767	232	8.73 832	232	11.26 168	9.99 935	52
9	8.73 997	230	8.74 063	231	11.25 937	9.99 934	51
10	8.74 226	229	8.74 292	229	11.25 708	9.99 934	50
11	8.74 454	228	8.74 521	229	11.25 479	9.99 933	49
12	8.74 680	226	8.74 748	227	11.25 252	9.99 932	48
13	8.74 906	226	8.74 974	226	11.25 026	9.99 932	47
14	8.75 130	224	8.75 199	225	11.24 801	9.99 931	46
15	8.75 353	223	8.75 423	224	11.24 577	9.99 930	45
16	8.75 575	222	8.75 645	222	11.24 355	9.99 929	44
17	8.75 795	220	8.75 867	222	11.24 133	9.99 929	43
18	8.76 015	220	8.76 087	220	11.23 913	9.99 928	42
19	8.76 234	219	8.76 306	219	11.23 694	9.99 927	41
20	8.76 451	217	8.76 525	219	11.23 475	9.99 926	40
21	8.76 667	216	8.76 742	217	11.23 258	9.99 926	39
22	8.76 883	216	8.76 958	216	11.23 042	9.99 925	38
23	8.77 097	214	8.77 173	215	11.22 827	9.99 924	37
24	8.77 310	213	8.77 387	214	11.22 613	9.99 923	36
25	8.77 522	212	8.77 600	213	11.22 400	9.99 923	35
26	8.77 733	211	8.77 811	211	11.22 189	9.99 922	34
27	8.77 943	210	8.78 022	211	11.21 978	9.99 921	33
28	8.78 152	209	8.78 232	210	11.21 768	9.99 920	32
29	8.78 360	208	8.78 441	209	11.21 559	9.99 920	31
30	8.78 568	208	8.78 649	208	11.21 351	9.99 919	30
31	8.78 774	206	8.78 855	206	11.21 145	9.99 918	29
32	8.78 979	205	8.79 061	206	11.20 939	9.99 917	28
33	8.79 183	204	8.79 266	205	11.20 734	9.99 917	27
34	8.79 386	203	8.79 470	204	11.20 530	9.99 916	26
35	8.79 588	202	8.79 673	203	11.20 327	9.99 915	25
36	8.79 789	201	8.79 875	202	11.20 125	9.99 914	24
37	8.79 990	201	8.80 076	201	11.19 924	9.99 913	23
38	8.80 189	199	8.80 277	201	11.19 723	9.99 913	22
39	8.80 388	199	8.80 476	199	11.19 524	9.99 912	21
40	8.80 585	197	8.80 674	198	11.19 326	9.99 911	20
41	8.80 782	197	8.80 872	198	11.19 128	9.99 910	19
42	8.80 978	196	8.81 068	196	11.18 932	9.99 909	18
43	8.81 173	195	8.81 264	196	11.18 736	9.99 909	17
44	8.81 367	194	8.81 459	195	11.18 541	9.99 908	16
45	8.81 560	193	8.81 653	194	11.18 347	9.99 907	15
46	8.81 752	192	8.81 846	193	11.18 154	9.99 906	14
47	8.81 944	192	8.82 038	192	11.17 962	9.99 905	13
48	8.82 134	190	8.82 230	192	11.17 770	9.99 904	12
49	8.82 324	190	8.82 420	190	11.17 580	9.99 904	11
50	8.82 513	189	8.82 610	190	11.17 390	9.99 903	10
51	8.82 701	188	8.82 799	189	11.17 201	9.99 902	9
52	8.82 888	187	8.82 987	188	11.17 013	9.99 901	8
53	8.83 075	187	8.83 175	188	11.16 825	9.99 900	7
54	8.83 261	186	8.83 361	186	11.16 639	9.99 899	6
55	8.83 446	185	8.83 547	186	11.16 453	9.99 898	5
56	8.83 630	184	8.83 732	185	11.16 268	9.99 898	4
57	8.83 813	183	8.83 916	184	11.16 084	9.99 897	3
58	8.83 996	183	8.84 100	184	11.15 900	9.99 896	2
59	8.84 177	181	8.84 282	182	11.15 718	9.99 895	1
60	8.84 358	181	8.84 464	182	11.15 536	9.99 894	0
	L. Cos.	d.	L. Cot.	c. d.	L. Tan.	L. Sin.	′

P. P.

	241	239	237	236	234
1	24.1	23.9	23.7	23.6	23.4
2	48.2	47.8	47.4	47.2	46.8
3	72.3	71.7	71.1	70.8	70.2
4	96.4	95.6	94.8	94.4	93.6
5	120.5	119.5	118.5	118.0	117.0
6	144.6	143.4	142.2	141.6	140.4
7	168.7	167.3	165.9	165.2	163.8
8	192.8	191.2	189.6	188.8	187.2
9	216.9	215.1	213.3	212.4	210.6

	232	231	229	227	226
1	23.2	23.1	22.9	22.7	22.6
2	46.4	46.2	45.8	45.4	45.2
3	69.6	69.3	68.7	68.1	67.8
4	92.8	92.4	91.6	90.8	90.4
5	116.0	115.5	114.5	113.5	113.0
6	139.2	138.6	137.4	136.2	135.6
7	162.4	161.7	160.3	158.9	158.2
8	185.6	184.8	183.2	181.6	180.8
9	208.8	207.9	206.1	204.3	203.4

	224	222	220	219	217
1	22.4	22.2	22.0	21.9	21.7
2	44.8	44.4	44.0	43.8	43.4
3	67.2	66.6	66.0	65.7	65.1
4	89.6	88.8	88.0	87.6	86.8
5	112.0	111.0	110.0	109.5	108.5
6	134.4	133.2	132.0	131.4	130.2
7	156.8	155.4	154.0	153.3	151.9
8	179.2	177.6	176.0	175.2	173.6
9	201.6	199.8	198.0	197.1	195.3

	216	214	213	211	209
1	21.6	21.4	21.3	21.1	20.9
2	43.2	42.8	42.6	42.2	41.8
3	64.8	64.2	63.9	63.3	62.7
4	86.4	85.6	85.2	84.4	83.6
5	108.0	107.0	106.5	105.5	104.5
6	129.6	128.4	127.8	126.6	125.4
7	151.2	149.8	149.1	147.7	146.3
8	172.8	171.2	170.4	168.8	167.2
9	194.4	192.6	191.7	189.9	188.1

	208	206	203	201	199
1	20.8	20.6	20.3	20.1	19.9
2	41.6	41.2	40.6	40.2	39.8
3	62.4	61.8	60.9	60.3	59.7
4	83.2	82.4	81.2	80.4	79.6
5	104.0	103.0	101.5	100.5	99.5
6	124.8	123.6	121.8	120.6	119.4
7	145.6	144.2	142.1	140.7	139.3
8	166.4	164.8	162.4	160.8	159.2
9	187.2	185.4	182.7	180.9	179.1

	198	196	194	192	190
1	19.8	19.6	19.4	19.2	19.0
2	39.6	39.2	38.8	38.4	38.0
3	59.4	58.8	58.2	57.6	57.0
4	79.2	78.4	77.6	76.8	76.0
5	99.0	98.0	97.0	96.0	95.0
6	118.8	117.6	116.4	115.2	114.0
7	138.6	137.2	135.8	134.4	133.0
8	158.4	156.8	155.2	153.6	152.0
9	178.2	176.4	174.6	172.8	171.0

	188	186	184	182	181
1	18.8	18.6	18.4	18.2	18.1
2	37.6	37.2	36.8	36.4	36.2
3	56.4	55.8	55.2	54.6	54.3
4	75.2	74.4	73.6	72.8	72.4
5	94.0	93.0	92.0	91.0	90.5
6	112.8	111.6	110.4	109.2	108.6
7	131.6	130.2	128.8	127.4	126.7
8	150.4	148.8	147.2	145.6	144.8
9	169.2	167.4	165.6	163.8	162.9

'	L. Sin.	d.	L. Tan.	c. d.	L. Cot.	L. Cos.	'
0	8.84 358		8.84 464		11.15 536	9.99 894	60
1	8.84 539	181	8.84 646	182	11.15 354	9.99 893	59
2	8.84 718	179	8.84 826	180	11.15 174	9.99 892	58
3	8.84 897	179	8.85 006	180	11.14 994	9.99 891	57
4	8.85 075	178	8.85 185	179	11.14 815	9.99 891	56
5	8.85 252	177	8.85 363	178	11.14 637	9.99 890	55
6	8.85 429	177	8.85 540	177	11.14 460	9.99 889	54
7	8.85 605	176	8.85 717	177	11.14 283	9.99 888	53
8	8.85 780	175	8.85 893	176	11.14 107	9.99 887	52
9	8.85 955	175	8.86 069	176	11.13 931	9.99 886	51
10	8.86 128	173	8.86 243	174	11.13 757	9.99 885	50
11	8.86 301	173	8.86 417	174	11.13 583	9.99 884	49
12	8.86 474	173	8.86 591	174	11.13 409	9.99 883	48
13	8.86 645	171	8.86 763	172	11.13 237	9.99 882	47
14	8.86 816	171	8.86 935	172	11.13 065	9.99 881	46
15	8.86 987	171	8.87 106	171	11.12 894	9.99 880	45
16	8.87 156	169	8.87 277	171	11.12 723	9.99 879	44
17	8.87 325	169	8.87 447	170	11.12 553	9.99 879	43
18	8.87 494	169	8.87 616	169	11.12 384	9.99 878	42
19	8.87 661	167	8.87 785	169	11.12 215	9.99 877	41
20	8.87 829	168	8.87 953	168	11.12 047	9.99 876	40
21	8.87 995	166	8.88 120	167	11.11 880	9.99 875	39
22	8.88 161	166	8.88 287	167	11.11 713	9.99 874	38
23	8.88 326	165	8.88 453	166	11.11 547	9.99 873	37
24	8.88 490	164	8.88 618	165	11.11 382	9.99 872	36
25	8.88 654	164	8.88 783	165	11.11 217	9.99 871	35
26	8.88 817	163	8.88 948	165	11.11 052	9.99 870	34
27	8.88 980	163	8.89 111	163	11.10 889	9.99 869	33
28	8.89 142	162	8.89 274	163	11.10 726	9.99 868	32
29	8.89 304	162	8.89 437	163	11.10 563	9.99 867	31
30	8.89 464	160	8.89 598	161	11.10 402	9.99 866	30
31	8.89 625	161	8.89 760	162	11.10 240	9.99 865	29
32	8.89 784	159	8.89 920	160	11.10 080	9.99 864	28
33	8.89 943	159	8.90 080	160	11.09 920	9.99 863	27
34	8.90 102	159	8.90 240	160	11.09 760	9.99 862	26
35	8.90 260	158	8.90 399	159	11.09 601	9.99 861	25
36	8.90 417	157	8.90 557	158	11.09 443	9.99 860	24
37	8.90 574	157	8.90 715	158	11.09 285	9.99 859	23
38	8.90 730	156	8.90 872	157	11.09 128	9.99 858	22
39	8.90 885	155	8.91 029	157	11.08 971	9.99 857	21
40	8.91 040	155	8.91 185	156	11.08 815	9.99 856	20
41	8.91 195	155	8.91 340	155	11.08 660	9.99 855	19
42	8.91 349	154	8.91 495	155	11.08 505	9.99 854	18
43	8.91 502	153	8.91 650	155	11.08 350	9.99 853	17
44	8.91 655	153	8.91 803	153	11.08 197	9.99 852	16
45	8.91 807	152	8.91 957	154	11.08 043	9.99 851	15
46	8.91 959	152	8.92 110	153	11.07 890	9.99 850	14
47	8.92 110	151	8.92 262	152	11.07 738	9.99 848	13
48	8.92 261	151	8.92 414	152	11.07 586	9.99 847	12
49	8.92 411	150	8.92 565	151	11.07 435	9.99 846	11
50	8.92 561	150	8.92 716	151	11.07 284	9.99 845	10
51	8.92 710	149	8.92 866	150	11.07 134	9.99 844	9
52	8.92 859	149	8.93 016	150	11.06 984	9.99 843	8
53	8.93 007	148	8.93 165	149	11.06 835	9.99 842	7
54	8.93 154	147	8.93 313	148	11.06 687	9.99 841	6
55	8.93 301	147	8.93 462	149	11.06 538	9.99 840	5
56	8.93 448	147	8.93 609	147	11.06 391	9.99 839	4
57	8.93 594	146	8.93 756	147	11.06 244	9.99 838	3
58	8.93 740	146	8.93 903	147	11.06 097	9.99 837	2
59	8.93 885	145	8.94 049	146	11.05 951	9.99 836	1
60	8.94 030	145	8.94 195	146	11.05 805	9.99 834	0
	L. Cos.	d.	L. Cot.	c. d.	L. Tan.	L. Sin.	'

P. P.

	182	181	180	179	178
1	18.2	18.1	18.0	17.9	17.8
2	36.4	36.2	36.0	35.8	35.6
3	54.6	54.3	54.0	53.7	53.4
4	72.8	72.4	72.0	71.6	71.2
5	91.0	90.5	90.0	89.5	89.0
6	109.2	108.6	108.0	107.4	106.8
7	127.4	126.7	126.0	125.3	124.6
8	145.6	144.8	144.0	143.2	142.4
9	163.8	162.9	162.0	161.1	160.2

	177	176	175	174	173
1	17.7	17.6	17.5	17.4	17.3
2	35.4	35.2	35.0	34.8	34.6
3	53.1	52.8	52.5	52.2	51.9
4	70.8	70.4	70.0	69.6	69.2
5	88.5	88.0	87.5	87.0	86.5
6	106.2	105.6	105.0	104.4	103.8
7	123.9	123.2	122.5	121.8	121.1
8	141.6	140.8	140.0	139.2	138.4
9	159.3	158.4	157.5	156.6	155.7

	172	171	170	169	168
1	17.2	17.1	17.0	16.9	16.8
2	34.4	34.2	34.0	33.8	33.6
3	51.6	51.3	51.0	50.7	50.4
4	68.8	68.4	68.0	67.6	67.2
5	86.0	85.5	85.0	84.5	84.0
6	103.2	102.6	102.0	101.4	100.8
7	120.4	119.7	119.0	118.3	117.6
8	137.6	136.8	136.0	135.2	134.4
9	154.8	153.9	153.0	152.1	151.2

	167	166	165	164	163
1	16.7	16.6	16.5	16.4	16.3
2	33.4	33.2	33.0	32.8	32.6
3	50.1	49.8	49.5	49.2	48.9
4	66.8	66.4	66.0	65.6	65.2
5	83.5	83.0	82.5	82.0	81.5
6	100.2	99.6	99.0	98.4	97.8
7	116.9	116.2	115.5	114.8	114.1
8	133.6	132.8	132.0	131.2	130.4
9	150.3	149.4	148.5	147.6	146.7

	162	161	160	159	158
1	16.2	16.1	16.0	15.9	15.8
2	32.4	32.2	32.0	31.8	31.6
3	48.6	48.3	48.0	47.7	47.4
4	64.8	64.4	64.0	63.6	63.2
5	81.0	80.5	80.0	79.5	79.0
6	97.2	96.6	96.0	95.4	94.8
7	113.4	112.7	112.0	111.3	110.6
8	129.6	128.8	128.0	127.2	126.4
9	145.8	144.9	144.0	143.1	142.2

	157	156	155	154	153
1	15.7	15.6	15.5	15.4	15.3
2	31.4	31.2	31.0	30.8	30.6
3	47.1	46.8	46.5	46.2	45.9
4	62.8	62.4	62.0	61.6	61.2
5	78.5	78.0	77.5	77.0	76.5
6	94.2	93.6	93.0	92.4	91.8
7	109.9	109.2	108.5	107.8	107.1
8	125.6	124.8	124.0	123.2	122.4
9	141.3	140.4	139.5	138.6	137.7

	152	151	150	149	148
1	15.2	15.1	15.0	14.9	14.8
2	30.4	30.2	30.0	29.8	29.6
3	45.6	45.3	45.0	44.7	44.4
4	60.8	60.4	60.0	59.6	59.2
5	76.0	75.5	75.0	74.5	74.0
6	91.2	90.6	90.0	89.4	88.8
7	106.4	105.7	105.0	104.3	103.6
8	121.6	120.8	120.0	119.2	118.4
9	136.8	135.9	135.0	134.1	133.2

P. P.

85°

'	L. Sin.	d.	L. Tan.	c. d.	L. Cot.	L. Cos.		P. P.
0	8.94 030	144	8.94 195	145	11.05 805	9.99 834	60	
1	8.94 174	143	8.94 340	145	11.05 660	9.99 833	59	
2	8.94 317	144	8.94 485	145	11.05 515	9.99 832	58	
3	8.94 461	142	8.94 630	143	11.05 370	9.99 831	57	
4	8.94 603	143	8.94 773	144	11.05 227	9.99 830	56	
5	8.94 746	141	8.94 917	143	11.05 083	9.99 829	55	
6	8.94 887	142	8.95 060	142	11.04 940	9.99 828	54	
7	8.95 029	141	8.95 202	142	11.04 798	9.99 827	53	
8	8.95 170	140	8.95 344	142	11.04 656	9.99 825	52	
9	8.95 310	140	8.95 486	141	11.04 514	9.99 824	51	
10	8.95 450	139	8.95 627	140	11.04 373	9.99 823	50	
11	8.95 589	139	8.95 767	141	11.04 233	9.99 822	49	
12	8.95 728	139	8.95 908	139	11.04 092	9.99 821	48	
13	8.95 867	138	8.96 047	140	11.03 953	9.99 820	47	
14	8.96 005	138	8.96 187	138	11.03 813	9.99 819	46	
15	8.96 143	137	8.96 325	139	11.03 675	9.99 817	45	
16	8.96 280	137	8.96 464	138	11.03 536	9.99 816	44	
17	8.96 417	136	8.96 602	137	11.03 398	9.99 815	43	
18	8.96 553	136	8.96 739	138	11.03 261	9.99 814	42	
19	8.96 689	136	8.96 877	136	11.03 123	9.99 813	41	
20	8.96 825	135	8.97 013	137	11.02 987	9.99 812	40	
21	8.96 960	135	8.97 150	135	11.02 850	9.99 810	39	
22	8.97 095	134	8.97 285	136	11.02 715	9.99 809	38	
23	8.97 229	134	8.97 421	135	11.02 579	9.99 808	37	
24	8.97 363	133	8.97 556	135	11.02 444	9.99 807	36	
25	8.97 496	133	8.97 691	134	11.02 309	9.99 806	35	
26	8.97 629	133	8.97 825	134	11.02 175	9.99 804	34	
27	8.97 762	132	8.97 959	133	11.02 041	9.99 803	33	
28	8.97 894	132	8.98 092	133	11.01 908	9.99 802	32	
29	8.98 026	131	8.98 225	133	11.01 775	9.99 801	31	
30	8.98 157	131	8.98 358	132	11.01 642	9.99 800	30	
31	8.98 288	131	8.98 490	132	11.01 510	9.99 798	29	
32	8.98 419	130	8.98 622	131	11.01 378	9.99 797	28	
33	8.98 549	130	8.98 753	131	11.01 247	9.99 796	27	
34	8.98 679	129	8.98 884	131	11.01 116	9.99 795	26	
35	8.98 808	129	8.99 015	130	11.00 985	9.99 793	25	
36	8.98 937	129	8.99 145	130	11.00 855	9.99 792	24	
37	8.99 066	128	8.99 275	130	11.00 725	9.99 791	23	
38	8.99 194	128	8.99 405	129	11.00 595	9.99 790	22	
39	8.99 322	128	8.99 534	128	11.00 466	9.99 788	21	
40	8.99 450	127	8.99 662	129	11.00 338	9.99 787	20	
41	8.99 577	127	8.99 791	128	11.00 209	9.99 786	19	
42	8.99 704	126	8.99 919	127	11.00 081	9.99 785	18	
43	8.99 830	126	9.00 046	128	10.99 954	9.99 783	17	
44	8.99 956	126	9.00 174	127	10.99 826	9.99 782	16	
45	9.00 082	125	9.00 301	126	10.99 699	9.99 781	15	
46	9.00 207	125	9.00 427	126	10.99 573	9.99 780	14	
47	9.00 332	124	9.00 553	126	10.99 447	9.99 778	13	
48	9.00 456	125	9.00 679	126	10.99 321	9.99 777	12	
49	9.00 581	123	9.00 805	125	10.99 195	9.99 776	11	
50	9.00 704	124	9.00 930	125	10.99 070	9.99 775	10	
51	9.00 828	123	9.01 055	124	10.98 945	9.99 773	9	
52	9.00 951	123	9.01 179	124	10.98 821	9.99 772	8	
53	9.01 074	122	9.01 303	124	10.98 697	9.99 771	7	
54	9.01 196	122	9.01 427	123	10.98 573	9.99 769	6	
55	9.01 318	122	9.01 550	123	10.98 450	9.99 768	5	
56	9.01 440	121	9.01 673	123	10.98 327	9.99 767	4	
57	9.01 561	121	9.01 796	122	10.98 204	9.99 765	3	
58	9.01 682	121	9.01 918	122	10.98 082	9.99 764	2	
59	9.01 803	120	9.02 040	122	10.97 960	9.99 763	1	
60	9.01 923		9.02 162		10.97 838	9.99 761	0	
	L. Cos.	d.	L. Cot.	c. d.	L. Tan.	L. Sin.	'	P. P.

P. P.

	147	146	145	144
1	14.7	14.6	14.5	14.4
2	29.4	29.2	29.0	28.8
3	44.1	43.8	43.5	43.2
4	58.8	58.4	58.0	57.6
5	73.5	73.0	72.5	72.0
6	88.2	87.6	87.0	86.4
7	102.9	102.2	101.5	100.8
8	117.6	116.8	116.0	115.2
9	132.3	131.4	130.5	129.6

	143	142	141	140
1	14.3	14.2	14.1	14.0
2	28.6	28.4	28.2	28.0
3	42.9	42.6	42.3	42.0
4	57.2	56.8	56.4	56.0
5	71.5	71.0	70.5	70.0
6	85.8	85.2	84.6	84.0
7	100.1	99.4	98.7	98.0
8	114.4	113.6	112.8	112.0
9	128.7	127.8	126.9	126.0

	139	138	137	136
1	13.9	13.8	13.7	13.6
2	27.8	27.6	27.4	27.2
3	41.7	41.4	41.1	40.8
4	55.6	55.2	54.8	54.4
5	69.5	69.0	68.5	68.0
6	83.4	82.8	82.2	81.6
7	97.3	96.6	95.9	95.2
8	111.2	110.4	109.6	108.8
9	125.1	124.2	123.3	122.4

	135	134	133	132
1	13.5	13.4	13.3	13.2
2	27.0	26.8	26.6	26.4
3	40.5	40.2	39.9	39.6
4	54.0	53.6	53.2	52.8
5	67.5	67.0	66.5	66.0
6	81.0	80.4	79.8	79.2
7	94.5	93.8	93.1	92.4
8	108.0	107.2	106.4	105.6
9	121.5	120.6	119.7	118.8

	131	130	129	128
1	13.1	13.0	12.9	12.8
2	26.2	26.0	25.8	25.6
3	39.3	39.0	38.7	38.4
4	52.4	52.0	51.6	51.2
5	65.5	65.0	64.5	64.0
6	78.6	78.0	77.4	76.8
7	91.7	91.0	90.3	89.6
8	104.8	104.0	103.2	102.4
9	117.9	117.0	116.1	115.2

	127	126	125	124
1	12.7	12.6	12.5	12.4
2	25.4	25.2	25.0	24.8
3	38.1	37.8	37.5	37.2
4	50.8	50.4	50.0	49.6
5	63.5	63.0	62.5	62.0
6	76.2	75.6	75.0	74.4
7	88.9	88.2	87.5	86.8
8	101.6	100.8	100.0	99.2
9	114.3	113.4	112.5	111.6

	123	122	121	120
1	12.3	12.2	12.1	12.0
2	24.6	24.4	24.2	24.0
3	36.9	36.6	36.3	36.0
4	49.2	48.8	48.4	48.0
5	61.5	61.0	60.5	60.0
6	73.8	73.2	72.6	72.0
7	86.1	85.4	84.7	84.0
8	98.4	97.6	96.8	96.0
9	110.7	109.8	108.9	108.1

′	L. Sin.	d.	L. Tan.	c. d.	L. Cot.	L. Cos.	′
0	9.01 923	120	9.02 162	121	10.97 838	9.99 761	60
1	9.02 043	120	9.02 283	121	10.97 717	9.99 760	59
2	9.02 163	120	9.02 404	121	10.97 596	9.99 759	58
3	9.02 283	119	9.02 525	120	10.97 475	9.99 757	57
4	9.02 402	118	9.02 645	121	10.97 355	9.99 756	56
5	9.02 520	119	9.02 766	119	10.97 234	9.99 755	55
6	9.02 639	118	9.02 885	120	10.97 115	9.99 753	54
7	9.02 757	117	9.03 005	119	10.96 995	9.99 752	53
8	9.02 874	118	9.03 124	118	10.96 876	9.99 751	52
9	9.02 992	117	9.03 242	119	10.96 758	9.99 749	51
10	9.03 109	117	9.03 361	118	10.96 639	9.99 748	50
11	9.03 226	116	9.03 479	118	10.96 521	9.99 747	49
12	9.03 342	116	9.03 597	117	10 96 403	9.99 745	48
13	9.03 458	116	9.03 714	118	10.96 286	9.99 744	47
14	9.03 574	116	9.03 832	116	10.96 168	9.99 742	46
15	9.03 690	115	9.03 948	117	10.96 052	9.99 741	45
16	9.03 805	115	9.04 065	116	10.95 935	9.99 740	44
17	9.03 920	114	9.04 181	116	10.95 819	9.99 738	43
18	9.04 034	115	9.04 297	116	10.95 703	9 99 737	42
19	9.04 149	113	9.04 413	115	10.95 587	9.99 736	41
20	9.04 262	114	9.04 528	115	10.95 472	9.99 734	40
21	9.04 376	114	9.04 643	115	10.95 357	9.99 733	39
22	9.04 490	113	9.04 758	115	10.95 242	9.99 731	38
23	9.04 603	112	9.04 873	114	10.95 127	9.99 730	37
24	9.04 715	113	9.04 987	114	10.95 013	9.99 728	36
25	9.04 828	112	9.05 101	113	10.94 899	9.99 727	35
26	9.04 940	112	9.05 214	114	10.94 786	9.99 726	34
27	9.05 052	112	9.05 328	113	10.94 672	9.99 724	33
28	9.05 164	111	9.05 441	112	10.94 559	9.99 723	32
29	9.05 275	111	9.05 553	113	10.94 447	9.99 721	31
30	9.05 386	111	9.05 666	112	10.94 334	9.99 720	30
31	9.05 497	110	9.05 778	112	10.94 222	9.99 718	29
32	9.05 607	110	9.05 890	112	10.94 110	9.99 717	28
33	9.05 717	110	9.06 002	111	10.93 998	9.99 716	27
34	9.05 827	110	9.06 113	111	10.93 887	9.99 714	26
35	9.05 937	109	9.06 224	111	10.93 776	9.99 713	25
36	9.06 046	109	9.06 335	110	10.93 665	9.99 711	24
37	9.06 155	109	9.06 445	111	10.93 555	9.99 710	23
38	9.06 264	108	9.06 556	110	10.93 444	9.99 708	22
39	9.06 372	109	9.06 666	109	10.93 334	9.99 707	21
40	9.06 481	108	9.06 775	110	10.93 225	9.99 705	20
41	9.06 589	107	9.06 885	109	10.93 115	9.99 704	19
42	9.06 696	108	9.06 994	109	10.93 006	9.99 702	18
43	9.06 804	107	9.07 103	108	10.92 897	9.99 701	17
44	9.06 911	107	9.07 211	109	10.92 789	9.99 699	16
45	9.07 018	106	9.07 320	108	10.92 680	9.99 698	15
46	9.07 124	107	9.07 428	108	10.92 572	9.99 696	14
47	9.07 231	106	9.07 536	107	10.92 464	9.99 695	13
48	9.07 337	105	9.07 643	108	10.92 357	9.99 693	12
49	9.07 442	106	9.07 751	107	10.92 249	9.99 692	11
50	9.07 548	105	9.07 858	106	10.92 142	9.99 690	10
51	9.07 653	105	9.07 964	107	10.92 036	9.99 689	9
52	9.07 758	105	9.08·071	106	10.91 929	9.99 687	8
53	9.07 863	105	9.08 177	106	10.91 823	9.99 686	7
54	9.07 968	104	9.08 283	106	10.91 717	9.99 684	6
55	9.08 072	104	9.08 389	106	10.91 611	9.99 683	5
56	9.08 176	104	9.08 495	105	10.91 505	9.99 681	4
57	9.08 280	103	9.08 600	105	10.91 400	9.99 680	3
58	9.08 383	103	9.08 705	105	10.91 295	9.99 678	2
59	9.08 486	103	9.08 810	104	10.91 190	9.99 677	1
60	9.08 589		9.08 914		10.91 086	9.99 675	0

| | L. Cos. | d. | L. Cot. | c. d. | L. Tan. | L. Sin. | ′ |

P. P.

	121	120	119	118
1	12.1	12.0	11.9	11.8
2	24.2	24.0	23.8	23.6
3	36.3	36.0	35.7	35.4
4	48.4	48.0	47.6	47.2
5	60.5	60.0	59.5	59.0
6	72.6	72.0	71.4	70.8
7	84.7	84.0	83.3	82.6
8	96.8	96.0	95.2	94.4
9	108.9	108.0	107.1	106.2

	117	116	115	114
1	11.7	11.6	11.5	11.4
2	23.4	23.2	23.0	22.8
3	35.1	34.8	34.5	34.2
4	46.8	46.4	46.0	45.6
5	58.5	58.0	57.5	57.0
6	70.2	69.6	69.0	68.4
7	81.9	81.2	80.5	79.8
8	93.6	92.8	92.0	91.2
9	105.3	104.4	103.5	102.6

	113	112	111	110
1	11.3	11.2	11.1	11.0
2	22.6	22.4	22.2	22.0
3	33.9	33.6	33.3	33.0
4	45.2	44.8	44.4	44.0
5	56.5	56.0	55.5	55.0
6	67.8	67.2	66.6	66.0
7	79.1	78.4	77.7	77.0
8	90.4	89.6	88.8	88.0
9	101.7	100.8	99.9	99.0

	109	108	107	106
1	10.9	10.8	10.7	10.6
2	21.8	21.6	21.4	21.2
3	32.7	32.4	32.1	31.8
4	43.6	43.2	42.8	42.4
5	54.5	54.0	53.5	53.0
6	65.4	64.8	64.2	63.6
7	76.3	75.6	74.9	74.2
8	87.2	86.4	85.6	84.8
9	98.1	97.2	96.3	95.4

	105	104	103
1	10.5	10.4	10.3
2	21.0	20.8	20.6
3	31.5	31.2	30.9
4	42.0	41.6	41.2
5	52.5	52.0	51.5
6	63.0	62.4	61.8
7	73.5	72.8	72.1
8	84.0	83.2	82.4
9	94.5	93.6	92.7

'	L. Sin.	d.	L. Tan.	c. d.	L. Cot.	L. Cos.		'
0	9.08 589	103	9.08 914	105	10.91 086	9.99 675		60
1	9.08 692	103	9.09 019	104	10.90 981	9.99 674		59
2	9.08 795	102	9.09 123	104	10.90 877	9.99 672		58
3	9.08 897	102	9.09 227	103	10.90 773	9.99 670		57
4	9.08 999	102	9.09 330	104	10.90 670	9.99 669		56
5	9.09 101	101	9.09 434	103	10.90 566	9.99 667		55
6	9.09 202	102	9.09 537	103	10.90 463	9.99 666		54
7	9.09 304	101	9.09 640	103	10.90 360	9.99 664		53
8	9.09 405	101	9.09 742	103	10.90 258	9.99 663		52
9	9.09 506	100	9.09 845	102	10.90 155	9.99 661		51
10	9.09 606	101	9.09 947	102	10.90 053	9.99 659		50
11	9.09 707	100	9.10 049	101	10.89 951	9.99 658		49
12	9.09 807	100	9.10 150	102	10.89 850	9.99 656		48
13	9.09 907	99	9.10 252	101	10.89 748	9.99 655		47
14	9.10 006	100	9.10 353	101	10.89 647	9.99 653		46
15	9.10 106	99	9.10 454	101	10.89 546	9.99 651		45
16	9.10 205	99	9.10 555	101	10.89 445	9.99 650		44
17	9.10 304	98	9.10 656	100	10.89 344	9.99 648		43
18	9.10 402	99	9.10 756	100	10.89 244	9.99 647		42
19	9.10 501	98	9.10 856	100	10.89 144	9.99 645		41
20	9.10 599	98	9.10 956	100	10.89 044	9.99 643		40
21	9.10 697	98	9.11 056	99	10.88 944	9.99 642		39
22	9.10 795	98	9.11 155	99	10.88 845	9.99 640		38
23	9.10 893	97	9.11 254	99	10.88 746	9.99 638		37
24	9.10 990	97	9.11 353	99	10.88 647	9.99 637		36
25	9.11 087	97	9.11 452	99	10.88 548	9.99 635		35
26	9.11 184	97	9.11 551	98	10.88 449	9.99 633		34
27	9.11 281	96	9.11 649	98	10.88 351	9.99 632		33
28	9.11 377	97	9.11 747	98	10.88 253	9.99 630		32
29	9.11 474	96	9.11 845	98	10.88 155	9.99 629		31
30	9.11 570	96	9.11 943	97	10.88 057	9.99 627		30
31	9.11 666	95	9.12 040	98	10.87 960	9.99 625		29
32	9.11 761	96	9.12 138	97	10.87 862	9.99 624		28
33	9.11 857	95	9.12 235	97	10.87 765	9.99 622		27
34	9.11 952	95	9.12 332	96	10.87 668	9.99 620		26
35	9.12 047	95	9.12 428	97	10.87 572	9.99 618		25
36	9.12 142	94	9.12 525	96	10.87 475	9.99 617		24
37	9.12 236	95	9.12 621	96	10.87 379	9.99 615		23
38	9.12 331	94	9.12 717	96	10.87 283	9.99 613		22
39	9.12 425	94	9.12 813	96	10.87 187	9.99 612		21
40	9.12 519	93	9.12 909	95	10.87 091	9.99 610		20
41	9.12 612	94	9.13 004	95	10.86 996	9.99 608		19
42	9.12 706	93	9.13 099	95	10.86 901	9.99 607		18
43	9.12 799	93	9.13 194	95	10.86 806	9.99 605		17
44	9.12 892	93	9.13 289	95	10.86 711	9.99 603		16
45	9.12 985	93	9.13 384	94	10.86 616	9.99 601		15
46	9.13 078	93	9.13 478	95	10.86 522	9.99 600		14
47	9.13 171	92	9.13 573	94	10.86 427	9.99 598		13
48	9.13 263	92	9.13 667	94	10.86 333	9.99 596		12
49	9.13 355	92	9.13 761	93	10.86 239	9.99 595		11
50	9.13 447	92	9.13 854	94	10.86 146	9.99 593		10
51	9.13 539	91	9.13 948	93	10.86 052	9.99 591		9
52	9.13 630	92	9.14 041	93	10.85 959	9.99 589		8
53	9.13 722	91	9.14 134	93	10.85 866	9.99 588		7
54	9.13 813	91	9.14 227	93	10.85 773	9.99 586		6
55	9.13 904	90	9.14 320	92	10.85 680	9.99 584		5
56	9.13 994	91	9.14 412	92	10.85 588	9.99 582		4
57	9.14 085	90	9.14 504	93	10.85 496	9.99 581		3
58	9.14 175	91	9.14 597	91	10.85 403	9.99 579		2
59	9.14 266	90	9.14 688	92	10.85 312	9.99 577		1
60	9.14 356		9.14 780		10.85 220	9.99 575		0

| L. Cos. | d. | L. Cot. | c. d. | L. Tan. | L. Sin. | ' | P. P. |

P. P.

	105	104	103
1	10.5	10.4	10.3
2	21.0	20.8	20.6
3	31.5	31:2	30.9
4	42.0	41.6	41.2
5	52.5	52.0	51.5
6	63.0	62.4	61.8
7	73.5	72.8	72.1
8	84.0	83.2	82.4
9	94.5	93.6	92.7

	102	101	99
1	10.2	10.1	9.9
2	20.4	20.2	19.8
3	30.6	30.3	29.7
4	40.8	40.4	39.6
5	51.0	50.5	49.5
6	61.2	60.6	59.4
7	71.4	70.7	69.3
8	81.6	80.8	79.2
9	91.8	90.9	89.1

	98	97	96
1	9.8	9.7	9.6
2	19.6	19.4	19.2
3	29.4	29.1	28.8
4	39.2	38.8	38.4
5	49.0	48.5	48.0
6	58.8	58.2	57.6
7	68.6	67.9	67.2
8	78.4	77.6	76.8
9	88.2	87.3	86.4

	95	94	93
1	9.5	9.4	9.3
2	19.0	18.8	18.6
3	28.5	28.2	27.9
4	38.0	37.6	37.2
5	47.5	47.0	46.5
6	57.0	56.4	55.8
7	66.5	65.8	65.1
8	76.0	75.2	74.4
9	85.5	84.6	83.7

	92	91	90
1	9.2	9.1	9.0
2	18.4	18.2	18.0
3	27.6	27.3	27.0
4	36.8	36.4	36.0
5	46.0	45.5	45.0
6	55.2	54.6	54.0
7	64.4	63.7	63.0
8	73.6	72.8	72.0
9	82.8	81.9	81.0

′	L. Sin.	d.	L. Tan.	c. d.	L. Cot.	L. Cos.		P. P.
0	9.14 356		9.14 780		10.85 220	9.99 575	60	
1	9.14 445	89	9.14 872	92	10.85 128	9.99 574	59	
2	9.14 535	90	9.14 963	91	10.85 037	9.99 572	58	
3	9.14 624	89	9.15 054	91	10.84 946	9.99 570	57	
4	9.14 714	90	9.15 145	91	10.84 855	9.99 568	56	
		89		91				
5	9.14 803	88	9.15 236	91	10.84 764	9.99 566	55	
6	9.14 891	89	9.15 327	90	10.84 673	9.99 565	54	
7	9.14 980	89	9.15 417	91	10.84 583	9.99 563	53	
8	9.15 069	88	9.15 508	90	10.84 492	9.99 561	52	
9	9.15 157	88	9.15 598	90	10.84 402	9.99 559	51	
10	9.15 245	88	9.15 688	89	10.84 312	9.99 557	50	
11	9.15 333	88	9.15 777	90	10.84 223	9.99 556	49	
12	9.15 421	87	9.15 867	89	10.84 133	9.99 554	48	
13	9.15 508	88	9.15 956	90	10.84 044	9.99 552	47	
14	9.15 596	87	9.16 046	89	10.83 954	9.99 550	46	
15	9.15 683	87	9.16 135	89	10.83 865	9.99 548	45	
16	9.15 770	87	9.16 224	88	10.83 776	9.99 546	44	
17	9.15 857	87	9.16 312	89	10.83 688	9.99 545	43	
18	9.15 944	86	9.16 401	88	10.83 599	9.99 543	42	
19	9.16 030	86	9.16 489	88	10.83 511	9.99 541	41	
20	9.16 116	87	9.16 577	88	10.83 423	9.99 539	40	
21	9.16 203	86	9.16 665	88	10.83 335	9.99 537	39	
22	9.16 289	85	9.16 753	88	10.83 247	9.99 535	38	
23	9.16 374	86	9.16 841	87	10.83 159	9.99 533	37	
24	9.16 460	85	9.16 928	88	10.83 072	9.99 532	36	
25	9.16 545	86	9.17 016	87	10.82 984	9.99 530	35	
26	9.16 631	85	9.17 103	87	10.82 897	9.99 528	34	
27	9.16 716	85	9.17 190	87	10.82 810	9.99 526	33	
28	9.16 801	85	9.17 277	86	10.82 723	9.99 524	32	
29	9.16 886	84	9.17 363	87	10.82 637	9.99 522	31	
30	9.16 970	85	9.17 450	86	10.82 550	9.99 520	30	
31	9.17 055	84	9.17 536	86	10.82 464	9.99 518	29	
32	9.17 139	84	9.17 622	86	10.82 378	9.99 517	28	
33	9.17 223	84	9.17 708	86	10.82 292	9.99 515	27	
34	9.17 307	84	9.17 794	86	10.82 206	9.99 513	26	
35	9.17 391	83	9.17 880	85	10.82 120	9.99 511	25	
36	9.17 474	84	9.17 965	86	10.82 035	9.99 509	24	
37	9.17 558	83	9.18 051	85	10.81 949	9.99 507	23	
38	9.17 641	83	9.18 136	85	10.81 864	9.99 505	22	
39	9.17 724	83	9.18 221	85	10.81 779	9.99 503	21	
40	9.17 807	83	9.18 306	85	10.81 694	9.99 501	20	
41	9.17 890	83	9.18 391	84	10.81 609	9.99 499	19	
42	9.17 973	82	9.18 475	85	10.81 525	9.99 497	18	
43	9.18 055	82	9.18 560	84	10.81 440	9.99 495	17	
44	9.18 137	83	9.18 644	84	10.81 356	9.99 494	16	
45	9.18 220	82	9.18 728	84	10.81 272	9.99 492	15	
46	9.18 302	81	9.18 812	84	10.81 188	9.99 490	14	
47	9.18 383	82	9.18 896	83	10.81 104	9.99 488	13	
48	9.18 465	82	9.18 979	84	10.81 021	9.99 486	12	
49	9.18 547	81	9.19 063	83	10.80 937	9.99 484	11	
50	9.18 628	81	9.19 146	83	10.80 854	9.99 482	10	
51	9.18 709	81	9.19 229	83	10.80 771	9.99 480	9	
52	9.18 790	81	9.19 312	83	10.80 688	9.99 478	8	
53	9.18 871	81	9.19 395	83	10.80 605	9.99 476	7	
54	9.18 952	81	9.19 478	83	10.80 522	9.99 474	6	
55	9.19 033	80	9.19 561	82	10.80 439	9.99 472	5	
56	9.19 113	80	9.19 643	82	10.80 357	9.99 470	4	
57	9.19 193	80	9.19 725	82	10.80 275	9.99 468	3	
58	9.19 273	80	9.19 807	82	10.80 193	9.99 466	2	
59	9.19 353	80	9.19 889	82	10.80 111	9.99 464	1	
60	9.19 433		9.19 971		10.80 029	9.99 462	0	
	L. Cos.	d.	L. Cot.	c. d.	L. Tan.	L. Sin.	′	P. P.

P. P.

	92	91	90
1	9.2	9.1	9.0
2	18.4	18.2	18.0
3	27.6	27.3	27.0
4	36.8	36.4	36.0
5	46.0	45.5	45.0
6	55.2	54.6	54.0
7	64.4	63.7	63.0
8	73.6	72.8	72.0
9	82.8	81.9	81.0

	89	88
1	8.9	8.8
2	17.8	17.6
3	26.7	26.4
4	35.6	35.2
5	44.5	44.0
6	53.4	52.8
7	62.3	61.6
8	71.2	70.4
9	80.1	79.2

	87	86	85
1	8.7	8.6	8.5
2	17.4	17.2	17.0
3	26.1	25.8	25.5
4	34.8	34.4	34.0
5	43.5	43.0	42.5
6	52.2	51.6	51.0
7	60.9	60.2	59.5
8	69.6	68.8	68.0
9	78.3	77.4	76.5

	84	83
1	8.4	8.3
2	16.8	16.6
3	25.2	24.9
4	33.6	33.2
5	42.0	41.5
6	50.4	49.8
7	58.8	58.1
8	67.2	66.4
9	75.6	74.7

	82	81	80
1	8.2	8.1	8.0
2	16.4	16.2	16.0
3	24.6	24.3	24.0
4	32.8	32.4	32.0
5	41.0	40.5	40.0
6	49.2	48.6	48.0
7	57.4	56.7	56.0
8	65.6	64.8	64.0
9	73.8	72.9	72.0

′	L. Sin.	d.	L. Tan.	c. d.	L. Cot.	L. Cos.		P. P.
0	9.19 433		9.19 971		10.80 029	9.99 462	60	
1	9.19 513	80	9.20 053	82	10.79 947	9.99 460	59	
2	9.19 592	79	9.20 134	81	10.79 866	9.99 458	58	
3	9.19 672	80	9.20 216	82	10.79 784	9.99 456	57	
4	9.19 751	79	9.20 297	81	10.79 703	9.99 454	56	
5	9.19 830	79	9.20 378	81	10.79 622	9.99 452	55	**82** **81** **80**
6	9.19 909	79	9.20 459	81	10.79 541	9.99 450	54	
7	9.19 988	79	9.20 540	81	10.79 460	9.99 448	53	1 8.2 8.1 8.0
8	9.20 067	79	9.20 621	81	10.79 379	9.99 446	52	2 16.4 16.2 16.0
9	9.20 145	78	9.20 701	80	10.79 299	9.99 444	51	3 24.6 24.3 24.0
		78		81				4 32.8 32.4 32.0
10	9.20 223	79	9.20 782	80	10.79 218	9.99 442	50	5 41.0 40.5 40.0
11	9.20 302	78	9.20 862	80	10.79 138	9.99 440	49	6 49.2 48.6 48.0
12	9.20 380	78	9.20 942	80	10.79 058	9.99 438	48	7 57.4 56.7 56.0
13	9.20 458	77	9.21 022	80	10.78 978	9.99 436	47	8 65.6 64.8 64.0
14	9.20 535	78	9.21 102	80	10.78 898	9.99 434	46	9 73.8 72.9 72.0
15	9.20 613	78	9.21 182	79	10.78 818	9.99 432	45	
16	9.20 691	77	9.21 261	80	10.78 739	9.99 429	44	**79** **78** **77**
17	9.20 768	77	9.21 341	79	10.78 659	9.99 427	43	
18	9.20 845	77	9.21 420	79	10.78 580	9.99 425	42	1 7.9 7.8 7.7
19	9.20 922	77	9.21 499	79	10.78 501	9.99 423	41	2 15.8 15.6 15.4
		77		79				3 23.7 23.4 23.1
20	9.20 999	77	9.21 578	79	10.78 422	9.99 421	40	4 31.6 31.2 30.8
21	9.21 076	77	9.21 657	79	10.78 343	9.99 419	39	5 39.5 39.0 38.5
22	9.21 153	76	9.21 736	78	10.78 264	9.99 417	38	6 47.4 46.8 46.2
23	9.21 229	77	9.21 814	79	10.78 186	9.99 415	37	7 55.3 54.6 53.9
24	9.21 306	76	9.21 893	78	10.78 107	9.99 413	36	8 63.2 62.4 61.6
25	9.21 382	76	9.21 971	78	10.78 029	9.99 411	35	9 71.1 70.2 69.3
26	9.21 458	76	9.22 049	78	10.77 951	9.99 409	34	
27	9.21 534	76	9.22 127	78	10.77 873	9.99 407	33	**76** **75** **74**
28	9.21 610	75	9.22 205	78	10.77 795	9.99 404	32	
29	9.21 685	76	9.22 283	78	10.77 717	9.99 402	31	1 7.6 7.5 7.4
30	9.21 761	75	9.22 361	77	10.77 639	9.99 400	30	2 15.2 15.0 14.8
31	9.21 836	76	9.22 438	78	10.77 562	9.99 398	29	3 22.8 22.5 22.2
32	9.21 912	75	9.22 516	77	10.77 484	9.99 396	28	4 30.4 30.0 29.6
33	9.21 987	75	9.22 593	77	10.77 407	9.99 394	27	5 38.0 37.5 37.0
34	9.22 062	75	9.22 670	77	10.77 330	9.99 392	26	6 45.6 45.0 44.4
35	9.22 137	74	9.22 747	77	10.77 253	9.99 390	25	7 53.2 52.5 51.8
36	9.22 211	75	9.22 824	77	10.77 176	9.99 388	24	8 60.8 60.0 59.2
37	9.22 286	75	9.22 901	76	10.77 099	9.99 385	23	9 68.4 67.5 66.6
38	9.22 361	74	9.22 977	77	10.77 023	9.99 383	22	
39	9.22 435	74	9.23 054	76	10.76 946	9.99 381	21	**73** **72** **71**
40	9.22 509	74	9.23 130	76	10.76 870	9.99 379	20	
41	9.22 583	74	9.23 206	77	10.76 794	9.99 377	19	1 7.3 7.2 7.1
42	9.22 657	74	9.23 283	76	10.76 717	9.99 375	18	2 14.6 14.4 14.2
43	9.22 731	74	9.23 359	76	10.76 641	9.99 372	17	3 21.9 21.6 21.3
44	9.22 805	73	9.23 435	75	10.76 565	9.99 370	16	4 29.2 28.8 28.4
45	9.22 878	74	9.23 510	76	10.76 490	9.99 368	15	5 36.5 36.0 35.5
46	9.22 952	73	9.23 586	75	10.76 414	9.99 366	14	6 43.8 43.2 42.6
47	9.23 025	73	9.23 661	76	10.76 339	9.99 364	13	7 51.1 50.4 49.7
48	9.23 098	73	9.23 737	75	10.76 263	9.99 362	12	8 58.4 57.6 56.8
49	9.23 171	73	9.23 812	75	10.76 188	9.99 359	11	9 65.7 64.8 63.9
50	9.23 244	73	9.23 887	75	10.76 113	9.99 357	10	
51	9.23 317	73	9.23 962	75	10.76 038	9.99 355	9	
52	9.23 390	72	9.24 037	75	10.75 963	9.99 353	8	**3** **2**
53	9.23 462	73	9.24 112	74	10.75 888	9.99 351	7	1 0.3 0.2
54	9.23 535	72	9.24 186	75	10.75 814	9.99 348	6	2 0.6 0.4
55	9.23 607	72	9.24 261	74	10.75 739	9.99 346	5	3 0.9 0.6
56	9.23 679	73	9.24 335	75	10.75 665	9.99 344	4	4 1.2 0.8
57	9.23 752	71	9.24 410	74	10.75 590	9.99 342	3	5 1.5 1.0
58	9.23 823	72	9.24 484	74	10.75 516	9.99 340	2	6 1.8 1.2
59	9.23 895	72	9.24 558	74	10.75 442	9.99 337	1	7 2.1 1.4
60	9.23 967		9.24 632		10.75 368	9.99 335	0	8 2.4 1.6
								9 2.7 1.8
	L. Cos.	d.	L. Cot.	c. d.	L. Tan.	L. Sin.	′	P. P.

10°

'	L. Sin.	d.	L. Tan.	c. d.	L. Cot.	L. Cos.	d.		P. P.		
0	9.23 967	72	9.24 632	74	10.75 368	9.99 335	2	60			
1	9.24 039	71	9.24 706	73	10.75 294	9.99 333	2	59			
2	9.24 110	71	9.24 779	74	10.75 221	9.99 331	3	58			
3	9.24 181	72	9.24 853	73	10.75 147	9.99 328	2	57			
4	9.24 253	71	9.24 926	74	10.75 074	9.99 326	2	56	74	73	72
5	9.24 324	71	9.25 000	73	10.75 000	9.99 324	2	55	1 7.4	7.3	7.2
6	9.24 395	71	9.25 073	73	10.74 927	9.99 322	3	54	2 14.8	14.6	14.4
7	9.24 466	70	9.25 146	73	10.74 854	9.99 319	2	53	3 22.2	21.9	21.6
8	9.24 536	71	9.25 219	73	10.74 781	9.99 317	2	52	4 29.6	29.2	28.8
9	9.24 607	70	9.25 292	73	10.74 708	9.99 315	2	51	5 37.0	36.5	36.0
10	9.24 677	71	9.25 365	72	10.74 635	9.99 313	3	50	6 44.4	43.8	43.2
11	9.24 748	70	9.25 437	73	10.74 563	9.99 310	2	49	7 51.8	51.1	50.4
12	9.24 818	70	9.25 510	72	10.74 490	9.99 308	2	48	8 59.2	58.4	57.6
13	9.24 888	70	9.25 582	73	10.74 418	9.99 306	2	47	9 66.6	65.7	64.8
14	9.24 958	70	9.25 655	72	10.74 345	9.99 304	3	46			
15	9.25 028	70	9.25 727	72	10.74 273	9.99 301	2	45			
16	9.25 098	70	9.25 799	72	10.74 201	9.99 299	2	44			
17	9.25 168	69	9.25 871	72	10.74 129	9.99 297	3	43			
18	9.25 237	70	9.25 943	72	10.74 057	9.99 294	2	42			
19	9.25 307	69	9.26 015	71	10.73 985	9.99 292	2	41	71	70	69
20	9.25 376	69	9.26 086	72	10.73 914	9.99 290	2	40	1 7.1	7.0	6.9
21	9.25 445	69	9.26 158	71	10.73 842	9.99 288	3	39	2 14.2	14.0	13.8
22	9.25 514	69	9.26 229	72	10.73 771	9.99 285	2	38	3 21.3	21.0	20.7
23	9.25 583	69	9.26 301	71	10.73 699	9.99 283	2	37	4 28.4	28.0	27.6
24	9.25 652	69	9.26 372	71	10.73 628	9.99 281	3	36	5 35.5	35.0	34.5
25	9.25 721	69	9.26 443	71	10.73 557	9.99 278	2	35	6 42.6	42.0	41.4
26	9.25 790	68	9.26 514	71	10.73 486	9.99 276	2	34	7 49.7	49.0	48.3
27	9.25 858	69	9.26 585	70	10.73 415	9.99 274	3	33	8 56.8	56.0	55.2
28	9.25 927	68	9.26 655	71	10.73 345	9.99 271	2	32	9 63.9	63.0	62.1
29	9.25 995	68	9.26 726	71	10.73 274	9.99 269	2	31			
30	9.26 063	68	9.26 797	70	10.73 203	9.99 267	3	30			
31	9.26 131	68	9.26 867	70	10.73 133	9.99 264	2	29			
32	9.26 199	68	9.26 937	71	10.73 063	9.99 262	2	28			
33	9.26 267	68	9.27 008	70	10.72 992	9.99 260	3	27	68	67	66
34	9.26 335	68	9.27 078	70	10.72 922	9.99 257	2	26			
35	9.26 403	67	9.27 148	70	10.72 852	9.99 255	3	25	1 6.8	6.7	6.6
36	9.26 470	68	9.27 218	70	10.72 782	9.99 252	2	24	2 13.6	13.4	13.2
37	9.26 538	67	9.27 288	69	10.72 712	9.99 250	2	23	3 20.4	20.1	19.8
38	9.26 605	67	9.27 357	70	10.72 643	9.99 248	3	22	4 27.2	26.8	26.4
39	9.26 672	67	9.27 427	69	10.72 573	9.99 245	2	21	5 34.0	33.5	33.0
40	9.26 739	67	9.27 496	70	10.72 504	9.99 243	2	20	6 40.8	40.2	39.6
41	9.26 806	67	9.27 566	69	10.72 434	9.99 241	3	19	7 47.6	46.9	46.2
42	9.26 873	67	9.27 635	69	10.72 365	9.99 238	2	18	8 54.4	53.6	52.8
43	9.26 940	67	9.27 704	69	10.72 296	9.99 236	2	17	9 61.2	60.3	59.4
44	9.27 007	66	9.27 773	69	10.72 227	9.99 233	2	16			
45	9.27 073	67	9.27 842	69	10.72 158	9.99 231	2	15			
46	9.27 140	66	9.27 911	69	10.72 089	9.99 229	3	14			
47	9.27 206	67	9.27 980	69	10.72 020	9.99 226	2	13			
48	9.27 273	66	9.28 049	68	10.71 951	9.99 224	3	12	65	3	
49	9.27 339	66	9.28 117	69	10.71 883	9.99 221	2	11			
50	9.27 405	66	9.28 186	68	10.71 814	9.99 219	2	10	1 6.5	0.3	
51	9.27 471	66	9.28 254	69	10.71 746	9.99 217	3	9	2 13.0	0.6	
52	9.27 537	65	9.28 323	68	10.71 677	9.99 214	2	8	3 19.5	0.9	
53	9.27 602	66	9.28 391	68	10.71 609	9.99 212	3	7	4 26.0	1.2	
54	9.27 668	66	9.28 459	68	10.71 541	9.99 209	2	6	5 32.5	1.5	
55	9.27 734	65	9.28 527	68	10.71 473	9.99 207	3	5	6 39.0	1.8	
56	9.27 799	65	9.28 595	67	10.71 405	9.99 204	2	4	7 45.5	2.1	
57	9.27 864	66	9.28 662	68	10.71 338	9.99 202	2	3	8 52.0	2.4	
58	9.27 930	65	9.28 730	68	10.71 270	9.99 200	3	2	9 58.5	2.7	
59	9.27 995	65	9.28 798	67	10.71 202	9.99 197	2	1			
60	9.28 060		9.28 865		10.71 135	9.99 195		0			

| | L. Cos. | d. | L. Cot. | c. d. | L. Tan. | L. Sin. | d. | ' | P. P. | | |

79°

258

'	L. Sin.	d.	L. Tan.	c. d.	L. Cot.	L. Cos.	d.	'
0	9.28 060	65	9.28 865	68	10.71 135	9.99 195	3	60
1	9.28 125	65	9.28 933	67	10.71 067	9.99 192	2	59
2	9.28 190	64	9.29 000	67	10.71 000	9.99 190	2	58
3	9.28 254	65	9.29 067	67	10.70 933	9.99 187	3	57
4	9.28 319	65	9.29 134	67	10.70 866	9.99 185	3	56
5	9.28 384	64	9.29 201	67	10.70 799	9.99 182	2	55
6	9.28 448	64	9.29 268	67	10.70 732	9.99 180	3	54
7	9.28 512	65	9.29 335	67	10.70 665	9.99 177	2	53
8	9.28 577	64	9.29 402	66	10.70 598	9.99 175	3	52
9	9.28 641	64	9.29 468	67	10.70 532	9.99 172	2	51
10	9.28 705	64	9.29 535	66	10.70 465	9.99 170	3	50
11	9.28 769	64	9.29 601	67	10.70 399	9.99 167	2	49
12	9.28 833	63	9.29 668	66	10.70 332	9.99 165	3	48
13	9.28 896	64	9.29 734	66	10.70 266	9.99 162	2	47
14	9.28 960	64	9.29 800	66	10.70 200	9.99 160	3	46
15	9.29 024	63	9.29 866	66	10.70 134	9.99 157	2	45
16	9.29 087	63	9.29 932	66	10.70 068	9.99 155	3	44
17	9.29 150	64	9.29 998	66	10.70 002	9.99 152	2	43
18	9.29 214	63	9.30 064	66	10.69 936	9.99 150	3	42
19	9.29 277	63	9.30 130	65	10.69 870	9.99 147	2	41
20	9.29 340	63	9.30 195	66	10.69 805	9.99 145	3	40
21	9.29 403	63	9.30 261	65	10.69 739	9.99 142	2	39
22	9.29 466	63	9.30 326	65	10.69 674	9.99 140	3	38
23	9.29 529	62	9.30 391	66	10.69 609	9.99 137	2	37
24	9.29 591	63	9.30 457	65	10.69 543	9.99 135	3	36
25	9.29 654	62	9.30 522	65	10.69 478	9.99 132	2	35
26	9.29 716	63	9.30 587	65	10.69 413	9.99 130	3	34
27	9.29 779	62	9.30 652	65	10.69 348	9.99 127	3	33
28	9.29 841	62	9.30 717	65	10.69 283	9.99 124	2	32
29	9.29 903	63	9.30 782	64	10.69 218	9.99 122	3	31
30	9.29 966	62	9.30 846	65	10.69 154	9.99 119	2	30
31	9.30 028	62	9.30 911	64	10.69 089	9.99 117	3	29
32	9.30 090	61	9.30 975	65	10.69 025	9.99 114	2	28
33	9.30 151	62	9.31 040	64	10.68 960	9.99 112	3	27
34	9.30 213	62	9.31 104	64	10.68 896	9.99 109	3	26
35	9.30 275	61	9.31 168	65	10.68 832	9.99 106	2	25
36	9.30 336	62	9.31 233	64	10.68 767	9.99 104	3	24
37	9.30 398	61	9.31 297	64	10.68 703	9.99 101	2	23
38	9.30 459	62	9.31 361	64	10.68 639	9.99 099	3	22
39	9.30 521	61	9.31 425	64	10.68 575	9.99 096	3	21
40	9.30 582	61	9.31 489	63	10.68 511	9.99 093	2	20
41	9.30 643	61	9.31 552	64	10.68 448	9.99 091	3	19
42	9.30 704	61	9.31 616	63	10.68 384	9.99 088	2	18
43	9.30 765	61	9.31 679	64	10.68 321	9.99 086	3	17
44	9.30 826	61	9.31 743	63	10.68 257	9.99 083	3	16
45	9.30 887	60	9.31 806	64	10.68 194	9.99 080	2	15
46	9.30 947	61	9.31 870	63	10.68 130	9.99 078	3	14
47	9.31 008	60	9.31 933	63	10.68 067	9.99 075	3	13
48	9.31 068	61	9.31 996	63	10.68 004	9.99 072	2	12
49	9.31 129	60	9.32 059	63	10.67 941	9.99 070	3	11
50	9.31 189	61	9.32 122	63	10.67 878	9.99 067	3	10
51	9.31 250	60	9.32 185	63	10.67 815	9.99 064	2	9
52	9.31 310	60	9.32 248	63	10.67 752	9.99 062	3	8
53	9.31 370	60	9.32 311	62	10.67 689	9.99 059	3	7
54	9.31 430	60	9.32 373	63	10.67 627	9.99 056	2	6
55	9.31 490	59	9.32 436	62	10.67 564	9.99 054	3	5
56	9.31 549	60	9.32 498	63	10.67 502	9.99 051	3	4
57	9.31 609	60	9.32 561	62	10.67 439	9.99 048	2	3
58	9.31 669	59	9.32 623	62	10.67 377	9.99 046	3	2
59	9.31 728	60	9.32 685	62	10.67 315	9.99 043	3	1
60	9.31 788		9.32 747		10.67 253	9.99 040		0
	L. Cos.	d.	L. Cot.	c. d.	L. Tan.	L. Sin.	d.	'

P. P.

	68	67	66
1	6.8	6.7	6.6
2	13.6	13.4	13.2
3	20.4	20.1	19.8
4	27.2	26.8	26.4
5	34.0	33.5	33.0
6	40.8	40.2	39.6
7	47.6	46.9	46.2
8	54.4	53.6	52.8
9	61.2	60.3	59.4

	65	64	63
1	6.5	6.4	6.3
2	13.0	12.8	12.6
3	19.5	19.2	18.9
4	26.0	25.6	25.2
5	32.5	32.0	31.5
6	39.0	38.4	37.8
7	45.5	44.8	44.1
8	52.0	51.2	50.4
9	58.5	57.6	56.7

	62	61	60
1	6.2	6.1	6.0
2	12.4	12.2	12.0
3	18.6	18.3	18.0
4	24.8	24.4	24.0
5	31.0	30.5	30.0
6	37.2	36.6	36.0
7	43.4	42.7	42.0
8	49.6	48.8	48.0
9	55.8	54.9	54.0

	59	3	2
1	5.9	0.3	0.2
2	11.8	0.6	0.4
3	17.7	0.9	0.6
4	23.6	1.2	0.8
5	29.5	1.5	1.0
6	35.4	1.8	1.2
7	41.3	2.1	1.4
8	47.2	2.4	1.6
9	53.1	2.7	1.8

78°

12°

'	L. Sin.	d.	L. Tan.	c. d.	L. Cot.	L. Cos.	d.	
0	9.31 788	59	9.32 747	63	10.67 253	9.99 040	2	60
1	9.31 847	60	9.32 810	62	10.67 190	9.99 038	3	59
2	9.31 907	59	9.32 872	61	10.67 128	9.99 035	3	58
3	9.31 966	59	9.32 933	62	10.67 067	9.99 032	3	57
4	9.32 025	59	9.32 995	62	10.67 005	9.99 030	2	56
5	9.32 084	59	9.33 057	62	10.66 943	9.99 027	3	55
6	9.32 143	59	9.33 119	61	10.66 881	9.99 024	3	54
7	9.32 202	59	9.33 180	62	10.66 820	9.99 022	2	53
8	9.32 261	58	9.33 242	61	10.66 758	9.99 019	3	52
9	9.32 319	59	9.33 303	62	10.66 697	9.99 016	3	51
10	9.32 378	59	9.33 365	61	10.66 635	9.99 013	2	50
11	9.32 437	58	9.33 426	61	10.66 574	9.99 011	3	49
12	9.32 495	58	9.33 487	61	10.66 513	9.99 008	3	48
13	9.32 553	59	9.33 548	61	10.66 452	9.99 005	3	47
14	9.32 612	58	9.33 609	61	10.66 391	9.99 002	2	46
15	9.32 670	58	9.33 670	61	10.66 330	9.99 000	3	45
16	9.32 728	58	9.33 731	61	10.66 269	9.98 997	3	44
17	9.32 786	58	9.33 792	61	10.66 208	9.98 994	3	43
18	9.32 844	58	9.33 853	60	10.66 147	9.98 991	2	42
19	9.32 902	58	9.33 913	61	10.66 087	9.98 989	3	41
20	9.32 960	58	9.33 974	60	10.66 026	9.98 986	3	40
21	9.33 018	57	9.34 034	61	10.65 966	9.98 983	3	39
22	9.33 075	58	9.34 095	60	10.65 905	9.98 980	2	38
23	9.33 133	57	9.34 155	60	10.65 845	9.98 978	3	37
24	9.33 190	58	9.34 215	61	10.65 785	9.98 975	3	36
25	9.33 248	57	9.34 276	60	10.65 724	9.98 972	3	35
26	9.33 305	57	9.34 336	60	10.65 664	9.98 969	2	34
27	9.33 362	58	9.34 396	60	10.65 604	9.98 967	3	33
28	9.33 420	57	9.34 456	60	10.65 544	9.98 964	3	32
29	9.33 477	57	9.34 516	60	10.65 484	9.98 961	3	31
30	9.33 534	57	9.34 576	59	10.65 424	9.98 958	3	30
31	9.33 591	56	9.34 635	60	10.65 365	9.98 955	2	29
32	9.33 647	57	9.34 695	60	10.65 305	9.98 953	3	28
33	9.33 704	57	9.34 755	59	10.65 245	9.98 950	3	27
34	9.33 761	57	9.34 814	60	10.65 186	9.98 947	3	26
35	9.33 818	56	9.34 874	59	10.65 126	9.98 944	3	25
36	9.33 874	57	9.34 933	59	10.65 067	9.98 941	3	24
37	9.33 931	56	9.34 992	59	10.65 008	9.98 938	2	23
38	9.33 987	56	9.35 051	60	10.64 949	9.98 936	3	22
39	9.34 043	57	9.35 111	59	10.64 889	9.98 933	3	21
40	9.34 100	56	9.35 170	59	10.64 830	9.98 930	3	20
41	9.34 156	56	9.35 229	59	10.64 771	9.98 927	3	19
42	9.34 212	56	9.35 288	59	10.64 712	9.98 924	3	18
43	9.34 268	56	9.35 347	58	10.64 653	9.98 921	2	17
44	9.34 324	56	9.35 405	59	10.64 595	9.98 919	3	16
45	9.34 380	56	9.35 464	59	10.64 536	9.98 916	3	15
46	9.34 436	55	9.35 523	58	10.64 477	9.98 913	3	14
47	9.34 491	56	9.35 581	59	10.64 419	9.98 910	3	13
48	9.34 547	55	9.35 640	58	10.64 360	9.98 907	3	12
49	9.34 602	56	9.35 698	59	10.64 302	9.98 904	3	11
50	9.34 658	55	9.35 757	58	10.64 243	9.98 901	3	10
51	9.34 713	56	9.35 815	58	10.64 185	9.98 898	2	9
52	9.34 769	55	9.35 873	58	10.64 127	9.98 896	3	8
53	9.34 824	55	9.35 931	58	10.64 069	9.98 893	3	7
54	9.34 879	55	9.35 989	58	10.64 011	9.98 890	3	6
55	9.34 934	55	9.36 047	58	10.63 953	9.98 887	3	5
56	9.34 989	55	9.36 105	58	10.63 895	9.98 884	3	4
57	9.35 044	55	9.36 163	58	10.63 837	9.98 881	3	3
58	9.35 099	55	9.36 221	58	10.63 779	9.98 878	3	2
59	9.35 154	55	9.36 279	57	10.63 721	9.98 875	3	1
60	9.35 209		9.36 336		10.63 664	9.98 872		0

L. Cos.	d.	L. Cot.	c. d.	L. Tan.	L. Sin.	d.	'

P. P.

	63	62	61
1	6.3	6.2	6.1
2	12.6	12.4	12.2
3	18.9	18.6	18.3
4	25.2	24.8	24.4
5	31.5	31.0	30.5
6	37.8	37.2	36.6
7	44.1	43.4	42.7
8	50.4	49.6	48.8
9	56.7	55.8	54.9

	60	59
1	6.0	5.9
2	12.0	11.8
3	18.0	17.7
4	24.0	23.6
5	30.0	29.5
6	36.0	35.4
7	42.0	41.3
8	48.0	47.2
9	54.0	53.1

	58	57
1	5.8	5.7
2	11.6	11.4
3	17.4	17.1
4	23.2	22.8
5	29.0	28.5
6	34.8	34.2
7	40.6	39.9
8	46.4	45.6
9	52.2	51.3

	56	55	3
1	5.6	5.5	0.3
2	11.2	11.0	0.6
3	16.8	16.5	0.9
4	22.4	22.0	1.2
5	28.0	27.5	1.5
6	33.6	33.0	1.8
7	39.2	38.5	2.1
8	44.8	44.0	2.4
9	50.4	49.5	2.7

P. P.

77°

′	L. Sin.	d.	L. Tan.	c. d.	L. Cot.	L. Cos.	d.	′	P. P.		
0	9.35 209	54	9.36 336	58	10.63 664	9.98 872	3	60			
1	9.35 263	55	9.36 394	58	10.63 606	9.98 869	2	59			
2	9.35 318	55	9.36 452	57	10.63 548	9.98 867	3	58			
3	9.35 373	54	9.36 509	57	10.63 491	9.98 864	3	57			
4	9.35 427	54	9.36 566	58	10.63 434	9 98 861	3	56			
									58	57	56
5	9.35 481	55	9.36 624	57	10.63 376	9.98 858	3	55			
6	9.35 536	54	9.36 681	57	10.63 319	9.98 855	3	54	1	5.8	5.7	5.6
7	9.35 590	54	9.36 738	57	10.63 262	9.98 852	3	53	2	11.6	11.4	11.2
8	9.35 644	54	9.36 795	57	10.63 205	9.98 849	3	52	3	17.4	17.1	16.8
9	9.35 698	54	9.36 852	57	10.63 148	9.98 846	3	51	4	23.3	22.8	22.4
									5	29.0	28.5	28.0
10	9.35 752	54	9.36 909	57	10.63 091	9.98 843	3	50	6	34.8	34.2	33.6
11	9.35 806	54	9.36 966	57	10.63 034	9.98 840	3	49	7	40.6	39.9	39.2
12	9.35 860	54	9.37 023	57	10.62 977	9.98 837	3	48	8	46.4	45.6	44.8
13	9.35 914	54	9.37 080	57	10.62 920	9.98 834	3	47	9	52.2	51.3	50.4
14	9.35 968	54	9.37 137	56	10.62 863	9.98 831	3	46			
15	9.36 022	53	9.37 193	57	10.62 807	9.98 828	3	45			
16	9.36 075	54	9.37 250	56	10.62 750	9.98 825	3	44			
17	9.36 129	53	9.37 306	57	10.62 694	9.08 822	3	43			
18	9.36 182	54	9.37 363	56	10.62 637	9.98 819	3	42			
19	9.36 236	53	9.37 419	57	10.62 581	9.98 816	3	41			
									55	54	53
20	9.36 289	53	9.37 476	56	10.62 524	9.98 813	3	40			
21	9.36 342	53	9.37 532	56	10.62 468	9.98 810	3	39	1	5.5	5.4	5.3
22	9.36 395	54	9.37 588	56	10.62 412	9.98 807	3	38	2	11.0	10.8	10.6
23	9.36 449	53	9.37 644	56	10.62 356	9.98 804	3	37	3	16.5	16.2	15.9
24	9.36 502	53	9.37 700	56	10.62 300	9.98 801	3	36	4	22.0	21.6	21.2
									5	27.5	27.0	26.5
25	9.36 555	53	9.37 756	56	10.62 244	9.98 798	3	35	6	33.0	32.4	31.8
26	9.36 608	52	9.37 812	56	10.62 188	9.98 795	3	34	7	38.5	37.8	37.1
27	9.36 660	53	9.37 868	56	10.62 132	9.98 792	3	33	8	44.0	43.2	42.4
28	9.36 713	53	9.37 924	56	10.62 076	9.98 789	3	32	9	49.5	48.6	47.7
29	9.36 766	53	9.37 980	55	10.62 020	9.98 786	3	31			
30	9.36 819	52	9.38 035	56	10.61 965	9.98 783	3	30			
31	9.36 871	53	9.38 091	56	10.61 909	9.98 780	3	29			
32	9.36 924	52	9.38 147	55	10.61 853	9.98 777	3	28			
33	9.36 976	52	9.38 202	55	10.61 798	9.98 774	3	27			
34	9.37 028	53	9.38 257	56	10.61 743	9.98 771	3	26			
										52	51
35	9.37 081	52	9.38 313	55	10.61 687	9.98 768	3	25			
36	9.37 133	52	9.38 368	55	10.61 632	9.98 765	3	24	1	5.2	5.1
37	9.37 185	52	9.38 423	56	10.61 577	9.98 762	3	23	2	10.4	10.2
38	9.37 237	52	9.38 479	55	10.61 521	9.98 759	3	22	3	15.6	15.3
39	9.37 289	52	9.38 534	55	10.61 466	9.98 756	3	21	4	20.8	20.4
									5	26.0	25.5
40	9.37 341	52	9.38 589	55	10.61 411	9.98 753	3	20	6	31.2	30.6
41	9.37 393	52	9.38 644	55	10.61 356	9.98 750	4	19	7	36.4	35.7
42	9.37 445	52	9.38 699	55	10.61 301	9.98 746	3	18	8	41.6	40.8
43	9.37 497	52	9.38 754	54	10.61 246	9.98 743	3	17	9	46.8	45.9
44	9.37 549	51	9.38 808	55	10.61 192	9.98 740	3	16			
45	9.37 600	52	9.38 863	55	10.61 137	9.98 737	3	15			
46	9.37 652	51	9.38 918	54	10.61 082	9.98 734	3	14			
47	9.37 703	52	9.38 972	55	10.61 028	9.98 731	3	13			
48	9.37 755	51	9.39 027	55	10.60 973	9.98 728	3	12			
49	9.37 806	52	9.39 082	54	10.60 918	9.98 725	3	11	4	3	2
50	9.37 858	51	9.39 136	54	10.60 864	9.98 722	3	10	1	0.4	0.3	0.2
51	9.37 909	51	9.39 190	55	10.60 810	9.98 719	4	9	2	0.8	0.6	0.4
52	9.37 960	51	9.39 245	54	10.60 755	9.98 715	3	8	3	1.2	0.9	0.6
53	9.38 011	51	9.39 299	54	10.60 701	9.98 712	3	7	4	1.6	1.2	0.8
54	9.38 062	51	9.39 353	54	10.60 647	9.98 709	3	6	5	2.0	1.5	1.0
55	9.38 113	51	9.39 407	54	10.60 593	9.98 706	3	5	6	2.4	1.8	1.2
56	9.38 164	51	9.39 461	54	10.60 539	9.98 703	3	4	7	2.8	2.1	1.4
57	9.38 215	51	9.39 515	54	10.60 485	9.98 700	3	3	8	3.2	2.4	1.6
58	9.38 266	51	9.39 569	54	10.60 431	9.98 697	3	2	9	3.6	2.7	1.8
59	9.38 317	51	9.39 623	54	10.60 377	9.98 694	4	1			
60	9.38 368		9.39 677		10.60 323	9.98 690		0			
	L. Cos.	d.	L. Cot.	c. d.	L. Tan.	L. Sin.	d.	′	P. P.		

′	L. Sin.	d.	L. Tan.	c. d.	L. Cot.	L. Cos.	d.		P. P.
0	9.38 368	50	9.39 677	54	10.60 323	9.98 690	3	60	
1	9.38 418	51	9.39 731	54	10.60 269	9.98 687	3	59	
2	9.38 469	50	9.39 785	53	10.60 215	9.98 684	3	58	
3	9.38 519	51	9.39 838	54	10.60 162	9.98 681	3	57	
4	9.38 570	50	9.39 892	53	10.60 108	9.98 678	3	56	**54** **53**
5	9.38 620	50	9.39 945	54	10.60 055	9.98 675	4	55	1 5.4 5.3
6	9.38 670	51	9.39 999	53	10.60 001	9.98 671	3	54	2 10.8 10.6
7	9.38 721	50	9.40 052	54	10.59 948	9.98 668	3	53	3 16.2 15.9
8	9.38 771	50	9.40 106	53	10.59 894	9.98 665	3	52	4 21.6 21.2
9	9.38 821	50	9.40 159	53	10.59 841	9.98 662	3	51	5 27.0 26.5
10	9.38 871	50	9.40 212	54	10.59 788	9.98 659	3	50	6 32.4 31.8 7 37.8 37.1
11	9.38 921	50	9.40 266	53	10.59 734	9.98 656	4	49	8 43.2 42.4
12	9.38 971	50	9.40 319	53	10.59 681	9.98 652	3	48	9 48.6 47.7
13	9.39 021	50	9.40 372	53	10.59 628	9.98 649	3	47	
14	9.39 071	50	9.40 425	53	10.59 575	9.98 646	3	46	
15	9.39 121	49	9.40 478	53	10.59 522	9.98 643	3	45	
16	9.39 170	50	9.40 531	53	10.59 469	9.98 640	4	44	
17	9.39 220	50	9.40 584	52	10.59 416	9.98 636	3	43	
18	9.39 270	49	9.40 636	53	10.59 364	9.98 633	3	42	
19	9.39 319	50	9.40 689	53	10.59 311	9.98 630	3	41	**52** **51** **50**
20	9.39 369	49	9.40 742	53	10.59 258	9.98 627	4	40	1 5.2 5.1 5.0
21	9.39 418	49	9.40 795	52	10.59 205	9.98 623	3	39	2 10.4 10.2 10.0
22	9.39 467	50	9.40 847	53	10.59 153	9.98 620	3	38	3 15.6 15.3 15.0
23	9.39 517	49	9.40 900	52	10.59 100	9.98 617	3	37	4 20.8 20.4 20.0
24	9.39 566	49	9.40 952	53	10.59 048	9.98 614	4	36	5 26.0 25.5 25.0
25	9.39 615	49	9.41 005	52	10.58 995	9.98 610	3	35	6 31.2 30.6 30.0
26	9.39 664	49	9.41 057	52	10.58 943	9.98 607	3	34	7 36.4 35.7 35.0
27	9.39 713	49	9.41 109	52	10.58 891	9.98 604	3	33	8 41.6 40.8 40.0
28	9.39 762	49	9.41 161	53	10.58 839	9.98 601	4	32	9 46.8 45.9 45.0
29	9.39 811	49	9.41 214	52	10.58 786	9.98 597	3	31	
30	9.39 860	49	9.41 266	52	10.58 734	9.98 594	3	30	
31	9.39 909	49	9.41 318	52	10.58 682	9.98 591	3	29	
32	9.39 958	48	9.41 370	52	10.58 630	9.98 588	4	28	
33	9.40 006	49	9.41 422	52	10.58 578	9.98 584	3	27	
34	9.40 055	48	9.41 474	52	10.58 526	9.98 581	3	26	**49** **48** **47**
35	9.40 103	49	9.41 526	52	10.58 474	9.98 578	4	25	1 4.9 4.8 4.7
36	9.40 152	48	9.41 578	51	10.58 422	9.98 574	3	24	2 9.8 9.6 9.4
37	9.40 200	49	9.41 629	52	10.58 371	9.98 571	3	23	3 14.7 14.4 14.1
38	9.40 249	48	9.41 681	52	10.58 319	9.98 568	3	22	4 19.6 19.2 18.8
39	9.40 297	49	9.41 733	51	10.58 267	9.98 565	4	21	5 24.5 24.0 23.5
40	9.40 346	48	9.41 784	52	10.58 216	9.98 561	3	20	6 29.4 28.8 28.2
41	9.40 394	48	9.41 836	51	10.58 164	9.98 558	3	19	7 34.3 33.6 32.9
42	9.40 442	48	9.41 887	52	10.58 113	9.98 555	4	18	8 39.2 38.4 37.6
43	9.40 490	48	9.41 939	51	10.58 061	9.98 551	3	17	9 44.1 43.2 42.3
44	9.40 538	48	9.41 990	51	10.58 010	9.98 548	3	16	
45	9.40 586	48	9.42 041	52	10.57 959	9.98 545	4	15	
46	9.40 634	48	9.42 093	51	10.57 907	9.98 541	3	14	
47	9.40 682	48	9.42 144	51	10.57 856	9.98 538	3	13	
48	9.40 730	48	9.42 195	51	10.57 805	9.98 535	4	12	**4** **3**
49	9.40 778	47	9.42 246	51	10.57 754	9.98 531	3	11	
50	9.40 825	48	9.42 297	51	10.57 703	9.98 528	3	10	1 0.4 0.3
51	9.40 873	48	9.42 348	51	10.57 652	9.98 525	4	9	2 0.8 0.6
52	9.40 921	47	9.42 399	51	10.57 601	9.98 521	3	8	3 1.2 0.9
53	9.40 968	48	9.42 450	51	10.57 550	9.98 518	3	7	4 1.6 1.2
54	9.41 016	47	9.42 501	51	10.57 499	9.98 515	4	6	5 2.0 1.5
55	9.41 063	48	9.42 552	51	10.57 448	9.98 511	3	5	6 2.4 1.8 7 2.8 2.1
56	9.41 111	47	9.42 603	50	10.57 397	9.98 508	3	4	8 3.2 2.4
57	9.41 158	47	9.42 653	51	10.57 347	9.98 505	4	3	9 3.6 2.7
58	9.41 205	47	9.42 704	51	10.57 296	9.98 501	3	2	
59	9.41 252	48	9.42 755	50	10.57 245	9.98 498	4	1	
60	9.41 300		9.42 805		10.57 195	9.98 494		0	
	L. Cos.	d.	L. Cot.	c. d.	L. Tan.	L. Sin.	d.	′	P. P.

'	L. Sin.	d.	L. Tan.	c. d.	L. Cot.	L. Cos.	d.		P. P.
0	9.41 300	47	9.42 805	51	10.57 195	9.98 494	3	60	
1	9.41 347	47	9.42 856	50	10.57 144	9.98 491	3	59	
2	9.41 394	47	9.42 906	51	10.57 094	9.98 488	4	58	
3	9.41 441	47	9.42 957	50	10.57 043	9.98 484	3	57	
4	9.41 488	47	9.43 007	50	10.56 993	9.98 481	4	56	51 50 49
5	9.41 535	47	9.43 057	51	10.56 943	9.98 477	3	55	1 5.1 5.0 4.9
6	9.41 582	46	9.43 108	50	10.56 892	9.98 474	3	54	2 10.2 10.0 9.8
7	9.41 628	47	9.43 158	50	10.56 842	9.98 471	4	53	3 15.3 15.0 14.7
8	9.41 675	47	9.43 208	50	10.56 792	9.98 467	3	52	4 20.4 20.0 19.6
9	9.41 722	46	9.43 258	50	10.56 742	9.98 464	4	51	5 25.5 25.0 24.5
10	9.41 768	47	9.43 308	50	10.56 692	9.98 460	3	50	6 30.6 30.0 29.4
11	9.41 815	46	9.43 358	50	10.56 642	9.98 457	4	49	7 35.7 35.0 34.3
12	9.41 861	47	9.43 408	50	10.56 592	9.98 453	3	48	8 40.8 40.0 39.2
13	9.41 908	46	9.43 458	50	10.56 542	9.98 450	3	47	9 45.9 45.0 44.1
14	9.41 954	47	9.43 508	50	10.56 492	9.98 447	4	46	
15	9.42 001	46	9.43 558	49	10.56 442	9.98 443	3	45	
16	9.42 047	46	9.43 607	50	10.56 393	9.98 440	4	44	
17	9.42 093	47	9.43 657	50	10.56 343	9.98 436	3	43	
18	9.42 140	46	9.43 707	49	10.56 293	9.98 433	4	42	48 47 46
19	9.42 186	46	9.43 756	50	10.56 244	9.98 429	3	41	1 4.8 4.7 4.6
20	9.42 232	46	9.43 806	49	10.56 194	9.98 426	4	40	2 9.6 9.4 9.2
21	9.42 278	46	9.43 855	50	10.56 145	9.98 422	3	39	3 14.4 14.1 13.8
22	9.42 324	46	9.43 905	49	10.56 095	9.98 419	4	38	4 19.2 18.8 18.4
23	9.42 370	46	9.43 954	50	10.56 046	9.98 415	3	37	5 24.0 23.5 23.0
24	9.42 416	45	9.44 004	49	10.55 996	9.98 412	3	36	6 28.8 28.2 27.6
25	9.42 461	46	9.44 053	49	10.55 947	9.98 409	4	35	7 33.6 32.9 32.3
26	9.42 507	46	9.44 102	49	10.55 898	9.98 405	3	34	8 38.4 37.6 36.8
27	9.42 553	46	9.44 151	50	10.55 849	9.98 402	4	33	9 43.2 42.3 41.4
28	9.42 599	45	9.44 201	49	10.55 799	9.98 398	3	32	
29	9.42 644	46	9.44 250	49	10.55 750	9.98 395	4	31	
30	9.42 690	45	9.44 299	49	10.55 701	9.98 391	3	30	
31	9.42 735	46	9.44 348	49	10.55 652	9.98 388	4	29	
32	9.42 781	45	9.44 397	49	10.55 603	9.98 384	3	28	
33	9.42 826	46	9.44 446	49	10.55 554	9.98 381	4	27	45 44
34	9.42 872	45	9.44 495	49	10.55 505	9.98 377	4	26	
35	9.42 917	45	9.44 544	48	10.55 456	9.98 373	3	25	1 4.5 4.4
36	9.42 962	46	9.44 592	49	10.55 408	9.98 370	4	24	2 9.0 8.8
37	9.43 008	45	9.44 641	49	10.55 359	9.98 366	3	23	3 13.5 13.2
38	9.43 053	45	9.44 690	48	10.55 310	9.98 363	4	22	4 18.0 17.6
39	9.43 098	45	9.44 738	49	10.55 262	9.98 359	3	21	5 22.5 22.0
40	9.43 143	45	9.44 787	49	10.55 213	9.98 356	4	20	6 27.0 26.4
41	9.43 188	45	9.44 836	48	10.55 164	9.98 352	3	19	7 31.5 30.8
42	9.43 233	45	9.44 884	49	10.55 116	9.98 349	4	18	8 36.0 35.2
43	9.43 278	45	9.44 933	48	10.55 067	9.98 345	3	17	9 40.5 39.6
44	9.43 323	44	9.44 981	48	10.55 019	9.98 342	4	16	
45	9.43 367	45	9.45 029	49	10.54 971	9.98 338	4	15	
46	9.43 412	45	9.45 078	48	10.54 922	9.98 334	3	14	
47	9.43 457	45	9.45 126	48	10.54 874	9.98 331	4	13	
48	9.43 502	44	9.45 174	48	10.54 826	9.98 327	3	12	4 3
49	9.43 546	45	9.45 222	49	10.54 778	9.98 324	4	11	
50	9.43 591	44	9.45 271	48	10.54 729	9.98 320	3	10	1 0.4 0.3
51	9.43 635	45	9.45 319	48	10.54 681	9.98 317	4	9	2 0.8 0.6
52	9.43 680	44	9.45 367	48	10.54 633	9.98 313	4	8	3 1.2 0.9
53	9.43 724	45	9.45 415	48	10.54 585	9.98 309	3	7	4 1.6 1.2
54	9.43 769	44	9.45 463	48	10.54 537	9.98 306	4	6	5 2.0 1.5
55	9.43 813	44	9.45 511	48	10.54 489	9.98 302	3	5	6 2.4 1.8
56	9.43 857	44	9.45 559	47	10.54 441	9.98 299	4	4	7 2.8 2.1
57	9.43 901	45	9.45 606	48	10.54 394	9.98 295	4	3	8 3.2 2.4
58	9.43 946	44	9.45 654	48	10.54 346	9.98 291	3	2	9 3.6 2.7
59	9.43 990	44	9.45 702	48	10.54 298	9.98 288	4	1	
60	9.44 034		9.45 750		10.54 250	9.98 284		0	
	L. Cos.	d.	L. Cot.	c. d.	L. Tan.	L. Sin.	d.	'	P. P.

'	L. Sin.	d.	L. Tan.	c. d.	L. Cot.	L. Cos.	d.	
0	9.44 034	44	9.45 750	47	10.54 250	9.98 284	3	60
1	9.44 078	44	9.45 797	48	10.54 203	9.98 281	4	59
2	9.44 122	44	9.45 845	47	10.54 155	9.98 277	4	58
3	9.44 166	44	9.45 892	48	10.54 108	9.98 273	3	57
4	9.44 210	43	9.45 940	47	10.54 060	9.98 270	4	56
5	9.44 253	44	9.45 987	48	10.54 013	9.98 266	4	55
6	9.44 297	44	9.46 035	47	10.53 965	9.98 262	3	54
7	9.44 341	44	9.46 082	48	10.53 918	9.98 259	4	53
8	9.44 385	43	9.46 130	47	10.53 870	9.98 255	4	52
9	9.44 428	44	9.46 177	47	10.53 823	9.98 251	3	51
10	9.44 472	44	9.46 224	47	10.53 776	9.98 248	4	50
11	9.44 516	43	9.46 271	48	10.53 729	9.98 244	4	49
12	9.44 559	43	9.46 319	47	10.53 681	9.98 240	3	48
13	9.44 602	44	9.46 366	47	10.53 634	9.98 237	4	47
14	9.44 646	43	9.46 413	47	10.53 587	9.98 233	4	46
15	9.44 689	44	9.46 460	47	10.53 540	9.98 229	3	45
16	9.44 733	43	9.46 507	47	10.53 493	9.98 226	4	44
17	9.44 776	43	9.46 554	47	10.53 446	9.98 222	4	43
18	9.44 819	43	9.46 601	47	10.53 399	9.98 218	3	42
19	9.44 862	43	9.46 648	46	10.53 352	9.98 215	4	41
20	9.44 905	43	9.46 694	47	10.53 306	9.98 211	4	40
21	9.44 948	44	9.46 741	47	10.53 259	9.98 207	3	39
22	9.44 992	43	9.46 788	47	10.53 212	9.98 204	4	38
23	9.45 035	42	9.46 835	46	10.53 165	9.98 200	4	37
24	9.45 077	43	9.46 881	47	10.53 119	9.98 196	4	36
25	9.45 120	43	9.46 928	47	10.53 072	9.98 192	3	35
26	9.45 163	43	9.46 975	46	10.53 025	9.98 189	4	34
27	9.45 206	43	9.47 021	47	10.52 979	9.98 185	4	33
28	9.45 249	43	9.47 068	46	10.52 932	9.98 181	4	32
29	9.45 292	42	9.47 114	46	10.52 886	9.98 177	3	31
30	9.45 334	43	9.47 160	47	10.52 840	9.98 174	4	30
31	9.45 377	42	9.47 207	46	10.52 793	9.98 170	4	29
32	9.45 419	43	9.47 253	46	10.52 747	9.98 166	4	28
33	9.45 462	42	9.47 299	47	10.52 701	9.98 162	3	27
34	9.45 504	43	9.47 346	46	10.52 654	9.98 159	4	26
35	9.45 547	42	9.47 392	46	10.52 608	9.98 155	4	25
36	9.45 589	43	9.47 438	46	10.52 562	9.98 151	4	24
37	9.45 632	42	9.47 484	46	10.52 516	9.98 147	3	23
38	9.45 674	42	9.47 530	46	10.52 470	9.98 144	4	22
39	9.45 716	42	9.47 576	46	10.52 424	9.98 140	4	21
40	9.45 758	43	9.47 622	46	10.52 378	9.98 136	4	20
41	9.45 801	42	9.47 668	46	10.52 332	9.98 132	3	19
42	9.45 843	42	9.47 714	46	10.52 286	9.98 129	4	18
43	9.45 885	42	9.47 760	46	10.52 240	9.98 125	4	17
44	9.45 927	42	9.47 806	46	10.52 194	9.98 121	4	16
45	9.45 969	42	9.47 852	45	10.52 148	9.98 117	4	15
46	9.46 011	42	9.47 897	46	10.52 103	9.98 113	3	14
47	9.46 053	42	9.47 943	46	10.52 057	9.98 110	4	13
48	9.46 095	41	9.47 989	46	10.52 011	9.98 106	4	12
49	9.46 136	42	9.48 035	45	10.51 965	9.98 102	4	11
50	9.46 178	42	9.48 080	46	10.51 920	9.98 098	4	10
51	9.46 220	42	9.48 126	45	10.51 874	9.98 094	4	9
52	9.46 262	41	9.48 171	46	10.51 829	9.98 090	3	8
53	9.46 303	42	9.48 217	45	10.51 783	9.98 087	4	7
54	9.46 345	41	9.48 262	45	10.51 738	9.98 083	4	6
55	9.46 386	42	9.48 307	46	10.51 693	9.98 079	4	5
56	9.46 428	41	9.48 353	45	10.51 647	9.98 075	4	4
57	9.46 469	42	9.48 398	45	10.51 602	9.98 071	4	3
58	9.46 511	41	9.48 443	46	10.51 557	9.98 067	4	2
59	9.46 552	42	9.48 489	45	10.51 511	9.98 063	3	1
60	9.46 594		9.48 534		10.51 466	9.98 060		0
	L. Cos.	d.	L. Cot.	c. d.	L. Tan.	L. Sin.	d.	'

P. P.

	48	47	46
1	4.8	4.7	4.6
2	9.6	9.4	9.2
3	14.4	14.1	13.8
4	19.2	18.8	18.4
5	24.0	23.5	23.0
6	28.8	28.2	27.6
7	33.6	32.9	32.2
8	38.4	37.6	36.8
9	43.2	42.3	41.4

	45	44	43
1	4.5	4.4	4.3
2	9.0	8.8	8.6
3	13.5	13.2	12.9
4	18.0	17.6	17.2
5	22.5	22.0	21.5
6	27.0	26.4	25.8
7	31.5	30.8	30.1
8	36.0	35.2	34.4
9	40.5	39.6	38.7

	42	41
1	4.2	4.1
2	8.4	8.2
3	12.6	12.3
4	16.8	16.4
5	21.0	20.5
6	25.2	24.6
7	29.4	28.7
8	33.6	32.8
9	37.8	36.9

	4	3
1	0.4	0.3
2	0.8	0.6
3	1.2	0.9
4	1.6	1.2
5	2.0	1.5
6	2.4	1.8
7	2.8	2.1
8	3.2	2.4
9	3.6	2.7

P. P.

73°

′	L. Sin.	d.	L. Tan.	c. d.	L. Cot.	L. Cos.	d.		P. P.
0	9.46 594		9.48 534		10.51 466	9.98 060		60	
1	9.46 635	41	9.48 579	45	10.51 421	9.98 056	4	59	
2	9.46 676	41	9.48 624	45	10.51 376	9.98 052	4	58	
3	9.46 717	41	9.48 669	45	10.51 331	9.98 048	4	57	
4	9.46 758	41	9.48 714	45	10.51 286	9.98 044	4	56	
		42		45			4		
5	9.46 800		9.48 759		10.51 241	9.98 040		55	
6	9.46 841	41	9.48 804	45	10.51 196	9.98 036	4	54	
7	9.46 882	41	9.48 849	45	10.51 151	9.98 032	4	53	
8	9.46 923	41	9.48 894	45	10.51 106	9.98 029	3	52	
9	9.46 964	41	9.48 939	45	10.51 061	9.98 025	4	51	**45** **44** **43**
		41		45			4		
10	9.47 005		9.48 984		10.51 016	9.98 021		50	1| 4.5 4.4 4.3
11	9.47 045	40	9.49 029	45	10.50 971	9.98 017	4	49	2| 9.0 8.8 8.6
12	9.47 086	41	9.49 073	44	10.50 927	9.98 013	4	48	3| 13.5 13.2 12.9
13	9.47 127	41	9.49 118	45	10.50 882	9.98 009	4	47	4| 18.0 17.6 17.2
14	9.47 168	41	9.49 163	45	10.50 837	9.98 005	4	46	5| 22.5 22.0 21.5
		41		44			4		6| 27.0 26.4 25.8
15	9.47 209		9.49 207		10.50 793	9.98 001		45	7| 31.5 30.8 30.1
16	9.47 249	40	9.49 252	45	10.50 748	9.97 997	4	44	8| 36.0 35.2 34.4
17	9.47 290	41	9.49 296	44	10.50 704	9.97 993	4	43	9| 40.5 39.6 38.7
18	9.47 330	40	9.49 341	45	10.50 659	9.97 989	4	42	
19	9.47 371	41	9.49 385	44	10.50 615	9.97 986	3	41	
		40		45			4		
20	9.47 411		9.49 430		10.50 570	9.97 982		40	
21	9.47 452	41	9.49 474	44	10.50 526	9.97 978	4	39	
22	9.47 492	40	9.49 519	45	10.50 481	9.97 974	4	38	
23	9.47 533	41	9.49 563	44	10.50 437	9.97 970	4	37	
24	9.47 573	40	9.49 607	44	10.50 393	9.97 966	4	36	
		40		45			4		
25	9.47 613		9.49 652		10.50 348	9.97 962		35	
26	9.47 654	41	9.49 696	44	10.50 304	9.97 958	4	34	**42** **41** **40**
27	9.47 694	40	9.49 740	44	10.50 260	9.97 954	4	33	
28	9.47 734	40	9.49 784	44	10.50 216	9.97 950	4	32	1| 4.2 4.1 4.0
29	9.47 774	40	9.49 828	44	10.50 172	9.97 946	4	31	2| 8.4 8.2 8.0
		40		44			4		3| 12.6 12.3 12.0
30	9.47 814		9.49 872		10.50 128	9.97 942		30	4| 16.8 16.4 16.0
31	9.47 854	40	9.49 916	44	10.50 084	9.97 938	4	29	5| 21.0 20.5 20.0
32	9.47 894	40	9.49 960	44	10.50 040	9.97 934	4	28	6| 25.2 24.6 24.0
33	9.47 934	40	9.50 004	44	10.49 996	9.97 930	4	27	7| 29.4 28.7 28.0
34	9.47 974	40	9.50 048	44	10.49 952	9.97 926	4	26	8| 33.6 32.8 32.0
		40		44			4		9| 37.8 36.9 36.0
35	9.48 014		9.50 092		10.49 908	9.97 922		25	
36	9.48 054	40	9.50 136	44	10.49 864	9.97 918	4	24	
37	9.48 094	40	9.50 180	44	10.49 820	9.97 914	4	23	
38	9.48 133	39	9.50 223	43	10.49 777	9.97 910	4	22	
39	9.48 173	40	9.50 267	44	10.49 733	9.97 906	4	21	
		40		44			4		
40	9.48 213		9.50 311		10.49 689	9.97 902		20	
41	9.48 252	39	9.50 355	44	10.49 645	9.97 898	4	19	
42	9.48 292	40	9.50 398	43	10.49 602	9.97 894	4	18	
43	9.48 332	40	9.50 442	44	10.49 558	9.97 890	4	17	**39** **5** **4** **3**
44	9.48 371	39	9.50 485	43	10.49 515	9.97 886	4	16	
		40		44			4		1| 3.9 0.5 0.4 0.3
45	9.48 411		9.50 529		10.49 471	9.97 882		15	2| 7.8 1.0 0.8 0.6
46	9.48 450	39	9.50 572	43	10.49 428	9.97 878	4	14	3| 11.7 1.5 1.2 0.9
47	9.48 490	40	9.50 616	44	10.49 384	9.97 874	4	13	4| 15.6 2.0 1.6 1.2
48	9.48 529	39	9.50 659	43	10.49 341	9.97 870	4	12	5| 19.5 2.5 2.0 1.5
49	9.48 568	39	9.50 703	44	10.49 297	9.97 866	5	11	6| 23.4 3.0 2.4 1.8
		39		43					7| 27.3 3.5 2.8 2.1
50	9.48 607		9.50 746		10.49 254	9.97 861		10	8| 31.2 4.0 3.2 2.4
51	9.48 647	40	9.50 789	43	10.49 211	9.97 857	4	9	9| 35.1 4.5 3.6 2.7
52	9.48 686	39	9.50 833	44	10.49 167	9.97 853	4	8	
53	9.48 725	39	9.50 876	43	10.49 124	9.97 849	4	7	
54	9.48 764	39	9.50 919	43	10.49 081	9.97 845	4	6	
		39		43			4		
55	9.48 803		9.50 962		10.49 038	9.97 841		5	
56	9.48 842	39	9.51 005	.43	10.48 995	9.97 837	4	4	
57	9.48 881	39	9.51 048	43	10.48 952	9.97 833	4	3	
58	9.48 920	39	9.51 092	44	10.48 908	9.97 829	4	2	
59	9.48 959	39	9.51 135	43	10.48 865	9.97 825	4	1	
		39		43			4		
60	9.48 998		9.51 178		10.48 822	9.97 821		0	
	L. Cos.	d.	L. Cot.	c. d.	L. Tan.	L. Sin.	d.	′	P. P.

′	L. Sin.	d.	L. Tan.	c. d.	L. Cot.	L. Cos.	d.		P. P.		
0	9.48 998		9.51 178		10.48 822	9.97 821		60			
1	9.49 037	39	9.51 221	43	10.48 779	9.97 817	4	59			
2	9.49 076	39	9.51 264	43	10.48 736	9.97 812	5	58			
3	9.49 115	39	9.51 306	42	10.48 694	9.97 808	4	57			
4	9.49 153	38	9.51 349	43	10.48 651	9.97 804	4	56			
5	9.49 192	39	9.51 392	43	10.48 608	9.97 800	4	55			
6	9.49 231	39	9.51 435	43	10.48 565	9.97 796	4	54			
7	9.49 269	38	9.51 478	43	10.48 522	9.97 792	4	53			
8	9.49 308	39	9.51 520	42	10.48 480	9.97 788	4	52			
9	9.49 347	39	9.51 563	43	10.48 437	9.97 784	4	51		43	42	41
10	9.49 385	38	9.51 606	43	10.48 394	9.97 779	5	50			
11	9.49 424	39	9.51 648	42	10.48 352	9.97 775	4	49	**1**	4.3	4.2	4.1
12	9.49 462	38	9.51 691	43	10.48 309	9.97 771	4	48	**2**	8.6	8.4	8.2
13	9.49 500	38	9.51 734	43	10.48 266	9.97 767	4	47	**3**	12.9	12.6	12.3
14	9.49 539	39	9.51 776	42	10.48 224	9.97 763	4	46	**4**	17.2	16.8	16.4
15	9.49 577	38	9.51 819	43	10.48 181	9.97 759	4	45	**5**	21.5	21.0	20.5
16	9.49 615	38	9.51 861	42	10.48 139	9.97 754	5	44	**6**	25.8	25.2	24.6
17	9.49 654	39	9.51 903	42	10.48 097	9.97 750	4	43	**7**	30.1	29.4	28.7
18	9.49 692	38	9.51 946	43	10.48 054	9.97 746	4	42	**8**	34.4	33.6	32.8
19	9.49 730	38	9.51 988	42	10.48 012	9.97 742	4	41	**9**	38.7	37.8	36.9
20	9.49 768	38	9.52 031	43	10.47 969	9.97 738	4	40			
21	9.49 806	38	9.52 073	42	10.47 927	9.97 734	4	39			
22	9.49 844	38	9.52 115	42	10.47 885	9.97 729	5	38			
23	9.49 882	38	9.52 157	42	10.47 843	9.97 725	4	37			
24	9.49 920	38	9.52 200	43	10.47 800	9.97 721	4	36			
25	9.49 958	38	9.52 242	42	10.47 758	9.97 717	4	35			
26	9.49 996	38	9.52 284	42	10.47 716	9.97 713	4	34		39	38	37
27	9.50 034	38	9.52 326	42	10.47 674	9.97 708	5	33			
28	9.50 072	38	9.52 368	42	10.47 632	9.97 704	4	32	**1**	3.9	3.8	3.7
29	9.50 110	38	9.52 410	42	10.47 590	9.97 700	4	31	**2**	7.8	7.6	7.4
30	9.50 148	38	9.52 452	42	10.47 548	9.97 696	4	30	**3**	11.7	11.4	11.1
31	9.50 185	37	9.52 494	42	10.47 506	9.97 691	5	29	**4**	15.6	15.2	14.8
32	9.50 223	38	9.52 536	42	10.47 464	9.97 687	4	28	**5**	19.5	19.0	18.5
33	9.50 261	38	9.52 578	42	10.47 422	9.97 683	4	27	**6**	23.4	22.8	22.2
34	9.50 298	37	9.52 620	41	10.47 380	9.97 679	4	26	**7**	27.3	26.6	25.9
35	9.50 336	38	9.52 661	42	10.47 339	9.97 674	5	25	**8**	31.2	30.4	29.6
36	9.50 374	38	9.52 703	42	10.47 297	9.97 670	4	24	**9**	35.1	34.2	33.3
37	9.50 411	37	9.52 745	42	10.47 255	9.97 666	4	23			
38	9.50 449	38	9.52 787	42	10.47 213	9.97 662	4	22			
39	9.50 486	37	9.52 829	41	10.47 171	9.97 657	5	21			
40	9.50 523	37	9.52 870	42	10.47 130	9.97 653	4	20			
41	9.50 561	38	9.52 912	41	10.47 088	9.97 649	4	19			
42	9.50 598	37	9.52 953	42	10.47 047	9.97 645	4	18			
43	9.50 635	37	9.52 995	42	10.47 005	9.97 640	5	17			
44	9.50 673	38	9.53 037	41	10.46 963	9.97 636	4	16		36	5	4
45	9.50 710	37	9.53 078	42	10.46 922	9.97 632	4	15			
46	9.50 747	37	9.53 120	41	10.46 880	9.97 628	4	14	**1**	3.6	0.5	0.4
47	9.50 784	37	9.53 161	41	10.46 839	9.97 623	5	13	**2**	7.2	1.0	0.8
48	9.50 821	37	9.53 202	42	10.46 798	9.97 619	4	12	**3**	10.8	1.5	1.2
49	9.50 858	37	9.53 244	41	10.46 756	9.97 615	4	11	**4**	14.4	2.0	1.6
50	9.50 896	38	9.53 285	42	10.46 715	9.97 610	5	10	**5**	18.0	2.5	2.0
51	9.50 933	37	9.53 327	41	10.46 673	9.97 606	4	9	**6**	21.6	3.0	2.4
52	9.50 970	37	9.53 368	41	10.46 632	9.97 602	4	8	**7**	25.2	3.5	2.8
53	9.51 007	37	9.53 409	41	10.46 591	9.97 597	5	7	**8**	28.8	4.0	3.2
54	9.51 043	36	9.53 450	42	10.46 550	9.97 593	4	6	**9**	32.4	4.5	3.6
55	9.51 080	37	9.53 492	41	10.46 508	9.97 589	4	5			
56	9.51 117	37	9.53 533	41	10.46 467	9.97 584	5	4			
57	9.51 154	37	9.53 574	41	10.46 426	9.97 580	4	3			
58	9.51 191	37	9.53 615	41	10.46 385	9.97 576	4	2			
59	9.51 227	36	9.53 656	41	10.46 344	9.97 571	5	1			
60	9.51 264	37	9.53 697	41	10.46 303	9.97 567	4	0			
	L. Cos.	d.	L. Cot.	c. d.	L. Tan.	L. Sin.	d.	′		P. P.		

′	L. Sin.	d.	L. Tan.	c. d.	L. Cot.	L. Cos.	d.	
0	9.51 264	37	9.53 697	41	10.46 303	9.97 567	4	60
1	9.51 301	37	9.53 738	41	10.46 262	9.97 563	5	59
2	9.51 338	36	9.53 779	41	10.46 221	9.97 558	4	58
3	9.51 374	37	9.53 820	41	10.46 180	9.97 554	4	57
4	9.51 411	36	9.53 861	41	10.46 139	9.97 550	5	56
5	9.51 447	37	9.53 902	41	10.46 098	9.97 545	4	55
6	9.51 484	36	9.53 943	41	10.46 057	9.97 541	5	54
7	9.51 520	37	9.53 984	41	10.46 016	9.97 536	4	53
8	9.51 557	36	9.54 025	40	10.45 975	9.97 532	4	52
9	9.51 593	36	9.54 065	41	10.45 935	9.97 528	5	51
10	9.51 629	37	9.54 106	41	10.45 894	9.97 523	4	50
11	9.51 666	36	9.54 147	40	10.45 853	9.97 519	4	49
12	9.51 702	36	9.54 187	41	10.45 813	9.97 515	5	48
13	9.51 738	36	9.54 228	41	10.45 772	9.97 510	4	47
14	9.51 774	37	9.54 269	40	10.45 731	9.97 506	5	46
15	9.51 811	36	9.54 309	41	10.45 691	9.97 501	4	45
16	9.51 847	36	9.54 350	40	10.45 650	9.97 497	5	44
17	9.51 883	36	9.54 390	41	10.45 610	9.97 492	4	43
18	9.51 919	36	9.54 431	40	10.45 569	9.97 488	4	42
19	9.51 955	36	9.54 471	41	10.45 529	9.97 484	5	41
20	9.51 991	36	9.54 512	40	10.45 488	9.97 479	4	40
21	9.52 027	36	9.54 552	41	10.45 448	9.97 475	5	39
22	9.52 063	36	9.54 593	40	10.45 407	9.97 470	4	38
23	9.52 099	36	9.54 633	40	10.45 367	9.97 466	5	37
24	9.52 135	36	9.54 673	41	10.45 327	9.97 461	4	36
25	9.52 171	36	9.54 714	40	10.45 286	9.97 457	4	35
26	9.52 207	35	9.54 754	40	10.45 246	9.97 453	5	34
27	9.52 242	36	9.54 794	41	10.45 206	9.97 448	4	33
28	9.52 278	36	9.54 835	40	10.45 165	9.97 444	5	32
29	9.52 314	36	9.54 875	40	10.45 125	9.97 439	4	31
30	9.52 350	35	9.54 915	40	10.45 085	9.97 435	5	30
31	9.52 385	36	9.54 955	40	10.45 045	9.97 430	4	29
32	9.52 421	35	9.54 995	40	10.45 005	9.97 426	5	28
33	9.52 456	36	9.55 035	40	10.44 965	9.97 421	4	27
34	9.52 492	35	9.55 075	40	10.44 925	9.97 417	5	26
35	9.52 527	36	9.55 115	40	10.44 885	9.97 412	4	25
36	9.52 563	35	9.55 155	40	10.44 845	9.97 408	5	24
37	9.52 598	36	9.55 195	40	10.44 805	9.97 403	4	23
38	9.52 634	35	9.55 235	40	10.44 765	9.97 399	5	22
39	9.52 669	36	9.55 275	40	10.44 725	9.97 394	4	21
40	9.52 705	35	9.55 315	40	10.44 685	9.97 390	5	20
41	9.52 740	35	9.55 355	40	10.44 645	9.97 385	4	19
42	9.52 775	36	9.55 395	39	10.44 605	9.97 381	5	18
43	9.52 811	35	9.55 434	40	10.44 566	9.97 376	4	17
44	9.52 846	35	9.55 474	40	10.44 526	9.97 372	5	16
45	9.52 881	35	9.55 514	40	10.44 486	9.97 367	4	15
46	9.52 916	35	9.55 554	39	10.44 446	9.97 363	5	14
47	9.52 951	35	9.55 593	40	10.44 407	9.97 358	5	13
48	9.52 986	35	9.55 633	40	10.44 367	9.97 353	4	12
49	9.53 021	35	9.55 673	39	10.44 327	9.97 349	5	11
50	9.53 056	36	9.55 712	40	10.44 288	9.97 344	4	10
51	9.53 092	34	9.55 752	39	10.44 248	9.97 340	5	9
52	9.53 126	35	9.55 791	40	10.44 209	9.97 335	4	8
53	9.53 161	35	9.55 831	39	10.44 169	9.97 331	5	7
54	9.53 196	35	9.55 870	40	10.44 130	9.97 326	4	6
55	9.53 231	35	9.55 910	39	10.44 090	9.97 322	5	5
56	9.53 266	35	9.55 949	40	10.44 051	9.97 317	5	4
57	9.53 301	35	9.55 989	39	10.44 011	9.97 312	5	3
58	9.53 336	34	9.56 028	39	10.43 972	9.97 308	5	2
59	9.53 370	35	9.56 067	40	10.43 933	9.97 303	5	1
60	9.53 405		9.56 107		10.43 893	9.97 299	4	0
	L. Cos.	d.	L. Cot.	c. d.	L. Tan.	L. Sin.	d.	′

P. P.

	41	40	39
1	4.1	4.0	3.9
2	8.2	8.0	7.8
3	12.3	12.0	11.7
4	16.4	16.0	15.6
5	20.5	20.0	19.5
6	24.6	24.0	23.4
7	28.7	28.0	27.3
8	32.8	32.0	31.2
9	36.9	36.0	35.1

	37	36	35
1	3.7	3.6	3.5
2	7.4	7.2	7.0
3	11.1	10.8	10.5
4	14.8	14.4	14.0
5	18.5	18.0	17.5
6	22.2	21.6	21.0
7	25.9	25.2	24.5
8	29.6	28.8	28.0
9	33.3	32.4	31.5

	34	5	4
1	3.4	0.5	0.4
2	6.8	1.0	0.8
3	10.2	1.5	1.2
4	13.6	2.0	1.6
5	17.0	2.5	2.0
6	20.4	3.0	2.4
7	23.8	3.5	2.8
8	27.2	4.0	3.2
9	30.6	4.5	3.6

′	L. Sin.	d.	L. Tan.	c. d.	L. Cot.	L. Cos.	d.				P. P.	
0	9.53 405	35	9.56 107	39	10.43 893	9.97 299	5	60				
1	9.53 440	35	9.56 146	39	10.43 854	9.97 294	5	59				
2	9.53 475	34	9.56 185	39	10.43 815	9.97 289	4	58				
3	9.53 509	35	9.56 224	40	10.43 776	9.97 285	5	57				
4	9.53 544	34	9.56 264	39	10.43 736	9.97 280	4	56				
5	9.53 578	35	9.56 303	39	10.43 697	9.97 276	5	55				
6	9.53 613	34	9.56 342	39	10.43 658	9.97 271	5	54				
7	9.53 647	35	9.56 381	39	10.43 619	9.97 266	4	53				
8	9.53 682	34	9.56 420	39	10.43 580	9.97 262	5	52				
9	9.53 716	35	9.56 459	39	10.43 541	9.97 257	5	51		40	39	38
10	9.53 751	34	9.56 498	39	10.43 502	9.97 252	4	50				
11	9.53 785	34	9.56 537	39	10.43 463	9.97 248	5	49	1	4.0	3.9	3.8
12	9.53 819	35	9.56 576	39	10.43 424	9.97 243	5	48	2	8.0	7.8	7.6
13	9.53 854	34	9.56 615	39	10.43 385	9.97 238	4	47	3	12.0	11.7	11.4
14	9.53 888	34	9.56 654	39	10.43 346	9.97 234	5	46	4	16.0	15.6	15.2
15	9.53 922	35	9.56 693	39	10.43 307	9.97 229	5	45	5	20.0	19.5	19.0
16	9.53 957	34	9.56 732	39	10.43 268	9.97 224	4	44	6	24.0	23.4	22.8
17	9.53 991	34	9.56 771	39	10.43 229	9.97 220	5	43	7	28.0	27.3	26.6
18	9.54 025	34	9.56 810	39	10.43 190	9.97 215	5	42	8	32.0	31.2	30.4
19	9.54 059	34	9.56 849	38	10.43 151	9.97 210	4	41	9	36.0	35.1	34.2
20	9.54 093	34	9.56 887	39	10.43 113	9.97 206	5	40				
21	9.54 127	34	9.56 926	39	10.43 074	9.97 201	5	39				
22	9.54 161	34	9.56 965	39	10.43 035	9.97 196	4	38				
23	9.54 195	34	9.57 004	38	10.42 996	9.97 192	5	37				
24	9.54 229	34	9.57 042	39	10.42 958	9.97 187	5	36				
25	9.54 263	34	9.57 081	39	10.42 919	9.97 182	4	35				
26	9.54 297	34	9.57 120	38	10.42 880	9.97 178	5	34		37	35	34
27	9.54 331	34	9.57 158	39	10.42 842	9.97 173	5	33				
28	9.54 365	34	9.57 197	38	10.42 803	9.97 168	5	32	1	3.7	3.5	3.4
29	9.54 399	34	9.57 235	39	10.42 765	9.97 163	4	31	2	7.4	7.0	6.8
30	9.54 433	33	9.57 274	38	10.42 726	9.97 159	5	30	3	11.1	10.5	10.2
31	9.54 466	34	9.57 312	39	10.42 688	9.97 154	5	29	4	14.8	14.0	13.6
32	9.54 500	34	9.57 351	38	10.42 649	9.97 149	4	28	5	18.5	17.5	17.0
33	9.54 534	33	9.57 389	39	10.42 611	9.97 145	5	27	6	22.2	21.0	20.4
34	9.54 567	34	9.57 428	38	10.42 572	9.97 140	5	26	7	25.9	24.5	23.8
35	9.54 601	34	9.57 466	38	10.42 534	9.97 135	5	25	8	29.6	28.0	27.2
36	9.54 635	33	9.57 504	39	10.42 496	9.97 130	4	24	9	33.3	31.5	30.6
37	9.54 668	34	9.57 543	38	10.42 457	9.97 126	5	23				
38	9.54 702	33	9.57 581	38	10.42 419	9.97 121	5	22				
39	9.54 735	34	9.57 619	39	10.42 381	9.97 116	5	21				
40	9.54 769	33	9.57 658	38	10.42 342	9.97 111	4	20				
41	9.54 802	34	9.57 696	38	10.42 304	9.97 107	5	19				
42	9.54 836	33	9.57 734	38	10.42 266	9.97 102	5	18				
43	9.54 869	34	9.57 772	38	10.42 228	9.97 097	5	17		33	5	4
44	9.54 903	33	9.57 810	39	10.42 190	9.97 092	5	16				
45	9.54 936	33	9.57 849	38	10.42 151	9.97 087	4	15	1	3.3	0.5	0.4
46	9.54 969	34	9.57 887	38	10.42 113	9.97 083	5	14	2	6.6	1.0	0.8
47	9.55 003	33	9.57 925	38	10.42 075	9.97 078	5	13	3	9.9	1.5	1.2
48	9.55 036	33	9.57 963	38	10.42 037	9.97 073	5	12	4	13.2	2.0	1.6
49	9.55 069	33	9.58 001	38	10.41 999	9.97 068	5	11	5	16.5	2.5	2.0
50	9.55 102	34	9.58 039	38	10.41 961	9.97 063	4	10	6	19.8	3.0	2.4
51	9.55 136	33	9.58 077	38	10.41 923	9.97 059	5	9	7	23.1	3.5	2.8
52	9.55 169	33	9.58 115	38	10.41 885	9.97 054	5	8	8	26.4	4.0	3.2
53	9.55 202	33	9.58 153	38	10.41 847	9.97 049	5	7	9	29.7	4.5	3.6
54	9.55 235	33	9.58 191	38	10.41 809	9.97 044	5	6				
55	9.55 268	33	9.58 229	38	10.41 771	9.97 039	4	5				
56	9.55 301	33	9.58 267	37	10.41 733	9.97 035	5	4				
57	9.55 334	33	9.58 304	38	10.41 696	9.97 030	5	3				
58	9.55 367	33	9.58 342	38	10.41 658	9.97 025	5	2				
59	9.55 400	33	9.58 380	38	10.41 620	9.97 020	5	1				
60	9.55 433		9.58 418		10.41 582	9.97 015		0				
	L. Cos.	d.	L. Cot.	c. d.	L. Tan.	L. Sin.	d.	′			P. P.	

′	L. Sin.	d.	L. Tan.	c. d.	L. Cot.	L. Cos.	d.		P. P.		
0	9.55 433	33	9.58 418	37	10.41 582	9.97 015	5	60			
1	9.55 466	33	9.58 455	38	10.41 545	9.97 010	5	59			
2	9.55 499	33	9.58 493	38	10.41 507	9.97 005	4	58			
3	9.55 532	32	9.58 531	38	10.41 469	9.97 001	5	57			
4	9.55 564	33	9.58 569	37	10.41 431	9.96 996	5	56			
5	9.55 597	33	9.58 606	38	10.41 394	9.96 991	5	55			
6	9.55 630	33	9.58 644	37	10.41 356	9.96 986	5	54			
7	9.55 663	32	9.58 681	38	10.41 319	9.96 981	5	53			
8	9.55 695	33	9.58 719	38	10.41 281	9.96 976	5	52			
9	9.55 728	33	9.58 757	37	10.41 243	9.96 971	5	51	**38**	**37**	**36**
10	9.55 761	32	9.58 794	38	10.41 206	9.96 966	4	50	1 3.8	3.7	3.6
11	9.55 793	33	9.58 832	37	10.41 168	9.96 962	5	49	2 7.6	7.4	7.2
12	9.55 826	32	9.58 869	38	10.41 131	9.96 957	5	48	3 11.4	11.1	10.8
13	9.55 858	33	9.58 907	37	10.41 093	9.96 952	5	47	4 15.2	14.8	14.4
14	9.55 891	32	9.58 944	37	10.41 056	9.96 947	5	46	5 19.0	18.5	18.0
15	9.55 923	33	9.58 981	38	10.41 019	9.96 942	5	45	6 22.8	22.2	21.6
16	9.55 956	32	9.59 019	37	10.40 981	9.96 937	5	44	7 26.6	25.9	25.2
17	9.55 988	33	9.59 056	38	10.40 944	9.96 932	5	43	8 30.4	29.6	28.8
18	9.56 021	32	9.59 094	37	10.40 906	9.96 927	5	42	9 34.2	33.3	32.4
19	9.56 053	32	9.59 131	37	10.40 869	9.96 922	5	41			
20	9.56 085	33	9.59 168	37	10.40 832	9.96 917	5	40			
21	9.56 118	32	9.59 205	38	10.40 795	9.96 912	5	39			
22	9.56 150	32	9.59 243	37	10.40 757	9.96 907	4	38			
23	9.56 182	33	9.59 280	37	10.40 720	9.96 903	5	37			
24	9.56 215	32	9.59 317	37	10.40 683	9.96 898	5	36			
25	9.56 247	32	9.59 354	37	10.40 646	9.96 893	5	35			
26	9.56 279	32	9.59 391	38	10.40 609	9.96 888	5	34	**33**	**32**	**31**
27	9.56 311	32	9.59 429	37	10.40 571	9.96 883	5	33	1 3.3	3.2	3.1
28	9.56 343	32	9.59 466	37	10.40 534	9.96 878	5	32	2 6.6	6.4	6.2
29	9.56 375	33	9.59 503	37	10.40 497	9.96 873	5	31	3 9.9	9.6	9.3
30	9.56 408	32	9.59 540	37	10.40 460	9.96 868	5	30	4 13.2	12.8	12.4
31	9.56 440	32	9.59 577	37	10.40 423	9.96 863	5	29	5 16.5	16.0	15.5
32	9.56 472	32	9.59 614	37	10.40 386	9.96 858	5	28	6 19.8	19.2	18.6
33	9.56 504	32	9.59 651	37	10.40 349	9.96 853	5	27	7 23.1	22.4	21.7
34	9.56 536	32	9.59 688	37	10.40 312	9.96 848	5	26	8 26.4	25.6	24.8
35	9.56 568	31	9.59 725	37	10.40 275	9.96 843	5	25	9 29.7	28.8	27.9
36	9.56 599	32	9.59 762	37	10.40 238	9.96 838	5	24			
37	9.56 631	32	9.59 799	36	10.40 201	9.96 833	5	23			
38	9.56 663	32	9.59 835	37	10.40 165	9.96 828	5	22			
39	9.56 695	32	9.59 872	37	10.40 128	9.96 823	5	21			
40	9.56 727	32	9.59 909	37	10.40 091	9.96 818	5	20			
41	9.56 759	31	9.59 946	37	10.40 054	9.96 813	5	19			
42	9.56 790	32	9.59 983	36	10.40 017	9.96 808	5	18			
43	9.56 822	32	9.60 019	37	10.39 981	9.96 803	5	17	**6**	**5**	**4**
44	9.56 854	32	9.60 056	37	10.39 944	9.96 798	5	16			
45	9.56 886	31	9.60 093	37	10.39 907	9.96 793	5	15	1 0.6	0.5	0.4
46	9.56 917	32	9.60 130	36	10.39 870	9.96 788	5	14	2 1.2	1.0	0.8
47	9.56 949	31	9.60 166	37	10.39 834	9.96 783	5	13	3 1.8	1.5	1.2
48	9.56 980	32	9.60 203	37	10.39 797	9.96 778	6	12	4 2.4	2.0	1.6
49	9.57 012	32	9.60 240	36	10.39 760	9.96 772	5	11	5 3.0	2.5	2.0
50	9.57 044	31	9.60 276	37	10.39 724	9.96 767	5	10	6 3.6	3.0	2.4
51	9.57 075	32	9.60 313	36	10.39 687	9.96 762	5	9	7 4.2	3.5	2.8
52	9.57 107	31	9.60 349	37	10.39 651	9.96 757	5	8	8 4.8	4.0	3.2
53	9.57 138	31	9.60 386	36	10.39 614	9.96 752	5	7	9 5.4	4.5	3.6
54	9.57 169	32	9.60 422	37	10.39 578	9.96 747	5	6			
55	9.57 201	31	9.60 459	36	10.39 541	9.96 742	5	5			
56	9.57 232	32	9.60 495	37	10.39 505	9.96 737	5	4			
57	9.57 264	31	9.60 532	36	10.39 468	9.96 732	5	3			
58	9.57 295	31	9.60 568	37	10.39 432	9.96 727	5	2			
59	9.57 326	32	9.60 605	36	10.39 395	9.96 722	5	1			
60	9.57 358		9.60 641		10.39 359	9.96 717		0			
	L. Cos.	d.	L. Cot.	c. d.	L. Tan.	L. Sin.	d.	′	P. P.		

′	L. Sin.	d.	L. Tan.	c. d.	L. Cot.	L. Cos.	d.		P. P.		
0	9.57 358	31	9.60 641	36	10.39 359	9.96 717	6	**60**			
1	9.57 389	31	9.60 677	37	10.39 323	9.96 711	5	59			
2	9.57 420	31	9.60 714	36	10.39 286	9.96 706	5	58			
3	9.57 451	31	9.60 750	36	10.39 250	9.96 701	5	57			
4	9.57 482	32	9.60 786	37	10.39 214	9.96 696	5	56			
5	9.57 514	31	9.60 823	36	10.39 177	9.96 691	5	**55**			
6	9.57 545	31	9.60 859	36	10.39 141	9.96 686	5	54			
7	9.57 576	31	9.60 895	36	10.39 105	9.96 681	5	53			
8	9.57 607	31	9.60 931	36	10.39 069	9.96 676	6	52			
9	9.57 638	31	9.60 967	37	10.39 033	9.96 670	5	51	**37**	**36**	**35**
10	9.57 669	31	9.61 004	36	10.38 996	9.96 665	5	**50**			
11	9.57 700	31	9.61 040	36	10.38 960	9.96 660	5	49	1 3.7	3.6	3.5
12	9.57 731	31	9.61 076	36	10.38 924	9.96 655	5	48	2 7.4	7.2	7.0
13	9.57 762	31	9.61 112	36	10.38 888	9.96 650	5	47	3 11.1	10.8	10.5
14	9.57 793	31	9.61 148	36	10.38 852	9.96 645	5	46	4 14.8	14.4	14.0
15	9.57 824	31	9.61 184	36	10.38 816	9.96 640	6	**45**	5 18.5	18.0	17.5
16	9.57 855	30	9.61 220	36	10.38 780	9.96 634	5	44	6 22.2	21.6	21.0
17	9.57 885	31	9.61 256	36	10.38 744	9.96 629	5	43	7 25.9	25.2	24.5
18	9.57 916	31	9.61 292	36	10.38 708	9.96 624	5	42	8 29.6	28.8	28.0
19	9.57 947	31	9.61 328	36	10.38 672	9.96 619	5	41	9 33.3	32.4	31.5
20	9.57 978	30	9.61 364	36	10.38 636	9.96 614	6	**40**			
21	9.58 008	31	9.61 400	36	10.38 600	9.96 608	5	39			
22	9.58 039	31	9.61 436	36	10.38 564	9.96 603	5	38			
23	9.58 070	31	9.61 472	36	10.38 528	9.96 598	5	37			
24	9.58 101	30	9.61 508	36	10.38 492	9.96 593	5	36			
25	9.58 131	31	9.61 544	35	10.38 456	9.96 588	6	**35**			
26	9.58 162	30	9.61 579	36	10.38 421	9.96 582	5	34	**32**	**31**	**30**
27	9.58 192	31	9.61 615	36	10.38 385	9.96 577	5	33			
28	9.58 223	30	9.61 651	36	10.38 349	9.96 572	5	32	1 3.2	3.1	3.0
29	9.58 253	31	9.61 687	35	10.38 313	9.96 567	5	31	2 6.4	6.2	6.0
30	9.58 284	30	9.61 722	36	10.38 278	9.96 562	6	**30**	3 9.6	9.3	9.0
31	9.58 314	31	9.61 758	36	10.38 242	9.96 556	5	29	4 12.8	12.4	12.0
32	9.58 345	30	9.61 794	36	10.38 206	9.96 551	5	28	5 16.0	15.5	15.0
33	9.58 375	31	9.61 830	35	10.38 170	9.96 546	5	27	6 19.2	18.6	18.0
34	9.58 406	30	9.61 865	36	10.38 135	9.96 541	6	26	7 22.4	21.7	21.0
35	9.58 436	31	9.61 901	35	10.38 099	9.96 535	5	**25**	8 25.6	24.8	24.0
36	9.58 467	30	9.61 936	36	10.38 064	9.96 530	5	24	9 28.8	27.9	27.0
37	9.58 497	30	9.61 972	36	10.38 028	9.96 525	5	23			
38	9.58 527	30	9.62 008	35	10.37 992	9.96 520	6	22			
39	9.58 557	31	9.62 043	36	10.37 957	9.96 514	5	21			
40	9.58 588	30	9.62 079	35	10.37 921	9.96 509	5	**20**			
41	9.58 618	30	9.62 114	36	10.37 886	9.96 504	6	19			
42	9.58 648	30	9.62 150	35	10.37 850	9.96 498	5	18			
43	9.58 678	31	9.62 185	36	10.37 815	9.96 493	5	17	**29**	**6**	**5**
44	9.58 709	30	9.62 221	35	10.37 779	9.96 488	5	16			
45	9.58 739	30	9.62 256	36	10.37 744	9.96 483	6	**15**	1 2.9	0.6	0.5
46	9.58 769	30	9.62 292	35	10.37 708	9.96 477	5	14	2 5.8	1.2	1.0
47	9.58 799	30	9.62 327	35	10.37 673	9.96 472	5	13	3 8.7	1.8	1.5
48	9.58 829	30	9.62 362	36	10.37 638	9.96 467	6	12	4 11.6	2.4	2.0
49	9.58 859	30	9.62 398	35	10.37 602	9.96 461	5	11	5 14.5	3.0	2.5
50	9.58 889	30	9.62 433	35	10.37 567	9.96 456	5	**10**	6 17.4	3.6	3.0
51	9.58 919	30	9.62 468	36	10.37 532	9.96 451	6	9	7 20.3	4.2	3.5
52	9.58 949	30	9.62 504	35	10.37 496	9.96 445	5	8	8 23.2	4.8	4.0
53	9.58 979	30	9.62 539	35	10.37 461	9.96 440	5	7	9 26.1	5.4	4.5
54	9.59 009	30	9.62 574	35	10.37 426	9.96 435	6	6			
55	9.59 039	30	9.62 609	36	10.37 391	9.96 429	5	**5**			
56	9.59 069	29	9.62 645	35	10.37 355	9.96 424	5	4			
57	9.59 098	30	9.62 680	35	10.37 320	9.96 419	6	3			
58	9.59 128	30	9.62 715	35	10.37 285	9.96 413	5	2			
59	9.59 158	30	9.62 750	35	10.37 250	9.96 408	5	1			
60	9.59 188		9.62 785		10.37 215	9.96 403		**0**			
	L. Cos.	d.	L. Cot.	c. d.	L. Tan.	L. Sin.	d.	′	P. P.		

′	L. Sin.	d.	L. Tan.	c. d.	L. Cot.	L. Cos.	d.		P. P.
0	9.59 188	30	9.62 785	35	10.37 215	9.96 403	6	60	
1	9.59 218	29	9.62 820	35	10.37 180	9.96 397	5	59	
2	9.59 247	30	9.62 855	35	10.37 145	9.96 392	5	58	
3	9.59 277	30	9.62 890	36	10.37 110	9.96 387	6	57	
4	9.59 307	29	9.62 926	35	10.37 074	9.96 381	5	56	
5	9.59 336	30	9.62 961	35	10.37 039	9.96 376	6	55	
6	9.59 366	30	9.62 996	35	10.37 004	9.96 370	5	54	
7	9.59 396	29	9.63 031	35	10.36 969	9.96 365	6	53	
8	9.59 425	30	9.63 066	35	10.36 934	9.96 360	6	52	
9	9.59 455	29	9.63 101	34	10.36 899	9.96 354	5	51	
10	9.59 484	30	9.63 135	35	10.36 865	9.96 349	6	50	**36** **35** **34**
11	9.59 514	29	9.63 170	35	10.36 830	9.96 343	5	49	1 3.6 3.5 3.4
12	9.59 543	30	9.63 205	35	10.36 795	9.96 338	5	48	2 7.2 7.0 6.8
13	9 59 573	29	9.63 240	35	10.36 760	9.96 333	6	47	3 10.8 10.5 10.2
14	9.59 602	30	9.63 275	35	10.36 725	9.96 327	5	46	4 14.4 14.0 13.6
15	9.59 632	29	9.63 310	35	10.36 690	9.96 322	6	45	5 18.0 17.5 17.0
16	9.59 661	29	9.63 345	34	10.36 655	9.96 316	5	44	6 21.6 21.0 20.4
17	9.59 690	30	9.63 379	35	10.36 621	9.96 311	6	43	7 25.2 24.5 23.8
18	9.59 720	29	9.63 414	35	10.36 586	9.96 305	5	42	8 28.8 28.0 27.2
19	9.59 749	29	9.63 449	35	10.36 551	9.96 300	6	41	9 32.4 31.5 30.6
20	9.59 778	30	9.63 484	35	10.36 516	9.96 294	5	40	
21	9.59 808	29	9.63 519	34	10.36 481	9.96 289	5	39	
22	9.59 837	29	9.63 553	35	10.36 447	9.96 284	5	38	
23	9.59 866	29	9.63 588	35	10.36 412	9.96 278	5	37	
24	9.59 895	29	9.63 623	34	10.36 377	9.96 273	6	36	
25	9.59 924	30	9.63 657	35	10.36 343	9.96 267	5	35	
26	9.59 954	29	9.63 692	34	10.36 308	9.96 262	6	34	**30** **29** **28**
27	9.59 983	29	9.63 726	35	10.36 274	9.96 256	5	33	1 3.0 2.9 2.8
28	9.60 012	29	9.63 761	35	10.36 239	9.96 251	6	32	2 6.0 5.8 5.6
29	9.60 041	29	9.63 796	34	10.36 204	9.96 245	5	31	3 9.0 8.7 8.4
30	9.60 070	29	9.63 830	35	10.36 170	9.96 240	6	30	4 12.0 11.6 11.2
31	9.60 099	29	9.63 865	34	10.36 135	9.96 234	5	29	5 15.0 14.5 14.0
32	9.60 128	29	9.63 899	35	10.36 101	9.96 229	6	28	6 18.0 17.4 16.8
33	9.60 157	29	9.63 934	34	10.36 066	9.96 223	5	27	7 21.0 20.3 19.6
34	9.60 186	29	9.63 968	35	10.36 032	9.96 218	6	26	8 24.0 23.2 22.4
35	9.60 215	29	9.64 003	34	10.35 997	9.96 212	5	25	9 27.0 26.1 25.2
36	9.60 244	29	9.64 037	35	10.35 963	9.96 207	6	24	
37	9.60 273	29	9.64 072	34	10.35 928	9.96 201	5	23	
38	9.60 302	29	9.64 106	34	10.35 894	9.96 196	6	22	
39	9.60 331	28	9.64 140	35	10.35 860	9.96 190	5	21	
40	9.60 359	29	9.64 175	34	10.35 825	9.96 185	6	20	
41	9.60 388	29	9.64 209	34	10.35 791	9.96 179	5	19	
42	9.60 417	29	9.64 243	35	10.35 757	9.96 174	6	18	
43	9.60 446	28	9.64 278	34	10.35 722	9.96 168	6	17	**6** **5**
44	9.60 474	29	9.64 312	34	10.35 688	9.96 162	5	16	
45	9.60 503	29	9.64 346	35	10.35 654	9.96 157	6	15	1 0.6 0.5
46	9.60 532	29	9.64 381	34	10.35 619	9.96 151	5	14	2 1.2 1.0
47	9.60 561	28	9.64 415	34	10.35 585	9.96 146	6	13	3 1.8 1.5
48	9.60 589	29	9.64 449	34	10.35 551	9.96 140	5	12	4 2.4 2.0
49	9.60 618	28	9.64 483	34	10.35 517	9.96 135	6	11	5 3.0 2.5
50	9.60 646	29	9.64 517	35	10.35 483	9.96 129	6	10	6 3.6 3.0
51	9.60 675	29	9.64 552	34	10.35 448	9.96 123	5	9	7 4.2 3.5
52	9.60 704	28	9.64 586	34	10.35 414	9.96 118	6	8	8 4.8 4.0
53	9.60 732	29	9.64 620	34	10.35 380	9.96 112	6	7	9 5.4 4.5
54	9.60 761	28	9.64 654	34	10.35 346	9.96 107	5	6	
55	9.60 789	29	9.64 688	34	10.35 312	9.96 101	6	5	
56	9.60 818	28	9.64 722	34	10.35 278	9.96 095	5	4	
57	9.60 846	29	9.64 756	34	10.35 244	9.96 090	6	3	
58	9.60 875	28	9.64 790	34	10.35 210	9.96 084	5	2	
59	9.60 903	28	9.64 824	34	10.35 176	9.96 079	6	1	
60	9.60 931		9.64 858		10.35 142	9.96 073		0	
	L. Cos.	d.	L. Cot.	c. d.	L. Tan.	L. Sin.	d.	′	P. P.

′	L. Sin.	d.	L. Tan.	c. d.	L. Cot.	L. Cos.	d.	′
0	9.60 931	29	9.64 858	34	10.35 142	9.96 073	6	60
1	9.60 960	28	9.64 892	34	10.35 108	9.96 067		59
2	9.60 988	28	9.64 926	34	10.35 074	9.96 062	5	58
3	9.61 016	29	9.64 960	34	10.35 040	9.96 056	6	57
4	9.61 045	28	9.64 994	34	10.35 006	9.96 050	6	56
5	9.61 073	28	9.65 028	34	10.34 972	9.96 045	5	55
6	9.61 101	28	9.65 062	34	10.34 938	9.96 039	6	54
7	9.61 129	29	9.65 096	34	10.34 904	9.96 034	5	53
8	9.61 158	28	9.65 130	34	10.34 870	9.96 028	6	52
9	9.61 186	28	9.65 164	33	10.34 836	9.96 022	6	51
10	9.61 214	28	9.65 197	34	10.34 803	9.96 017	5	50
11	9.61 242	28	9.65 231	34	10.34 769	9.96 011	6	49
12	9.61 270	28	9.65 265	34	10.34 735	9.96 005	6	48
13	9.61 298	28	9.65 299	34	10.34 701	9.96 000	5	47
14	9.61 326	28	9.65 333	33	10.34 667	9.95 994	6	46
15	9.61 354	28	9.65 366	34	10.34 634	9.95 988	6	45
16	9.61 382	29	9.65 400	34	10.34 600	9.95 982	6	44
17	9.61 411	27	9.65 434	33	10.34 566	9.95 977	5	43
18	9.61 438	28	9.65 467	34	10.34 533	9.95 971	6	42
19	9.61 466	28	9.65 501	34	10.34 499	9.95 965	6	41
20	9.61 494	28	9.65 535	33	10.34 465	9.95 960	5	40
21	9.61 522	28	9.65 568	34	10.34 432	9.95 954	6	39
22	9.61 550	28	9.65 602	34	10.34 398	9.95 948	6	38
23	9.61 578	28	9.65 636	33	10.34 364	9.95 942	6	37
24	9.61 606	28	9.65 669	34	10.34 331	9.95 937	5	36
25	9.61 634	28	9.65 703	33	10.34 297	9.95 931	6	35
26	9.61 662	27	9.65 736	34	10.34 264	9.95 925	6	34
27	9.61 689	28	9.65 770	33	10.34 230	9.95 920	5	33
28	9.61 717	28	9.65 803	34	10.34 197	9.95 914	6	32
29	9.61 745	28	9.65 837	33	10.34 163	9.95 908	6	31
30	9.61 773	27	9.65 870	34	10.34 130	9.95 902	5	30
31	9.61 800	28	9.65 904	33	10.34 096	9.95 897	6	29
32	9.61 828	28	9.65 937	34	10.34 063	9.95 891	6	28
33	9.61 856	27	9.65 971	33	10.34 029	9.95 885	6	27
34	9.61 883	28	9.66 004	34	10.33 996	9.95 879	6	26
35	9.61 911	28	9.66 038	33	10.33 962	9.95 873	5	25
36	9.61 939	27	9.66 071	33	10.33 929	9.95 868	6	24
37	9.61 966	28	9.66 104	34	10.33 896	9.95 862	6	23
38	9.61 994	27	9.66 138	33	10.33 862	9.95 856	6	22
39	9.62 021	28	9.66 171	33	10.33 829	9.95 850	6	21
40	9.62 049	27	9.66 204	34	10.33 796	9.95 844	5	20
41	9.62 076	28	9.66 238	33	10.33 762	9.95 839	6	19
42	9.62 104	27	9.66 271	33	10.33 729	9.95 833	6	18
43	9.62 131	28	9.66 304	33	10.33 696	9.95 827	6	17
44	9.62 159	27	9.66 337	34	10.33 663	9.95 821	6	16
45	9.62 186	28	9.66 371	33	10.33 629	9.95 815	5	15
46	9.62 214	27	9.66 404	33	10.33 596	9.95 810	6	14
47	9.62 241	27	9.66 437	33	10.33 563	9.95 804	6	13
48	9.62 268	28	9.66 470	33	10.33 530	9.95 798	6	12
49	9.62 296	27	9.66 503	34	10.33 497	9.95 792	6	11
50	9.62 323	27	9.66 537	33	10.33 463	9.95 786	6	10
51	9.62 350	27	9.66 570	33	10.33 430	9.95 780	5	9
52	9.62 377	28	9.66 603	33	10.33 397	9.95 775	6	8
53	9.62 405	27	9.66 636	33	10.33 364	9.95 769	6	7
54	9.62 432	27	9.66 669	33	10.33 331	9.95 763	6	6
55	9.62 459	27	9.66 702	33	10.33 298	9.95 757	6	5
56	9.62 486	27	9.66 735	33	10.33 265	9.95 751	6	4
57	9.62 513	28	9.66 768	33	10.33 232	9.95 745	6	3
58	9.62 541	27	9.66 801	33	10.33 199	9.95 739	6	2
59	9.62 568	27	9.66 834	33	10.33 166	9.95 733	5	1
60	9.62 595		9.66 867		10.33 133	9.95 728		0

	L. Cos.	d	L. Cot.	c. d.	L. Tan.	L. Sin.	d.	′

P. P.

	34	33
1	3.4	3.3
2	6.8	6.6
3	10.2	9.9
4	13.6	13.2
5	17.0	16.5
6	20.4	19.8
7	23.8	23.1
8	27.2	26.4
9	30.6	29.7

	29	28	27
1	2.9	2.8	2.7
2	5.8	5.6	5.4
3	8.7	8.4	8.1
4	11.6	11.2	10.8
5	14.5	14.0	13.5
6	17.4	16.8	16.2
7	20.3	19.6	18.9
8	23.2	22.4	21.6
9	26.1	25.2	24.3

	6	5
1	0.6	0.5
2	1.2	1.0
3	1.8	1.5
4	2.4	2.0
5	3.0	2.5
6	3.6	3.0
7	4.2	3.5
8	4.8	4.0
9	5.4	4.5

25°

′	L. Sin.	d.	L. Tan.	c. d.	L. Cot.	L. Cos.	d.	′
0	9.62 595	27	9.66 867	33	10.33 133	9.95 728	6	60
1	9.62 622	27	9.66 900	33	10.33 100	9.95 722	6	59
2	9.62 649	27	9.66 933	33	10.33 067	9.95 716	6	58
3	9.62 676	27	9.66 966	33	10.33 034	9.95 710	6	57
4	9.62 703	27	9.66 999	33	10.33 001	9.95 704	6	56
5	9.62 730	27	9.67 032	33	10.32 968	9.95 698	6	55
6	9.62 757	27	9.67 065	33	10.32 935	9.95 692	6	54
7	9.62 784	27	9.67 098	33	10.32 902	9.95 686	6	53
8	9.62 811	27	9.67 131	32	10.32 869	9.95 680	6	52
9	9.62 838	27	9.67 163	33	10.32 837	9.95 674	6	51
10	9.62 865	27	9.67 196	33	10.32 804	9.95 668	5	50
11	9.62 892	26	9.67 229	33	10.32 771	9.95 663	6	49
12	9.62 918	27	9.67 262	33	10.32 738	9.95 657	6	48
13	9.62 945	27	9.67 295	32	10.32 705	9.95 651	6	47
14	9.62 972	27	9.67 327	33	10.32 673	9.95 645	6	46
15	9.62 999	27	9.67 360	33	10.32 640	9.95 639	6	45
16	9.63 026	26	9.67 393	33	10.32 607	9.95 633	6	44
17	9.63 052	27	9.67 426	32	10.32 574	9.95 627	6	43
18	9.63 079	27	9.67 458	33	10.32 542	9.95 621	6	42
19	9.63 106	27	9.67 491	33	10.32 509	9.95 615	6	41
20	9.63 133	26	9.67 524	32	10.32 476	9.95 609	6	40
21	9.63 159	27	9.67 556	33	10.32 444	9.95 603	6	39
22	9.63 186	27	9.67 589	33	10.32 411	9.95 597	6	38
23	9.63 213	26	9.67 622	32	10.32 378	9.95 591	6	37
24	9.63 239	27	9.67 654	33	10.32 346	9.95 585	6	36
25	9.63 266	26	9.67 687	32	10.32 313	9.95 579	6	35
26	9.63 292	27	9.67 719	33	10.32 281	9.95 573	6	34
27	9.63 319	26	9.67 752	33	10.32 248	9.95 567	6	33
28	9.63 345	27	9.67 785	32	10.32 215	9.95 561	6	32
29	9.63 372	26	9.67 817	33	10.32 183	9.95 555	6	31
30	9.63 398	27	9.67 850	32	10.32 150	9.95 549	6	30
31	9.63 425	26	9.67 882	33	10.32 118	9.95 543	6	29
32	9.63 451	27	9.67 915	32	10.32 085	9.95 537	6	28
33	9.63 478	26	9.67 947	33	10.32 053	9.95 531	6	27
34	9.63 504	27	9.67 980	32	10.32 020	9.95 525	6	26
35	9.63 531	26	9.68 012	32	10.31 988	9.95 519	6	25
36	9.63 557	26	9.68 044	33	10.31 956	9.95 513	6	24
37	9.63 583	27	9.68 077	32	10.31 923	9.95 507	7	23
38	9.63 610	26	9.68 109	33	10.31 891	9.95 500	6	22
39	9.63 636	26	9.68 142	32	10.31 858	9.95 494	6	21
40	9.63 662	27	9.68 174	32	10.31 826	9.95 488	6	20
41	9.63 689	26	9.68 206	33	10.31 794	9.95 482	6	19
42	9.63 715	26	9.68 239	32	10.31 761	9.95 476	6	18
43	9.63 741	26	9.68 271	32	10.31 729	9.95 470	6	17
44	9.63 767	27	9.68 303	33	10.31 697	9.95 464	6	16
45	9.63 794	26	9.68 336	32	10.31 664	9.95 458	6	15
46	9.63 820	26	9.68 368	32	10.31 632	9.95 452	6	14
47	9.63 846	26	9.68 400	32	10.31 600	9.95 446	6	13
48	9.63 872	26	9.68 432	33	10.31 568	9.95 440	6	12
49	9.63 898	26	9.68 465	32	10.31 535	9.95 434	7	11
50	9.63 924	26	9.68 497	32	10.31 503	9.95 427	6	10
51	9.63 950	26	9.68 529	32	10.31 471	9.95 421	6	9
52	9.63 976	26	9.68 561	32	10.31 439	9.95 415	6	8
53	9.64 002	26	9.68 593	33	10.31 407	9.95 409	6	7
54	9.64 028	26	9.68 626	32	10.31 374	9.95 403	6	6
55	9.64 054	26	9.68 658	32	10.31 342	9.95 397	7	5
56	9.64 080	26	9.68 690	32	10.31 310	9.95 391	6	4
57	9.64 106	26	9.68 722	32	10.31 278	9.95 384	6	3
58	9.64 132	26	9.68 754	32	10.31 246	9.95 378	6	2
59	9.64 158	26	9.68 786	32	10.31 214	9.95 372	6	1
60	9.64 184		9.68 818		10.31 182	9.95 366		0
	L. Cos.	d.	L. Cot.	c. d.	L. Tan.	L. Sin.	d.	′

P. P.

	33	32
1	3.3	3.2
2	6.6	6.4
3	9.9	9.6
4	13.2	12.8
5	16.5	16.0
6	19.8	19.2
7	23.1	22.4
8	26.4	25.6
9	29.7	28.8

	27	26
1	2.7	2.6
2	5.4	5.2
3	8.1	7.8
4	10.8	10.4
5	13.5	13.0
6	16.2	15.6
7	18.9	18.2
8	21.6	20.8
9	24.3	23.4

	7	6	5
1	0.7	0.6	0.5
2	1.4	1.2	1.0
3	2.1	1.8	1.5
4	2.8	2.4	2.0
5	3.5	3.0	2.5
6	4.2	3.6	3.0
7	4.9	4.2	3.5
8	5.6	4.8	4.0
9	6.3	5.4	4.5

64°

273

′	L. Sin.	d.	L. Tan.	c. d.	L. Cot.	L. Cos.	d.		P. P.
0	9.64 184	26	9.68 818	32	10.31 182	9.95 366	6	60	
1	9.64 210	26	9.68 850	32	10.31 150	9.95 360	6	59	
2	9.64 236	26	9.68 882	32	10.31 118	9.95 354	6	58	
3	9.64 262	26	9.68 914	32	10.31 086	9.95 348	7	57	
4	9.64 288	25	9.68 946	32	10.31 054	9.95 341	6	56	
5	9.64 313	26	9.68 978	32	10.31 022	9.95 335	6	55	
6	9.64 339	26	9.69 010	32	10.30 990	9.95 329	6	54	
7	9.64 365	26	9.69 042	32	10.30 958	9.95 323	6	53	
8	9.64 391	26	9.69 074	32	10.30 926	9.95 317	7	52	
9	9.64 417	25	9.69 106	32	10.30 894	9.95 310	6	51	32 31
10	9.64 442	26	9.69 138	32	10.30 862	9.95 304	6	50	
11	9.64 468	26	9.69 170	32	10.30 830	9.95 298	6	49	1 | 3.2 3.1
12	9.64 494	25	9.69 202	32	10.30 798	9.95 292	6	48	2 | 6.4 6.2
13	9.64 519	26	9.69 234	32	10.30 766	9.95 286	7	47	3 | 9.6 9.3
14	9.64 545	26	9.69 266	32	10.30 734	9.95 279	6	46	4 | 12.8 12.4
15	9.64 571	25	9.69 298	31	10.30 702	9.95 273	6	45	5 | 16.0 15.5
16	9.64 596	26	9.69 329	32	10.30 671	9.95 267	6	44	6 | 19.2 18.6
17	9.64 622	25	9.69 361	32	10.30 639	9.95 261	7	43	7 | 22.4 21.7
18	9.64 647	26	9.69 393	32	10.30 607	9.95 254	6	42	8 | 25.6 24.8
19	9.64 673	25	9.69 425	32	10.30 575	9.95 248	6	41	9 | 28.8 27.9
20	9.64 698	26	9.69 457	31	10.30 543	9.95 242	6	40	
21	9.64 724	25	9.69 488	32	10.30 512	9.95 236	7	39	
22	9.64 749	26	9.69 520	32	10.30 480	9.95 229	6	38	
23	9.64 775	25	9.69 552	32	10.30 448	9.95 223	6	37	
24	9.64 800	26	9.69 584	31	10.30 416	9.95 217	6	36	
25	9.64 826	25	9.69 615	32	10.30 385	9.95 211	7	35	
26	9.64 851	26	9.69 647	32	10.30 353	9.95 204	6	34	26 25 24
27	9.64 877	25	9.69 679	31	10.30 321	9.95 198	6	33	
28	9.64 902	25	9.69 710	32	10.30 290	9.95 192	7	32	1 | 2.6 2.5 2.4
29	9.64 927	26	9.69 742	32	10.30 258	9.95 185	6	31	2 | 5.2 5.0 4.8
30	9.64 953	25	9.69 774	31	10.30 226	9.95 179	6	30	3 | 7.8 7.5 7.2
31	9.64 978	25	9.69 805	32	10.30 195	9.95 173	6	29	4 | 10.4 10.0 9.6
32	9.65 003	26	9.69 837	31	10.30 163	9.95 167	7	28	5 | 13.0 12.5 12.0
33	9.65 029	25	9.69 868	32	10.30 132	9.95 160	6	27	6 | 15.6 15.0 14.4
34	9.65 054	25	9.69 900	32	10.30 100	9.95 154	6	26	7 | 18.2 17.5 16.8
35	9.65 079	25	9.69 932	31	10.30 068	9.95 148	7	25	8 | 20.8 20.0 19.2
36	9.65 104	26	9.69 963	32	10.30 037	9.95 141	6	24	9 | 23.4 22.5 21.6
37	9.65 130	25	9.69 995	31	10.30 005	9.95 135	6	23	
38	9.65 155	25	9.70 026	32	10.29 974	9.95 129	7	22	
39	9.65 180	25	9.70 058	31	10.29 942	9.95 122	6	21	
40	9.65 205	25	9.70 089	32	10.29 911	9.95 116	6	20	
41	9.65 230	25	9.70 121	31	10.29 879	9.95 110	7	19	
42	9.65 255	26	9.70 152	32	10.29 848	9.95 103	6	18	
43	9.65 281	25	9.70 184	31	10.29 816	9.95 097	7	17	
44	9.65 306	25	9.70 215	32	10.29 785	9.95 090	6	16	7 6
45	9.65 331	25	9.70 247	31	10.29 753	9.95 084	6	15	
46	9.65 356	25	9.70 278	31	10.29 722	9.95 078	7	14	1 | 0.7 0.6
47	9.65 381	25	9.70 309	32	10.29 691	9.95 071	6	13	2 | 1.4 1.2
48	9.65 406	25	9.70 341	31	10.29 659	9.95 065	6	12	3 | 2.1 1.8
49	9.65 431	25	9.70 372	32	10.29 628	9.95 059	7	11	4 | 2.8 2.4
50	9.65 456	25	9.70 404	31	10.29 596	9.95 052	6	10	5 | 3.5 3.0
51	9.65 481	25	9.70 435	31	10.29 565	9.95 046	7	9	6 | 4.2 3.6
52	9.65 506	25	9.70 466	32	10.29 534	9.95 039	6	8	7 | 4.9 4.2
53	9.65 531	25	9.70 498	31	10.29 502	9.95 033	6	7	8 | 5.6 4.8
54	9.65 556	24	9.70 529	31	10.29 471	9.95 027	7	6	9 | 6.3 5.4
55	9.65 580	25	9.70 560	32	10.29 440	9.95 020	6	5	
56	9.65 605	25	9.70 592	31	10.29 408	9.95 014	7	4	
57	9.65 630	25	9.70 623	31	10.29 377	9.95 007	6	3	
58	9.65 655	25	9.70 654	31	10.29 346	9.95 001	6	2	
59	9.65 680	25	9.70 685	32	10.29 315	9.94 995	7	1	
60	9.65 705		9.70 717		10.29 283	9.94 988		0	
	L. Cos.	d.	L. Cot.	c. d.	L. Tan.	L. Sin.	d.	′	P. P.

′	L. Sin.	d.	L. Tan.	c. d.	L. Cot.	L. Cos.	d.		P. P.		
0	9.65 705		9.70 717		10.29 283	9.94 988		60			
1	9.65 729	24	9.70 748	31	10.29 252	9.94 982	6	59			
2	9.65 754	25	9.70 779	31	10.29 221	9.94 975	7	58			
3	9.65 779	25	9.70 810	31	10.29 190	9.94 969	6	57			
4	9.65 804	25	9.70 841	31	10.29 159	9.94 962	7	56			
		24		32			6				
5	9.65 828	25	9.70 873	31	10.29 127	9.94 956	7	55			
6	9.65 853	25	9.70 904	31	10.29 096	9.94 949	6	54			
7	9.65 878	24	9.70 935	31	10.29 065	9.94 943	7	53			
8	9.65 902	25	9.70 966	31	10.29 034	9.94 936	6	52			
9	9.65 927	25	9.70 997	31	10.29 003	9.94 930	7	51		32	31	30
10	9.65 952	24	9.71 028	31	10.28 972	9.94 923	6	50			
11	9.65 976	25	9.71 059	31	10.28 941	9.94 917	6	49	1	3.2 3.1 3.0
12	9.66 001	24	9.71 090	31	10.28 910	9.94 911	7	48	2	6.4 6.2 6.0
13	9.66 025	25	9.71 121	32	10.28 879	9.94 904	6	47	3	9.6 9.3 9.0
14	9.66 050	25	9.71 153	31	10.28 847	9.94 898	7	46	4	12.8 12.4 12.0
15	9.66 075	24	9.71 184	31	10.28 816	9.94 891	6	45	5	16.0 15.5 15.0
16	9.66 099	25	9.71 215	31	10.28 785	9.94 885	7	44	6	19.2 18.6 18.0
17	9.66 124	24	9.71 246	31	10.28 754	9.94 878	7	43	7	22.4 21.7 21.0
18	9.66 148	25	9.71 277	31	10.28 723	9.94 871	6	42	8	25.6 24.8 24.0
19	9.66 173	24	9.71 308	31	10.28 692	9.94 865	7	41	9	28.8 27.9 27.0
20	9.66 197	24	9.71 339	31	10.28 661	9.94 858	6	40			
21	9.66 221	25	9.71 370	31	10.28 630	9.94 852	7	39			
22	9.66 246	24	9.71 401	30	10.28 599	9.94 845	6	38			
23	9.66 270	25	9.71 431	31	10.28 569	9.94 839	7	37			
24	9.66 295	24	9.71 462	31	10.28 538	9.94 832	6	36			
25	9.66 319	24	9.71 493	31	10.28 507	9.94 826	7	35			
26	9.66 343	25	9.71 524	31	10.28 476	9.94 819	6	34			
27	9.66 368	24	9.71 555	31	10.28 445	9.94 813	7	33		25	24	23
28	9.66 392	24	9.71 586	31	10.28 414	9.94 806	7	32			
29	9.66 416	25	9.71 617	31	10.28 383	9.94 799	6	31	1	2.5 2.4 2.3
30	9.66 441	24	9.71 648	31	10.28 352	9.94 793	7	30	2	5.0 4.8 4.6
31	9.66 465	24	9.71 679	30	10.28 321	9.94 786	6	29	3	7.5 7.2 6.9
32	9.66 489	24	9.71 709	31	10.28 291	9.94 780	7	28	4	10.0 9.6 9.2
33	9.66 513	24	9.71 740	31	10.28 260	9.94 773	6	27	5	12.5 12.0 11.5
34	9.66 537	25	9.71 771	31	10.28 229	9.94 767	7	26	6	15.0 14.4 13.8
35	9.66 562	24	9.71 802	31	10.28 198	9.94 760	7	25	7	17.5 16.8 16.1
36	9.66 586	24	9.71 833	30	10.28 167	9.94 753	6	24	8	20.0 19.2 18.4
37	9.66 610	24	9.71 863	31	10.28 137	9.94 747	7	23	9	22.5 21.6 20.7
38	9.66 634	24	9.71 894	31	10.28 106	9.94 740	6	22			
39	9.66 658	24	9.71 925	30	10.28 075	9.94 734	7	21			
40	9.66 682	24	9.71 955	31	10.28 045	9.94 727	7	20			
41	9.66 706	25	9.71 986	31	10.28 014	9.94 720	6	19			
42	9.66 731	24	9.72 017	31	10.27 983	9.94 714	7	18			
43	9.66 755	24	9.72 048	30	10.27 952	9.94 707	7	17			
44	9.66 779	24	9.72 078	31	10.27 922	9.94 700	6	16		7	6
45	9.66 803	24	9.72 109	31	10.27 891	9.94 694	7	15			
46	9.66 827	24	9.72 140	30	10.27 860	9.94 687	7	14	1	0.7 0.6
47	9.66 851	24	9.72 170	31	10.27 830	9.94 680	6	13	2	1.4 1.2
48	9.66 875	24	9.72 201	30	10.27 799	9.94 674	7	12	3	2.1 1.8
49	9.66 899	23	9.72 231	31	10.27 769	9.94 667	7	11	4	2.8 2.4
50	9.66 922	24	9.72 262	31	10.27 738	9.94 660	6	10	5	3.5 3.0
51	9.66 946	24	9.72 293	30	10.27 707	9.94 654	7	9	6	4.2 3.6
52	9.66 970	24	9.72 323	31	10.27 677	9.94 647	7	8	7	4.9 4.2
53	9.66 994	24	9.72 354	30	10.27 646	9.94 640	6	7	8	5.6 4.8
54	9.67 018	24	9.72 384	31	10.27 616	9.94 634	7	6	9	6.3 5.4
55	9.67 042	24	9.72 415	30	10.27 585	9.94 627	7	5			
56	9.67 066	24	9.72 445	31	10.27 555	9.94 620	6	4			
57	9.67 090	23	9.72 476	30	10.27 524	9.94 614	7	3			
58	9.67 113	24	9.72 506	31	10.27 494	9.94 607	7	2			
59	9.67 137	24	9.72 537	30	10.27 463	9.94 600	7	1			
60	9.67 161		9.72 567		10.27 433	9.94 593		0			
	L. Cos.	d.	L. Cot.	c. d.	L. Tan.	L. Sin.	d.	′		P. P.	

′	L. Sin.	d.	L. Tan.	c. d.	L. Cot.	L. Cos.	d.	
0	9.67 161	24	9.72 567	31	10.27 433	9.94 593	6	60
1	9.67 185	23	9.72 598	30	10.27 402	9.94 587	7	59
2	9.67 208	24	9.72 628	31	10.27 372	9.94 580	7	58
3	9.67 232	24	9.72 659	30	10.27 341	9.94 573	6	57
4	9.67 256	24	9.72 689	31	10.27 311	9.94 567	7	56
5	9.67 280	23	9.72 720	30	10.27 280	9.94 560	7	55
6	9.67 303	24	9.72 750	30	10.27 250	9.94 553	7	54
7	9.67 327	23	9.72 780	31	10.27 220	9.94 546	6	53
8	9.67 350	24	9.72 811	30	10.27 189	9.94 540	7	52
9	9.67 374	24	9.72 841	31	10.27 159	9.94 533	7	51
10	9.67 398	23	9.72 872	30	10.27 128	9.94 526	7	50
11	9.67 421	24	9.72 902	30	10.27 098	9.94 519	6	49
12	9.67 445	23	9.72 932	31	10.27 068	9.94 513	7	48
13	9.67 468	24	9.72 963	30	10.27 037	9.94 506	7	47
14	9.67 492	23	9.72 993	30	10.27 007	9.94 499	7	46
15	9.67 515	24	9.73 023	31	10.26 977	9.94 492	7	45
16	9.67 539	23	9.73 054	30	10.26 946	9.94 485	7	44
17	9.67 562	24	9.73 084	30	10.26 916	9.94 479	6	43
18	9.67 586	23	9.73 114	30	10.26 886	9.94 472	7	42
19	9.67 609	24	9.73 144	31	10.26 856	9.94 465	7	41
20	9.67 633	23	9.73 175	30	10.26 825	9.94 458	7	40
21	9.67 656	24	9.73 205	30	10.26 795	9.94 451	6	39
22	9.67 680	23	9.73 235	30	10.26 765	9.94 445	7	38
23	9.67 703	23	9.73 265	30	10.26 735	9.94 438	7	37
24	9.67 726	24	9.73 295	31	10.26 705	9.94 431	7	36
25	9.67 750	23	9.73 326	30	10.26 674	9.94 424	7	35
26	9.67 773	23	9.73 356	30	10.26 644	9.94 417	7	34
27	9.67 796	24	9.73 386	30	10.26 614	9.94 410	6	33
28	9.67 820	23	9.73 416	30	10.26 584	9.94 404	7	32
29	9.67 843	23	9.73 446	30	10.26 554	9.94 397	7	31
30	9.67 866	24	9.73 476	31	10.26 524	9.94 390	7	30
31	9.67 890	23	9.73 507	30	10.26 493	9.94 383	7	29
32	9.67 913	23	9.73 537	30	10.26 463	9.94 376	7	28
33	9.67 936	23	9.73 567	30	10.26 433	9.94 369	7	27
34	9.67 959	23	9.73 597	30	10.26 403	9.94 362	7	26
35	9.67 982	24	9.73 627	30	10.26 373	9.94 355	6	25
36	9.68 006	23	9.73 657	30	10.26 343	9.94 349	7	24
37	9.68 029	23	9.73 687	30	10.26 313	9.94 342	7	23
38	9.68 052	23	9.73 717	30	10.26 283	9.94 335	7	22
39	9.68 075	23	9.73 747	30	10.26 253	9.94 328	7	21
40	9.68 098	23	9.73 777	30	10.26 223	9.94 321	7	20
41	9.68 121	23	9.73 807	30	10.26 193	9.94 314	7	19
42	9.68 144	23	9.73 837	30	10.26 163	9.94 307	7	18
43	9.68 167	23	9.73 867	30	10.26 133	9.94 300	7	17
44	9.68 190	23	9.73 897	30	10.26 103	9.94 293	7	16
45	9.68 213	24	9.73 927	30	10.26 073	9.94 286	7	15
46	9.68 237	23	9.73 957	30	10.26 043	9.94 279	6	14
47	9.68 260	23	9.73 987	30	10.26 013	9.94 273	7	13
48	9.68 283	22	9.74 017	30	10.25 983	9.94 266	7	12
49	9.68 305	23	9.74 047	30	10.25 953	9.94 259	7	11
50	9.68 328	23	9.74 077	30	10.25 923	9.94 252	7	10
51	9.68 351	23	9.74 107	30	10.25 893	9.94 245	7	9
52	9.68 374	23	9.74 137	29	10.25 863	9.94 238	7	8
53	9.68 397	23	9.74 166	30	10.25 834	9.94 231	7	7
54	9.68 420	23	9.74 196	30	10.25 804	9.94 224	7	6
55	9.68 443	23	9.74 226	30	10.25 774	9.94 217	7	5
56	9.68 466	23	9.74 256	30	10.25 744	9.94 210	7	4
57	9.68 489	23	9.74 286	30	10.25 714	9.94 203	7	3
58	9.68 512	22	9.74 316	29	10.25 684	9.94 196	7	2
59	9.68 534	23	9.74 345	30	10.25 655	9.94 189	7	1
60	9.68 557		9.74 375		10.25 625	9.94 182		0

| | L. Cos. | d. | L. Cot. | c. d. | L. Tan. | L. Sin. | d. | ′ |

P. P.

	31	30	29
1	3.1	3.0	2.9
2	6.2	6.0	5.8
3	9.3	9.0	8.7
4	12.4	12.0	11.6
5	15.5	15.0	14.5
6	18.6	18.0	17.4
7	21.7	21.0	20.3
8	24.8	24.0	23.2
9	27.9	27.0	26.1

	24	23	22
1	2.4	2.3	2.2
2	4.8	4.6	4.4
3	7.2	6.9	6.6
4	9.6	9.2	8.8
5	12.0	11.5	11.0
6	14.4	13.8	13.2
7	16.8	16.1	15.4
8	19.2	18.4	17.6
9	21.6	20.7	19.8

	7	6
1	0.7	0.6
2	1.4	1.2
3	2.1	1.8
4	2.8	2.4
5	3.5	3.0
6	4.2	3.6
7	4.9	4.2
8	5.6	4.8
9	6.3	5.4

′	L. Sin.	d.	L. Tan.	c. d.	L. Cot.	L. Cos.	d.	′
0	9.68 557	23	9.74 375	30	10.25 625	9.94 182	7	60
1	9.68 580	23	9.74 405	30	10.25 595	9.94 175	7	59
2	9.68 603	22	9.74 435	30	10.25 565	9.94 168	7	58
3	9.68 625	23	9.74 465	29	10.25 535	9.94 161	7	57
4	9.68 648	23	9.74 494	30	10.25 506	9.94 154	7	56
5	9.68 671	23	9.74 524	30	10.25 476	9.94 147	7	55
6	9.68 694	22	9.74 554	29	10.25 446	9.94 140	7	54
7	9.68 716	23	9.74 583	30	10.25 417	9.94 133	7	53
8	9.68 739	23	9.74 613	30	10.25 387	9.94 126	7	52
9	9.68 762	22	9.74 643	30	10.25 357	9.94 119	7	51
10	9.68 784	23	9.74 673	29	10.25 327	9.94 112	7	50
11	9.68 807	22	9.74 702	30	10.25 298	9.94 105	7	49
12	9.68 829	23	9.74 732	30	10.25 268	9.94 098	8	48
13	9.68 852	23	9.74 762	29	10.25 238	9.94 090	7	47
14	9.68 875	22	9.74 791	30	10.25 209	9.94 083	7	46
15	9.68 897	23	9.74 821	30	10.25 179	9.94 076	7	45
16	9.68 920	22	9.74 851	29	10.25 149	9.94 069	7	44
17	9.68 942	23	9.74 880	30	10.25 120	9.94 062	7	43
18	9.68 965	22	9.74 910	29	10.25 090	9.94 055	7	42
19	9.68 987	23	9.74 939	30	10.25 061	9.94 048	7	41
20	9.69 010	22	9.74 969	29	10.25 031	9.94 041	7	40
21	9.69 032	23	9.74 998	30	10.25 002	9.94 034	7	39
22	9.69 055	22	9.75 028	30	10.24 972	9.94 027	7	38
23	9.69 077	23	9.75 058	29	10.24 942	9.94 020	8	37
24	9.69 100	22	9.75 087	30	10.24 913	9.94 012	7	36
25	9.69 122	22	9.75 117	29	10.24 883	9.94 005	7	35
26	9.69 144	23	9.75 146	30	10.24 854	9.93 998	7	34
27	9.69 167	22	9.75 176	29	10.24 824	9.93 991	7	33
28	9.69 189	23	9.75 205	30	10.24 795	9.93 984	7	32
29	9.69 212	22	9.75 235	29	10.24 765	9.93 977	7	31
30	9.69 234	22	9.75 264	30	10.24 736	9.93 970	7	30
31	9.69 256	23	9.75 294	29	10.24 706	9.93 963	8	29
32	9.69 279	22	9.75 323	30	10.24 677	9.93 955	7	28
33	9.69 301	22	9.75 353	29	10.24 647	9.93 948	7	27
34	9.69 323	22	9.75 382	29	10.24 618	9.93 941	7	26
35	9.69 345	23	9.75 411	30	10.24 589	9.93 934	7	25
36	9.69 368	22	9.75 441	29	10.24 559	9.93 927	7	24
37	9.69 390	22	9.75 470	30	10.24 530	9.93 920	8	23
38	9.69 412	22	9.75 500	29	10.24 500	9.93 912	7	22
39	9.69 434	22	9.75 529	29	10.24 471	9.93 905	7	21
40	9.69 456	23	9.75 558	30	10.24 442	9.93 898	7	20
41	9.69 479	22	9.75 588	29	10.24 412	9.93 891	7	19
42	9.69 501	22	9.75 617	30	10.24 383	9.93 884	8	18
43	9.69 523	22	9.75 647	29	10.24 353	9.93 876	7	17
44	9.69 545	22	9.75 676	29	10.24 324	9.93 869	7	16
45	9.69 567	22	9.75 705	30	10.24 295	9.93 862	7	15
46	9.69 589	22	9.75 735	29	10.24 265	9.93 855	8	14
47	9.69 611	22	9.75 764	29	10.24 236	9.93 847	7	13
48	9.69 633	22	9.75 793	29	10.24 207	9.93 840	7	12
49	9.69 655	22	9.75 822	30	10.24 178	9.93 833	7	11
50	9.69 677	22	9.75 852	29	10.24 148	9.93 826	7	10
51	9.69 699	22	9.75 881	29	10.24 119	9.93 819	8	9
52	9.69 721	22	9.75 910	29	10.24 090	9.93 811	7	8
53	9.69 743	22	9.75 939	30	10.24 061	9.93 804	7	7
54	9.69 765	22	9.75 969	29	10.24 031	9.93 797	8	6
55	9.69 787	22	9.75 998	29	10.24 002	9.93 789	7	5
56	9.69 809	22	9.76 027	29	10.23 973	9.93 782	7	4
57	9.69 831	22	9.76 056	30	10.23 944	9.93 775	7	3
58	9.69 853	22	9.76 086	29	10.23 914	9.93 768	8	2
59	9.69 875	22	9.76 115	29	10.23 885	9.93 760	7	1
60	9.69 897		9.76 144		10.23 856	9.93 753		0
	L. Cos.	d.	L. Cot.	c. d.	L. Tan.	L. Sin.	d.	′

P. P.

	30	29	23
1	3.0	2.9	2.3
2	6.0	5.8	4.6
3	9.0	8.7	6.9
4	12.0	11.6	9.2
5	15.0	14.5	11.5
6	18.0	17.4	13.8
7	21.0	20.3	16.1
8	24.0	23.2	18.4
9	27.0	26.1	20.7

	22	8	7
1	2.2	0.8	0.7
2	4.4	1.6	1.4
3	6.6	2.4	2.1
4	8.8	3.2	2.8
5	11.0	4.0	3.5
6	13.2	4.8	4.2
7	15.4	5.6	4.9
8	17.6	6.4	5.6
9	19.8	7.2	6.3

′	L. Sin.	d.	L. Tan.	c. d.	L. Cot.	L. Cos.	d.	′
0	9.69 897	22	9.76 144	29	10.23 856	9.93 753	7	60
1	9.69 919	22	9.76 173	29	10.23 827	9.93 746	8	59
2	9.69 941	22	9.76 202	29	10.23 798	9.93 738	8	58
3	9.69 963	21	9.76 231	30	10.23 769	9.93 731	7	57
4	9.69 984	22	9.76 261	29	10.23 739	9.93 724	7	56
5	9.70 006	22	9.76 290	29	10.23 710	9.93 717	7	55
6	9.70 028	22	9.76 319	29	10.23 681	9.93 709	8	54
7	9.70 050	22	9.76 348	29	10.23 652	9.93 702	7	53
8	9.70 072	21	9.76 377	29	10.23 623	9.93 695	7	52
9	9.70 093	22	9.76 406	29	10.23 594	9.93 687	8	51
10	9.70 115	22	9.76 435	29	10.23 565	9.93 680	7	50
11	9.70 137	22	9.76 464	29	10.23 536	9.93 673	7	49
12	9.70 159	21	9.76 493	29	10.23 507	9.93 665	8	48
13	9.70 180	22	9.76 522	29	10.23 478	9.93 658	7	47
14	9.70 202	22	9.76 551	29	10.23 449	9.93 650	8	46
15	9.70 224	21	9.76 580	29	10.23 420	9.93 643	7	45
16	9.70 245	22	9.76 609	30	10.23 391	9.93 636	7	44
17	9.70 267	21	9.76 639	29	10.23 361	9.93 628	8	43
18	9.70 288	22	9.76 668	29	10.23 332	9.93 621	7	42
19	9.70 310	22	9.76 697	28	10.23 303	9.93 614	7	41
20	9.70 332	21	9.76 725	29	10.23 275	9.93 606	8	40
21	9.70 353	22	9.76 754	29	10.23 246	9.93 599	7	39
22	9.70 375	21	9.76 783	29	10.23 217	9.93 591	8	38
23	9.70 396	22	9.76 812	29	10.23 188	9.93 584	7	37
24	9.70 418	21	9.76 841	29	10.23 159	9.93 577	7	36
25	9.70 439	22	9.76 870	29	10.23 130	9.93 569	8	35
26	9.70 461	21	9.76 899	29	10.23 101	9.93 562	7	34
27	9.70 482	22	9.76 928	29	10.23 072	9.93 554	8	33
28	9.70 504	21	9.76 957	29	10.23 043	9.93 547	7	32
29	9.70 525	22	9.76 986	29	10.23 014	9.93 539	8	31
30	9.70 547	21	9.77 015	29	10.22 985	9.93 532	7	30
31	9.70 568	22	9.77 044	29	10.22 956	9.93 525	8	29
32	9.70 590	21	9.77 073	28	10.22 927	9.93 517	7	28
33	9.70 611	22	9.77 101	29	10.22 899	9.93 510	8	27
34	9.70 633	21	9.77 130	29	10.22 870	9.93 502	7	26
35	9.70 654	21	9.77 159	29	10.22 841	9.93 495	8	25
36	9.70 675	22	9.77 188	29	10.22 812	9.93 487	7	24
37	9.70 697	21	9.77 217	29	10.22 783	9.93 480	8	23
38	9.70 718	21	9.77 246	28	10.22 754	9.93 472	7	22
39	9.70 739	22	9.77 274	29	10.22 726	9.93 465	8	21
40	9.70 761	21	9.77 303	29	10.22 697	9.93 457	7	20
41	9.70 782	21	9.77 332	29	10.22 668	9.93 450	8	19
42	9.70 803	21	9.77 361	29	10.22 639	9.93 442	7	18
43	9.70 824	22	9.77 390	28	10.22 610	9.93 435	8	17
44	9.70 846	21	9.77 418	29	10.22 582	9.93 427	7	16
45	9.70 867	21	9.77 447	29	10.22 553	9.93 420	8	15
46	9.70 888	21	9.77 476	29	10.22 524	9.93 412	7	14
47	9.70 909	22	9.77 505	28	10.22 495	9.93 405	8	13
48	9.70 931	21	9.77 533	29	10.22 467	9.93 397	7	12
49	9.70 952	21	9.77 562	29	10.22 438	9.93 390	8	11
50	9.70 973	21	9.77 591	28	10.22 409	9.93 382	7	10
51	9.70 994	21	9.77 619	29	10.22 381	9.93 375	8	9
52	9.71 015	21	9.77 648	29	10.22 352	9.93 367	7	8
53	9.71 036	22	9.77 677	29	10.22 323	9.93 360	8	7
54	9.71 058	21	9.77 706	28	10.22 294	9.93 352	8	6
55	9.71 079	21	9.77 734	29	10.22 266	9.93 344	7	5
56	9.71 100	21	9.77 763	28	10.22 237	9.93 337	8	4
57	9.71 121	21	9.77 791	29	10.22 209	9.93 329	7	3
58	9.71 142	21	9.77 820	29	10.22 180	9.93 322	8	2
59	9.71 163	21	9.77 849	28	10.22 151	9.93 314	7	1
60	9.71 184		9.77 877		10.22 123	9.93 307		0
	L. Cos.	d.	L. Cot.	c. d.	L. Tan.	L. Sin.	d.	′

P. P.

	30	29	28
1	3.0	2.9	2.8
2	6.0	5.8	5.6
3	9.0	8.7	8.4
4	12.0	11.6	11.2
5	15.0	14.5	14.0
6	18.0	17.4	16.8
7	21.0	20.3	19.6
8	24.0	23.2	22.4
9	27.0	26.1	25.2

	22	21
1	2.2	2.1
2	4.4	4.2
3	6.6	6.3
4	8.8	8.4
5	11.0	10.5
6	13.2	12.6
7	15.4	14.7
8	17.6	16.8
9	19.8	18.9

	8	7
1	0.8	0.7
2	1.6	1.4
3	2.4	2.1
4	3.2	2.8
5	4.0	3.5
6	4.8	4.2
7	5.6	4.9
8	6.4	5.6
9	7.2	6.3

′	L. Sin.	d.	L. Tan.	c. d.	L. Cot.	L. Cos.	d.	
0	9.71 184	21	9.77 877	29	10.22 123	9.93 307	8	60
1	9.71 205	21	9.77 906	29	10.22 094	9.93 299	8	59
2	9.71 226	21	9.77 935	28	10.22 065	9.93 291	7	58
3	9.71 247	21	9.77 963	29	10.22 037	9.93 284	8	57
4	9.71 268	21	9.77 992	28	10.22 008	9.93 276	7	56
5	9.71 289	21	9.78 020	29	10.21 980	9.93 269	8	55
6	9.71 310	21	9.78 049	28	10.21 951	9.93 261	8	54
7	9.71 331	21	9.78 077	29	10.21 923	9.93 253	7	53
8	9.71 352	21	9.78 106	29	10.21 894	9.93 246	8	52
9	9.71 373	20	9.78 135	28	10.21 865	9.93 238	8	51
10	9.71 393	21	9.78 163	29	10.21 837	9.93 230	7	50
11	9.71 414	21	9.78 192	28	10.21 808	9.93 223	8	49
12	9.71 435	21	9.78 220	29	10.21 780	9.93 215	8	48
13	9.71 456	21	9.78 249	28	10.21 751	9.93 207	7	47
14	9.71 477	21	9.78 277	29	10.21 723	9.93 200	8	46
15	9.71 498	21	9.78 306	28	10.21 694	9.93 192	8	45
16	9.71 519	20	9.78 334	29	10.21 666	9.93 184	7	44
17	9.71 539	21	9.78 363	28	10.21 637	9.93 177	8	43
18	9.71 560	21	9.78 391	28	10.21 609	9.93 169	8	42
19	9.71 581	21	9.78 419	29	10.21 581	9.93 161	7	41
20	9.71 602	20	9.78 448	28	10.21 552	9.93 154	8	40
21	9.71 622	21	9.78 476	29	10.21 524	9.93 146	8	39
22	9.71 643	21	9.78 505	28	10.21 495	9.93 138	7	38
23	9.71 664	21	9.78 533	29	10.21 467	9.93 131	8	37
24	9.71 685	20	9.78 562	28	10.21 438	9.93 123	8	36
25	9.71 705	21	9.78 590	28	10.21 410	9.93 115	7	35
26	9.71 726	21	9.78 618	29	10.21 382	9.93 108	8	34
27	9.71 747	20	9.78 647	28	10.21 353	9.93 100	8	33
28	9.71 767	21	9.78 675	29	10.21 325	9.93 092	8	32
29	9.71 788	21	9.78 704	28	10.21 296	9.93 084	7	31
30	9.71 809	20	9.78 732	28	10.21 268	9.93 077	8	30
31	9.71 829	21	9.78 760	29	10.21 240	9.93 069	8	29
32	9.71 850	20	9.78 789	28	10.21 211	9.93 061	8	28
33	9.71 870	21	9.78 817	28	10.21 183	9.93 053	7	27
34	9.71 891	20	9.78 845	29	10.21 155	9.93 046	8	26
35	9.71 911	21	9.78 874	28	10.21 126	9.93 038	8	25
36	9.71 932	20	9.78 902	28	10.21 098	9.93 030	8	24
37	9.71 952	21	9.78 930	29	10.21 070	9.93 022	8	23
38	9.71 973	21	9.78 959	28	10.21 041	9.93 014	7	22
39	9.71 994	20	9.78 987	28	10.21 013	9.93 007	8	21
40	9.72 014	20	9.79 015	28	10.20 985	9.92 999	8	20
41	9.72 034	21	9.79 043	29	10.20 957	9.92 991	8	19
42	9.72 055	20	9.79 072	28	10.20 928	9.92 983	7	18
43	9.72 075	21	9.79 100	28	10.20 900	9.92 976	8	17
44	9.72 096	20	9.79 128	28	10.20 872	9.92 968	8	16
45	9.72 116	21	9.79 156	29	10.20 844	9.92 960	8	15
46	9.72 137	20	9.79 185	28	10.20 815	9.92 952	8	14
47	9.72 157	20	9.79 213	28	10.20 787	9.92 944	7	13
48	9.72 177	21	9.79 241	28	10.20 759	9.92 936	8	12
49	9.72 198	20	9.79 269	28	10.20 731	9.92 929	8	11
50	9.72 218	20	9.79 297	29	10.20 703	9.92 921	8	10
51	9.72 238	21	9.79 326	28	10.20 674	9.92 913	8	9
52	9.72 259	20	9.79 354	28	10.20 646	9.92 905	8	8
53	9.72 279	20	9.79 382	28	10.20 618	9.92 897	8	7
54	9.72 299	21	9.79 410	28	10.20 590	9.92 889	8	6
55	9.72 320	20	9.79 438	28	10.20 562	9.92 881	7	5
56	9.72 340	20	9.79 466	29	10.20 534	9.92 874	8	4
57	9.72 360	20	9.79 495	28	10.20 505	9.92 866	8	3
58	9.72 381	21	9.79 523	28	10.20 477	9.92 858	8	2
59	9.72 401	20	9.79 551	28	10.20 449	9.92 850	8	1
60	9.72 421	20	9.79 579		10.20 421	9.92 842		0

| | L. Cos. | d. | L. Cot. | c. d. | L. Tan. | L. Sin. | d. | ′ |

P. P.

	29	28
1	2.9	2.8
2	5.8	5.6
3	8.7	8.4
4	11.6	11.2
5	14.5	14.0
6	17.4	16.8
7	20.3	19.6
8	23.2	22.4
9	26.1	25.2

	21	20
1	2.1	2.0
2	4.2	4.0
3	6.3	6.0
4	8.4	8.0
5	10.5	10.0
6	12.6	12.0
7	14.7	14.0
8	16.8	16.0
9	18.9	18.0

	8	7
1	0.8	0.7
2	1.6	1.4
3	2.4	2.1
4	3.2	2.8
5	4.0	3.5
6	4.8	4.2
7	5.6	4.9
8	6.4	5.6
9	7.2	6.3

′	L. Sin.	d.	L. Tan.	c. d.	L. Cot.	L. Cos.	d.		P. P.
0	9.72 421	20	9.79 579	28	10.20 421	9.92 842	8	60	
1	9.72 441	20	9.79 607	28	10.20 393	9.92 834	8	59	
2	9.72 461	21	9.79 635	28	10.20 365	9.92 826	8	58	
3	9.72 482	20	9.79 663	28	10.20 337	9.92 818	8	57	
4	9.72 502	20	9.79 691	28	10.20 309	9.92 810	7	56	
5	9.72 522	20	9.79 719	28	10.20 281	9.92 803	8	55	
6	9.72 542	20	9.79 747	29	10.20 253	9.92 795	8	54	
7	9.72 562	20	9.79 776	28	10.20 224	9.92 787	8	53	
8	9.72 582	20	9.79 804	28	10.20 196	9.92 779	8	52	
9	9.72 602	20	9.79 832	28	10.20 168	9.92 771	8	51	
10	9.72 622	21	9.79 860	28	10.20 140	9.92 763	8	50	
11	9.72 643	20	9.79 888	28	10.20 112	9.92 755	8	49	
12	9.72 663	20	9.79 916	28	10.20 084	9.92 747	8	48	
13	9.72 683	20	9.79 944	28	10.20 056	9.92 739	8	47	
14	9.72 703	20	9.79 972	28	10.20 028	9.92 731	8	46	
15	9.72 723	20	9.80 000	28	10.20 000	9.92 723	8	45	
16	9.72 743	20	9.80 028	28	10.19 972	9.92 715	8	44	
17	9.72 763	20	9.80 056	28	10.19 944	9.92 707	8	43	
18	9.72 783	20	9.80 084	28	10.19 916	9.92 699	8	42	
19	9.72 803	20	9.80 112	28	10.19 888	9.92 691	8	41	
20	9.72 823	20	9.80 140	28	10.19 860	9.92 683	8	40	
21	9.72 843	20	9.80 168	27	10.19 832	9.92 675	8	39	
22	9.72 863	20	9.80 195	28	10.19 805	9.92 667	8	38	
23	9.72 883	19	9.80 223	28	10.19 777	9.92 659	8	37	
24	9.72 902	20	9.80 251	28	10.19 749	9.92 651	8	36	
25	9.72 922	20	9.80 279	28	10.19 721	9.92 643	8	35	
26	9.72 942	20	9.80 307	28	10.19 693	9.92 635	8	34	
27	9.72 962	20	9.80 335	28	10.19 665	9.92 627	8	33	
28	9.72 982	20	9.80 363	28	10.19 637	9.92 619	8	32	
29	9.73 002	20	9.80 391	28	10.19 609	9.92 611	8	31	
30	9.73 022	19	9.80 419	28	10.19 581	9.92 603	8	30	
31	9.73 041	20	9.80 447	27	10.19 553	9.92 595	8	29	
32	9.73 061	20	9.80 474	28	10.19 526	9.92 587	8	28	
33	9.73 081	20	9.80 502	28	10.19 498	9.92 579	8	27	
34	9.73 101	20	9.80 530	28	10.19 470	9.92 571	8	26	
35	9.73 121	19	9.80 558	28	10.19 442	9.92 563	8	25	
36	9.73 140	20	9.80 586	28	10.19 414	9.92 555	9	24	
37	9.73 160	20	9.80 614	28	10.19 386	9.92 546	8	23	
38	9.73 180	20	9.80 642	27	10.19 358	9.92 538	8	22	
39	9.73 200	19	9.80 669	28	10.19 331	9.92 530	8	21	
40	9.73 219	20	9.80 697	28	10.19 303	9.92 522	8	20	
41	9.73 239	20	9.80 725	28	10.19 275	9.92 514	8	19	
42	9.73 259	19	9.80 753	28	10.19 247	9.92 506	8	18	
43	9.73 278	20	9.80 781	27	10.19 219	9.92 498	8	17	
44	9.73 298	20	9.80 808	28	10.19 192	9.92 490	8	16	
45	9.73 318	19	9.80 836	28	10.19 164	9.92 482	9	15	
46	9.73 337	20	9.80 864	28	10.19 136	9.92 473	8	14	
47	9.73 357	20	9.80 892	27	10.19 108	9.92 465	8	13	
48	9.73 377	19	9.80 919	28	10.19 081	9.92 457	8	12	
49	9.73 396	20	9.80 947	28	10.19 053	9.92 449	8	11	
50	9.73 416	19	9.80 975	28	10.19 025	9.92 441	8	10	
51	9.73 435	20	9.81 003	27	10.18 997	9.92 433	8	9	
52	9.73 455	19	9.81 030	28	10.18 970	9.92 425	9	8	
53	9.73 474	20	9.81 058	28	10.18 942	9.92 416	8	7	
54	9.73 494	19	9.81 086	27	10.18 914	9.92 408	8	6	
55	9.73 513	20	9.81 113	28	10.18 887	9.92 400	8	5	
56	9.73 533	19	9.81 141	28	10.18 859	9.92 392	8	4	
57	9.73 552	20	9.81 169	27	10.18 831	9.92 384	8	3	
58	9.73 572	19	9.81 196	28	10.18 804	9.92 376	9	2	
59	9.73 591	20	9.81 224	28	10.18 776	9.92 367	8	1	
60	9.73 611		9.81 252		10.18 748	9.92 359		0	
	L. Cos.	d.	L. Cot.	c. d.	L. Tan.	L. Sin.	d.	′	P. P.

P. P.

	29	28	27
1	2.9	2.8	2.7
2	5.8	5.6	5.4
3	8.7	8.4	8.1
4	11.6	11.2	10.8
5	14.5	14.0	13.5
6	17.4	16.8	16.2
7	20.3	19.6	18.9
8	23.2	22.4	21.6
9	26.1	25.2	24.3

	21	20	19
1	2.1	2.0	1.9
2	4.2	4.0	3.8
3	6.3	6.0	5.7
4	8.4	8.0	7.6
5	10.5	10.0	9.5
6	12.6	12.0	11.4
7	14.7	14.0	13.3
8	16.8	16.0	15.2
9	18.9	18.0	17.1

	9	8	7
1	0.9	0.8	0.7
2	1.8	1.6	1.4
3	2.7	2.4	2.1
4	3.6	3.2	2.8
5	4.5	4.0	3.5
6	5.4	4.8	4.2
7	6.3	5.6	4.9
8	7.2	6.4	5.6
9	8.1	7.2	6.3

′	L. Sin.	d.	L. Tan.	c. d.	L. Cot.	L. Cos.	d.		P. P.
0	9.73 611	19	9.81 252	27	10.18 748	9.92 359	8	60	
1	9.73 630	20	9.81 279	28	10.18 721	9.92 351	8	59	
2	9.73 650	19	9.81 307	28	10 18 693	9.92 343	8	58	
3	9.73 669	20	9.81 335	27	10.18 665	9.92 335	9	57	
4	9.73 689	19	9.81 362	28	10.18 638	9.92 326	8	56	
5	9.73 708	19	9.81 390	28	10.18 610	9.92 318	8	55	
6	9.73 727	20	9.81 418	27	10.18 582	9.92 310	8	54	
7	9.73 747	19	9.81 445	28	10.18 555	9.92 302	9	53	
8	9.73 766	19	9.81 473	27	10.18 527	9.92 293	8	52	
9	9.73 785	20	9.81 500	28	10.18 500	9.92 285	8	51	
10	9.73 805	19	9.81 528	28	10.18 472	9.92 277	8	50	**28 27**
11	9.73 824	19	9.81 556	27	10.18 444	9.92 269	9	49	1| 2.8 2.7
12	9.73 843	20	9.81 583	28	10.18 417	9.92 260	8	48	2| 5.6 5.4
13	9.73 863	19	9.81 611	27	10.18 389	9.92 252	8	47	3| 8.4 8.1
14	9.73 882	19	9.81 638	28	10.18 362	9.92 244	9	46	4|11.2 10.8
15	9.73 901	20	9.81 666	27	10.18 334	9.92 235	8	45	5|14.0 13.5
16	9.73 921	19	9.81 693	28	10.18 307	9.92 227	8	44	6|16.8 16.2
17	9.73 940	19	9.81 721	27	10.18 279	9.92 219	8	43	7|19.6 18.9
18	9.73 959	19	9.81 748	28	10.18 252	9.92 211	9	42	8|22.4 21.6
19	9.73 978	19	9.81 776	27	10.18 224	9.92 202	8	41	9|25.2 24.3
20	9.73 997	20	9.81 803	28	10.18 197	9.92 194	8	40	
21	9.74 017	19	9.81 831	27	10.18 169	9.92 186	9	39	
22	9.74 036	19	9.81 858	28	10.18 142	9.92 177	8	38	
23	9.74 055	19	9.81 886	27	10.18 114	9.92 169	8	37	
24	9.74 074	19	9.81 913	28	10.18 087	9.92 161	9	36	
25	9.74 093	20	9.81 941	27	10.18 059	9.92 152	8	35	
26	9.74 113	19	9.81 968	28	10.18 032	9.92 144	8	34	**20 19 18**
27	9.74 132	19	9.81 996	27	10.18 004	9.92 136	9	33	1| 2.0 1.9 1.8
28	9.74 151	19	9.82 023	28	10.17 977	9.92 127	8	32	2| 4.0 3.8 3.6
29	9.74 170	19	9.82 051	27	10.17 949	9.92 119	8	31	3| 6.0 5.7 5.4
30	9.74 189	19	9.82 078	28	10.17 922	9.92 111	9	30	4| 8.0 7.6 7.2
31	9.74 208	19	9.82 106	27	10.17 894	9.92 102	8	29	5|10.0 9.5 9.0
32	9.74 227	19	9.82 133	28	10.17 867	9.92 094	8	28	6|12.0 11.4 10.8
33	9.74 246	19	9.82 161	27	10.17 839	9.92 086	9	27	7|14.0 13.3 12.6
34	9.74 265	19	9.82 188	27	10.17 812	9.92 077	8	26	8|16.0 15.2 14.4
35	9.74 284	19	9.82 215	28	10.17 785	9.92 069	9	25	9|18.0 17.1 16.2
36	9.74 303	19	9.82 243	27	10.17 757	9.92 060	8	24	
37	9.74 322	19	9.82 270	28	10.17 730	9.92 052	8	23	
38	9.74 341	19	9.82 298	27	10.17 702	9.92 044	9	22	
39	9.74 360	19	9.82 325	27	10.17 675	9.92 035	8	21	
40	9.74 379	19	9.82 352	28	10.17 648	9.92 027	9	20	
41	9.74 398	19	9.82 380	27	10.17 620	9.92 018	8	19	
42	9.74 417	19	9.82 407	28	10.17 593	9.92 010	8	18	
43	9.74 436	19	9.82 435	27	10.17 565	9.92 002	9	17	
44	9.74 455	19	9.82 462	27	10.17 538	9.91 993	8	16	**9 8**
45	9.74 474	19	9.82 489	28	10.17 511	9.91 985	9	15	1| 0.9 0.8
46	9.74 493	19	9.82 517	27	10.17 483	9.91 976	8	14	2| 1.8 1.6
47	9.74 512	19	9.82 544	27	10.17 456	9.91 968	9	13	3| 2.7 2.4
48	9.74 531	18	9.82 571	28	10.17 429	9.91 959	8	12	4| 3.6 3.2
49	9.74 549	19	9.82 599	27	10.17 401	9.91 951	9	11	5| 4.5 4.0
50	9.74 568	19	9.82 626	27	10.17 374	9.91 942	8	10	6| 5.4 4.8
51	9.74 587	19	9.82 653	28	10.17 347	9.91 934	9	9	7| 6.3 5.6
52	9.74 606	19	9.82 681	27	10.17 319	9.91 925	8	8	8| 7.2 6.4
53	9.74 625	19	9.82 708	27	10.17 292	9.91 917	9	7	9| 8.1 7.2
54	9.74 644	18	9.82 735	27	10.17 265	9.91 908	8	6	
55	9.74 662	19	9.82 762	28	10.17 238	9.91 900	9	5	
56	9.74 681	19	9.82 790	27	10.17 210	9.91 891	8	4	
57	9.74 700	19	9.82 817	27	10 17 183	9.91 883	9	3	
58	9.74 719	18	9.82 844	27	10.17 156	9.91 874	8	2	
59	9.74 737	19	9.82 871	28	10.17 129	9.91 866	9	1	
60	9.74 756		9.82 899		10.17 101	9.91 857		0	
	L. Cos.	d.	L. Cot.	c. d.	L. Tan.	L. Sin.	d.	′	P. P.

'	L. Sin.	d.	L. Tan.	c. d.	L. Cot.	L. Cos.	d.	
0	9.74 756	19	9.82 899	27	10.17 101	9.91 857	8	60
1	9.74 775	19	9.82 926	27	10.17 074	9.91 849	9	59
2	9.74 794	18	9.82 953	27	10.17 047	9.91 840	8	58
3	9.74 812	19	9.82 980	28	10.17 020	9.91 832	9	57
4	9.74 831	19	9.83 008	27	10.16 992	9.91 823	8	56
5	9.74 850	18	9.83 035	27	10.16 965	9.91 815	9	55
6	9.74 868	19	9.83 062	27	10.16 938	9.91 806	8	54
7	9.74 887	19	9.83 089	28	10.16 911	9.91 798	9	53
8	9.74 906	18	9.83 117	27	10.16 883	9.91 789	8	52
9	9.74 924	19	9.83 144	27	10.16 856	9.91 781	9	51
10	9.74 943	18	9.83 171	27	10.16 829	9.91 772	9	50
11	9.74 961	19	9.83 198	27	10.16 802	9.91 763	8	49
12	9.74 980	19	9.83 225	27	10.16 775	9.91 755	9	48
13	9.74 999	18	9.83 252	28	10.16 748	9.91 746	8	47
14	9.75 017	19	9.83 280	27	10.16 720	9.91 738	9	46
15	9.75 036	18	9.83 307	27	10.16 693	9.91 729	9	45
16	9.75 054	19	9.83 334	27	10.16 666	9.91 720	8	44
17	9.75 073	18	9.83 361	27	10.16 639	9.91 712	9	43
18	9.75 091	19	9.83 388	27	10.16 612	9.91 703	8	42
19	9.75 110	18	9.83 415	27	10.16 585	9.91 695	9	41
20	9.75 128	19	9.83 442	28	10.16 558	9.91 686	9	40
21	9.75 147	18	9.83 470	27	10.16 530	9.91 677	8	39
22	9.75 165	19	9.83 497	27	10.16 503	9.91 669	9	38
23	9.75 184	18	9.83 524	27	10.16 476	9.91 660	9	37
24	9.75 202	19	9.83 551	27	10.16 449	9.91 651	8	36
25	9.75 221	18	9.83 578	27	10.16 422	9.91 643	9	35
26	9.75 239	19	9.83 605	27	10.16 395	9.91 634	9	34
27	9.75 258	18	9.83 632	27	10.16 368	9.91 625	8	33
28	9.75 276	18	9.83 659	27	10.16 341	9.91 617	9	32
29	9.75 294	19	9.83 686	27	10.16 314	9.91 608	9	31
30	9.75 313	18	9.83 713	27	10.16 287	9.91 599	8	30
31	9.75 331	19	9.83 740	28	10.16 260	9.91 591	9	29
32	9.75 350	18	9.83 768	27	10.16 232	9.91 582	9	28
33	9.75 368	18	9.83 795	27	10.16 205	9.91 573	8	27
34	9.75 386	19	9.83 822	27	10.16 178	9.91 565	9	26
35	9.75 405	18	9.83 849	27	10.16 151	9.91 556	9	25
36	9.75 423	18	9.83 876	27	10 16 124	9.91 547	9	24
37	9.75 441	18	9.83 903	27	10.16 097	9.91 538	8	23
38	9.75 459	19	9.83 930	27	10.16 070	9.91 530	9	22
39	9.75 478	18	9.83 957	27	10.16 043	9.91 521	9	21
40	9.75 496	18	9.83 984	27	10.16 016	9.91 512	8	20
41	9.75 514	19	9.84 011	27	10.15 989	9.91 504	9	19
42	9.75 533	18	9.84 038	27	10.15 962	9.91 495	9	18
43	9.75 551	18	9.84 065	27	10.15 935	9.91 486	9	17
44	9.75 569	18	9.84 092	27	10.15 908	9.91 477	8	16
45	9.75 587	18	9.84 119	27	10.15 881	9.91 469	9	15
46	9.75 605	19	9.84 146	27	10.15 854	9.91 460	9	14
47	9.75 624	18	9.84 173	27	10.15 827	9.91 451	9	13
48	9.75 642	18	9.84 200	27	10.15 800	9.91 442	9	12
49	9.75 660	18	9.84 227	27	10.15 773	9.91 433	8	11
50	9.75 678	18	9.84 254	26	10.15 746	9.91 425	9	10
51	9.75 696	18	9.84 280	27	10.15 720	9.91 416	9	9
52	9.75 714	19	9.84 307	27	10.15 693	9.91 40'	9	8
53	9.75 733	18	9.84 334	27	10.15 666	9.91 398	9	7
54	9.75 751	18	9.84 361	27	10.15 639	9.91 389	8	6
55	9.75 769	18	9.84 388	27	10.15 612	9.91 381	9	5
56	9.75 787	18	9.84 415	27	10.15 585	9.91 372	9	4
57	9.75 805	18	9.84 442	27	10.15 558	9.91 363	9	3
58	9.75 823	18	9.84 469	27	10.15 531	9.91 354	9	2
59	9.75 841	18	9.84 496	27	10.15 504	9.91 345	9	1
60	9.75 859		9.84 523		10.15 477	9.91 336		0
	L. Cos.	d.	L. Cot.	c. d.	L. Tan.	L. Sin.	d.	'

P. P.

	28	27	26
1	2.8	2.7	2.6
2	5.6	5.4	5.2
3	8.4	8.1	7.8
4	11.2	10.8	10.4
5	14.0	13.5	13.0
6	16.8	16.2	15.6
7	19.6	18.9	18.2
8	22.4	21.6	20.8
9	25.2	24.3	23.4

	19	18
1	1.9	1.8
2	3.8	3.6
3	5.7	5.4
4	7.6	7.2
5	9.5	9.0
6	11.4	10.8
7	13.3	12.6
8	15.2	14.4
9	17.1	16.2

	9	8
1	0.9	0.8
2	1.8	1.6
3	2.7	2.4
4	3.6	3.2
5	4.5	4.0
6	5.4	4.8
7	6.3	5.6
8	7.2	6.4
9	8.1	7.2

55°

′	L. Sin.	d.	L. Tan.	c. d.	L. Cot.	L. Cos.	d.	′
0	9.75 859		9.84 523		10.15 477	9.91 336		60
1	9.75 877	18	9.84 550	27	10.15 450	9.91 328	8	59
2	9.75 895	18	9.84 576	26	10.15 424	9.91 319	9	58
3	9.75 913	18	9.84 603	27	10.15 397	9.91 310	9	57
4	9.75 931	18	9.84 630	27	10.15 370	9.91 301	9	56
5	9.75 949	18	9.84 657	27	10.15 343	9.91 292	9	55
6	9.75 967	18	9.84 684	27	10.15 316	9.91 283	9	54
7	9.75 985	18	9.84 711	27	10.15 289	9.91 274	9	53
8	9.76 003	18	9.84 738	27	10.15 262	9.91 266	8	52
9	9.76 021	18	9.84 764	26	10.15 236	9.91 257	9	51
10	9.76 039	18	9.84 791	27	10.15 209	9.91 248	9	50
11	9.76 057	18	9.84 818	27	10.15 182	9.91 239	9	49
12	9.76 075	18	9.84 845	27	10.15 155	9.91 230	9	48
13	9.76 093	18	9.84 872	27	10.15 128	9.91 221	9	47
14	9.76 111	18	9.84 899	27	10.15 101	9.91 212	9	46
15	9.76 129	18	9.84 925	26	10.15 075	9.91 203	9	45
16	9.76 146	17	9.84 952	27	10.15 048	9.91 194	9	44
17	9.76 164	18	9.84 979	27	10.15 021	9.91 185	9	43
18	9.76 182	18	9.85 006	27	10.14 994	9.91 176	9	42
19	9.76 200	18	9.85 033	27	10.14 967	9.91 167	9	41
20	9.76 218	18	9.85 059	26	10.14 941	9.91 158	9	40
21	9.76 236	18	9.85 086	27	10.14 914	9.91 149	8	39
22	9.76 253	17	9.85 113	27	10.14 887	9.91 141	9	38
23	9.76 271	18	9.85 140	27	10.14 860	9.91 132	9	37
24	9.76 289	18	9.85 166	26	10.14 834	9.91 123	9	36
25	9.76 307	18	9.85 193	27	10.14 807	9.91 114	9	35
26	9.76 324	17	9.85 220	27	10.14 780	9.91 105	9	34
27	9.76 342	18	9.85 247	27	10.14 753	9.91 096	9	33
28	9.76 360	18	9.85 273	26	10.14 727	9.91 087	9	32
29	9.76 378	18	9.85 300	27	10.14 700	9.91 078	9	31
30	9.76 395	17	9.85 327	27	10.14 673	9.91 069	9	30
31	9.76 413	18	9.85 354	27	10.14 646	9.91 060	9	29
32	9.76 431	18	9.85 380	26	10.14 620	9.91 051	9	28
33	9.76 448	17	9.85 407	27	10.14 593	9.91 042	9	27
34	9.76 466	18	9.85 434	27	10.14 566	9.91 033	10	26
35	9.76 484	18	9.85 460	26	10.14 540	9.91 023	9	25
36	9.76 501	17	9.85 487	27	10.14 513	9.91 014	9	24
37	9.76 519	18	9.85 514	27	10.14 486	9.91 005	9	23
38	9.76 537	18	9.85 540	26	10.14 460	9.90 996	9	22
39	9.76 554	17	9.85 567	27	10.14 433	9.90 987	9	21
40	9.76 572	18	9.85 594	27	10.14 406	9.90 978	9	20
41	9.76 590	18	9.85 620	26	10.14 380	9.90 969	9	19
42	9.76 607	17	9.85 647	27	10.14 353	9.90 960	9	18
43	9.76 625	18	9.85 674	27	10.14 326	9.90 951	9	17
44	9.76 642	17	9.85 700	26	10.14 300	9.90 942	9	16
45	9.76 660	18	9.85 727	27	10.14 273	9.90 933	9	15
46	9.76 677	17	9.85 754	27	10.14 246	9.90 924	9	14
47	9.76 695	18	9.85 780	26	10.14 220	9.90 915	9	13
48	9.76 712	17	9.85 807	27	10.14 193	9.90 906	10	12
49	9.76 730	18	9.85 834	27	10.14 166	9.90 896	9	11
50	9.76 747	17	9.85 860	26	10.14 140	9.90 887	9	10
51	9.76 765	18	9.85 887	27	10.14 113	9.90 878	9	9
52	9.76 782	17	9.85 913	26	10.14 087	9.90 869	9	8
53	9.76 800	18	9.85 940	27	10.14 060	9.90 860	9	7
54	9.76 817	17	9.85 967	27	10.14 033	9.90 851	9	6
55	9.76 835	18	9.85 993	26	10.14 007	9.90 842	10	5
56	9.76 852	17	9.86 020	27	10.13 980	9.90 832	9	4
57	9.76 870	18	9.86 046	26	10.13 954	9.90 823	9	3
58	9.76 887	17	9.86 073	27	10.13 927	9.90 814	9	2
59	9.76 904	17	9.86 100	27	10.13 900	9.90 805	9	1
60	9.76 922	18	9.86 126	26	10.13 874	9.90 796	9	0
	L. Cos.	d.	L. Cot.	c. d.	L. Tan.	L. Sin.	d.	′

P. P.

	27	26
1	2.7	2.6
2	5.4	5.2
3	8.1	7.8
4	10.8	10.4
5	13.5	13.0
6	16.2	15.6
7	18.9	18.2
8	21.6	20.8
9	24.3	23.4

	18	17
1	1.8	1.7
2	3.6	3.4
3	5.4	5.1
4	7.2	6.8
5	9.0	8.5
6	10.8	10.2
7	12.6	11.9
8	14.4	13.6
9	16.2	15.3

	10	9	8
1	1.0	0.9	0.8
2	2.0	1.8	1.6
3	3.0	2.7	2.4
4	4.0	3.6	3.2
5	5.0	4.5	4.0
6	6.0	5.4	4.8
7	7.0	6.3	5.6
8	8.0	7.2	6.4
9	9.0	8.1	7.2

′	L. Sin.	d.	L. Tan.	c. d.	L. Cot.	L. Cos.	d.		P. P.
0	9.76 922	17	9.86 126	27	10.13 874	9.90 796	9	60	
1	9.76 939	18	9.86 153	26	10.13 847	9.90 787	10	59	
2	9.76 957	17	9.86 179	27	10.13 821	9.90 777	9	58	
3	9.76 974	17	9.86 206	26	10.13 794	9.90 768	9	57	
4	9.76 991	18	9.86 232	27	10.13 768	9.90 759	9	56	
5	9.77 009	17	9.86 259	26	10.13 741	9.90 750	9	55	
6	9.77 026	17	9.86 285	27	10.13 715	9.90 741	10	54	
7	9.77 043	18	9.86 312	26	10.13 688	9.90 731	9	53	
8	9.77 061	17	9.86 338	27	10.13 662	9.90 722	9	52	
9	9.77 078	17	9.86 365	27	10.13 635	9.90 713	9	51	27 26
10	9.77 095	17	9.86 392	26	10.13 608	9.90 704	10	50	
11	9.77 112	18	9.86 418	27	10.13 582	9.90 694	10	49	1|2.7 2.6
12	9.77 130	17	9.86 445	26	10.13 555	9.90 685	9	48	2|5.4 5.2
13	9.77 147	17	9.86 471	27	10.13 529	9.90 676	9	47	3|8.1 7.8
14	9.77 164	17	9.86 498	26	10.13 502	9.90 667	10	46	4|10.8 10.4
15	9.77 181	18	9.86 524	27	10.13 476	9.90 657	9	45	5|13.5 13.0
16	9.77 199	17	9.86 551	26	10.13 449	9.90 648	9	44	6|16.2 15.6
17	9.77 216	17	9.86 577	26	10.13 423	9.90 639	9	43	7|18.9 18.2
18	9.77 233	17	9.86 603	27	10.13 397	9.90 630	10	42	8|21.6 20.8
19	9.77 250	18	9.86 630	26	10.13 370	9.90 620	9	41	9|24.3 23.4
20	9.77 268	17	9.86 656	27	10.13 344	9.90 611	9	40	
21	9.77 285	17	9.86 683	26	10.13 317	9.90 602	10	39	
22	9.77 302	17	9.86 709	27	10.13 291	9.90 592	9	38	
23	9.77 319	17	9.86 736	26	10.13 264	9.90 583	9	37	
24	9.77 336	17	9.86 762	27	10.13 238	9.90 574	9	36	
25	9.77 353	17	9.86 789	26	10.13 211	9.90 565	10	35	
26	9.77 370	17	9.86 815	27	10.13 185	9.90 555	9	34	18 17 16
27	9.77 387	18	9.86 842	26	10.13 158	9.90 546	9	33	
28	9.77 405	17	9.86 868	26	10.13 132	9.90 537	10	32	1|1.8 1.7 1.6
29	9.77 422	17	9.86 894	27	10.13 106	9.90 527	9	31	2|3.6 3.4 3.2
30	9.77 439	17	9.86 921	26	10.13 079	9.90 518	9	30	3|5.4 5.1 4.8
31	9.77 456	17	9.86 947	27	10.13 053	9.90 509	10	29	4|7.2 6.8 6.4
32	9.77 473	17	9.86 974	26	10.13 026	9.90 499	9	28	5|9.0 8.5 8.0
33	9.77 490	17	9.87 000	27	10.13 000	9.90 490	10	27	6|10.8 10.2 9.6
34	9.77 507	17	9.87 027	26	10.12 973	9.90 480	9	26	7|12.6 11.9 11.2
35	9.77 524	17	9.87 053	26	10.12 947	9.90 471	9	25	8|14.4 13.6 12.8
36	9.77 541	17	9.87 079	27	10.12 921	9.90 462	10	24	9|16.2 15.3 14.4
37	9.77 558	17	9.87 106	26	10.12 894	9.90 452	9	23	
38	9.77 575	17	9.87 132	26	10.12 868	9.90 443	9	22	
39	9.77 592	17	9.87 158	27	10.12 842	9.90 434	10	21	
40	9.77 609	17	9.87 185	26	10.12 815	9.90 424	9	20	
41	9.77 626	17	9.87 211	27	10.12 789	9.90 415	10	19	
42	9.77 643	17	9.87 238	26	10.12 762	9.90 405	9	18	
43	9.77 660	17	9.87 264	26	10.12 736	9.90 396	10	17	
44	9.77 677	17	9.87 290	27	10.12 710	9.90 386	9	16	10 9
45	9.77 694	17	9.87 317	26	10.12 683	9.90 377	9	15	
46	9.77 711	17	9.87 343	26	10.12 657	9.90 368	10	14	1|1.0 0.9
47	9.77 728	16	9.87 369	27	10.12 631	9.90 358	9	13	2|2.0 1.8
48	9.77 744	17	9.87 396	26	10.12 604	9.90 349	10	12	3|3.0 2.7
49	9.77 761	17	9.87 422	26	10.12 578	9.90 339	9	11	4|4.0 3.6
50	9.77 778	17	9.87 448	27	10.12 552	9.90 330	10	10	5|5.0 4.5
51	9.77 795	17	9.87 475	26	10.12 525	9.90 320	9	9	6|6.0 5.4
52	9.77 812	17	9.87 501	26	10.12 499	9.90 311	10	8	7|7.0 6.3
53	9.77 829	17	9.87 527	27	10.12 473	9.90 301	9	7	8|8.0 7.2
54	9.77 846	16	9.87 554	26	10.12 446	9.90 292	10	6	9|9.0 8.1
55	9.77 862	17	9.87 580	26	10.12 420	9.90 282	9	5	
56	9.77 879	17	9.87 606	27	10.12 394	9.90 273	10	4	
57	9.77 896	17	9.87 633	26	10.12 367	9.90 263	9	3	
58	9.77 913	17	9.87 659	26	10.12 341	9.90 254	10	2	
59	9.77 930	16	9.87 685	26	10.12 315	9.90 244	9	1	
60	9.77 946		9.87 711		10.12 289	9.90 235		0	
	L. Cos.	d.	L. Cot.	c. d.	L. Tan.	L. Sin.	d.	′	P. P.

′	L. Sin.	d.	L. Tan.	c. d.	L. Cot.	L. Cos.	d.	′	P. P.
0	9.77 940	17	9.87 711	27	10.12 289	9.90 235	10	60	
1	9.77 963	17	9.87 738	26	10.12 262	9.90 225	9	59	
2	9.77 980	17	9.87 764	26	10.12 236	9.90 216	10	58	
3	9.77 997	16	9.87 790	27	10.12 210	9.90 206	9	57	
4	9.78 013	17	9.87 817	26	10.12 183	9.90 197	10	56	
5	9.78 030	17	9.87 843	26	10.12 157	9.90 187	9	55	
6	9.78 047	16	9.87 869	26	10.12 131	9.90 178	10	54	
7	9.78 063	17	9.87 895	27	10.12 105	9.90 168	9	53	
8	9.78 080	17	9.87 922	26	10.12 078	9.90 159	10	52	
9	9.78 097	16	9.87 948	26	10.12 052	9.90 149	10	51	**27** **26**
10	9.78 113	17	9.87 974	26	10.12 026	9.90 139	9	50	
11	9.78 130	17	9.88 000	27	10.12 000	9.90 130	10	49	1| 2.7 2.6
12	9.78 147	16	9.88 027	26	10.11 973	9.90 120	9	48	2| 5.4 5.2
13	9.78 163	17	9.88 053	26	10.11 947	9.90 111	10	47	3| 8.1 7.8
14	9.78 180	17	9.88 079	26	10.11 921	9.90 101	10	46	4| 10.8 10.4
15	9.78 197	16	9.88 105	26	10.11 895	9.90 091	9	45	5| 13.5 13.0
16	9.78 213	17	9.88 131	27	10.11 869	9.90 082	10	44	6| 16.2 15.6
17	9.78 230	16	9.88 158	26	10.11 842	9.90 072	9	43	7| 18.9 18.2
18	9.78 246	17	9.88 184	26	10.11 816	9.90 063	10	42	8| 21.6 20.8
19	9.78 263	17	9.88 210	26	10.11 790	9.90 053	10	41	9| 24.3 23.4
20	9.78 280	16	9.88 236	26	10.11 764	9.90 043	9	40	
21	9.78 296	17	9.88 262	27	10.11 738	9.90 034	10	39	
22	9.78 313	16	9.88 289	26	10.11 711	9.90 024	10	38	
23	9.78 329	17	9.88 315	26	10.11 685	9.90 014	9	37	
24	9.78 346	16	9.88 341	26	10.11 659	9.90 005	10	36	
25	9.78 362	17	9.88 367	26	10.11 633	9.89 995	10	35	
26	9.78 379	16	9.88 393	27	10.11 607	9.89 985	9	34	**17** **16**
27	9.78 395	17	9.88 420	26	10.11 580	9.89 976	10	33	
28	9.78 412	16	9.88 446	26	10.11 554	9.89 966	10	32	1| 1.7 1.6
29	9.78 428	17	9.88 472	26	10.11 528	9.89 956	9	31	2| 3.4 3.2
30	9.78 445	16	9.88 498	26	10.11 502	9.89 947	10	30	3| 5.1 4.8
31	9.78 461	17	9.88 524	26	10.11 476	9.89 937	10	29	4| 6.8 6.4
32	9.78 478	16	9.88 550	27	10.11 450	9.89 927	9	28	5| 8.5 8.0
33	9.78 494	16	9.88 577	26	10.11 423	9.89 918	10	27	6| 10.2 9.6
34	9.78 510	17	9.88 603	26	10.11 397	9.89 908	10	26	7| 11.9 11.2
35	9.78 527	16	9.88 629	26	10.11 371	9.89 898	10	25	8| 13.6 12.8
36	9.78 543	17	9.88 655	26	10.11 345	9.89 888	9	24	9| 15.3 14.4
37	9.78 560	16	9.88 681	26	10.11 319	9.89 879	10	23	
38	9.78 576	16	9.88 707	26	10.11 293	9.89 869	10	22	
39	9.78 592	17	9.88 733	26	10.11 267	9.89 859	10	21	
40	9.78 609	16	9.88 759	27	10.11 241	9.89 849	9	20	
41	9.78 625	17	9.88 786	26	10.11 214	9.89 840	10	19	
42	9.78 642	16	9.88 812	26	10.11 188	9.89 830	10	18	
43	9.78 658	16	9.88 838	26	10.11 162	9.89 820	10	17	**10** **9**
44	9.78 674	17	9.88 864	26	10.11 136	9.89 810	9	16	
45	9.78 691	16	9.88 890	26	10.11 110	9.89 801	10	15	1| 1.0 0.9
46	9.78 707	16	9.88 916	26	10.11 084	9.89 791	10	14	2| 2.0 1.8
47	9.78 723	16	9.88 942	26	10.11 058	9.89 781	10	13	3| 3.0 2.7
48	9.78 739	17	9.88 968	26	10.11 032	9.89 771	10	12	4| 4.0 3.6
49	9.78 756	16	9.88 994	26	10.11 006	9.89 761	9	11	5| 5.0 4.5
50	9.78 772	16	9.89 020	26	10.10 980	9.89 752	10	10	6| 6.0 5.4
51	9.78 788	17	9.89 046	27	10.10 954	9.89 742	10	9	7| 7.0 6.3
52	9.78 805	16	9.89 073	26	10.10 927	9.89 732	10	8	8| 8.0 7.2
53	9.78 821	16	9.89 099	26	10.10 901	9.89 722	10	7	9| 9.0 8.1
54	9.78 837	16	9.89 125	26	10.10 875	9.89 712	10	6	
55	9.78 853	16	9.89 151	26	10.10 849	9.89 702	9	5	
56	9.78 869	17	9.89 177	26	10.10 823	9.89 693	10	4	
57	9.78 886	16	9.89 203	26	10.10 797	9.89 683	10	3	
58	9.78 902	16	9.89 229	26	10.10 771	9.89 673	10	2	
59	9.78 918	16	9.89 255	26	10.10 745	9.89 663	10	1	
60	9.78 934		9.89 281		10.10 719	9.89 653		0	

	L. Cos.	d.	L. Cot.	c. d.	L. Tan.	L. Sin.	d.	′	P. P.

′	L. Sin.	d.	L. Tan.	c. d.	L. Cot.	L. Cos.	d.		P. P.
0	9.78 934	16	9.89 281	26	10.10 719	9.89 653	10	60	
1	9.78 950	17	9.89 307	26	10.10 693	9.89 643	10	59	
2	9.78 967	16	9.89 333	26	10.10 667	9.89 633	9	58	
3	9.78 983	16	9.89 359	26	10.10 641	9.89 624	10	57	
4	9.78 999	16	9.89 385	26	10.10 615	9.89 614	10	56	
5	9.79 015	16	9.89 411	26	10.10 589	9.89 604	10	55	
6	9.79 031	16	9.89 437	26	10.10 563	9.89 594	10	54	
7	9.79 047	16	9.89 463	26	10.10 537	9.89 584	10	53	
8	9.79 063	16	9.89 489	26	10.10 511	9.89 574	10	52	
9	9.79 079	16	9.89 515	26	10.10 485	9.89 564	10	51	26 25
10	9.79 095	16	9.89 541	26	10.10 459	9.89 554	10	50	
11	9.79 111	17	9.89 567	26	10.10 433	9.89 544	10	49	1 | 2.6 2.5
12	9.79 128	16	9.89 593	26	10.10 407	9.89 534	10	48	2 | 5.2 5.0
13	9.79 144	16	9.89 619	26	10.10 381	9.89 524	10	47	3 | 7.8 7.5
14	9.79 160	16	9.89 645	26	10.10 355	9.89 514	10	46	4 | 10.4 10.0
15	9.79 176	16	9.89 671	26	10.10 329	9.89 504	9	45	5 | 13.0 12.5
16	9.79 192	16	9.89 697	26	10.10 303	9.89 495	10	44	6 | 15.6 15.0
17	9.79 208	16	9.89 723	26	10.10 277	9.89 485	10	43	7 | 18.2 17.5
18	9.79 224	16	9.89 749	26	10.10 251	9.89 475	10	42	8 | 20.8 20.0
19	9.79 240	16	9.89 775	26	10.10 225	9.89 465	10	41	9 | 23.4 22.5
20	9.79 256	16	9.89 801	26	10.10 199	9.89 455	10	40	
21	9.79 272	16	9.89 827	26	10.10 173	9.89 445	10	39	
22	9.79 288	16	9.89 853	26	10.10 147	9.89 435	10	38	
23	9.79 304	15	9.89 879	26	10.10 121	9.89 425	10	37	
24	9.79 319	16	9.89 905	26	10.10 095	9.89 415	10	36	
25	9.79 335	16	9.89 931	26	10.10 069	9.89 405	10	35	
26	9.79 351	16	9.89 957	26	10.10 043	9.89 395	10	34	
27	9.79 367	16	9.89 983	26	10.10 017	9.89 385	10	33	17 16 15
28	9.79 383	16	9.90 009	26	10.09 991	9.89 375	11	32	
29	9.79 399	16	9.90 035	26	10.09 965	9.89 364	10	31	1 | 1.7 1.6 1.5
30	9.79 415	16	9.90 061	25	10.09 939	9.89 354	10	30	2 | 3.4 3.2 3.0
31	9.79 431	16	9.90 086	26	10.09 914	9.89 344	10	29	3 | 5.1 4.8 4.5
32	9.79 447	16	9.90 112	26	10.09 888	9.89 334	10	28	4 | 6.8 6.4 6.0
33	9.79 463	15	9.90 138	26	10.09 862	9.89 324	10	27	5 | 8.5 8.0 7.5
34	9.79 478	16	9.90 164	26	10.09 836	9.89 314	10	26	6 | 10.2 9.6 9.0
35	9.79 494	16	9.90 190	26	10.09 810	9.89 304	10	25	7 | 11.9 11.2 10.5
36	9.79 510	16	9.90 216	26	10.09 784	9.89 294	10	24	8 | 13.6 12.8 12.0
37	9.79 526	16	9.90 242	26	10.09 758	9.89 284	10	23	9 | 15.3 14.4 13.5
38	9.79 542	16	9.90 268	26	10.09 732	9.89 274	10	22	
39	9.79 558	15	9.90 294	26	10.09 706	9.89 264	10	21	
40	9.79 573	16	9.90 320	26	10.09 680	9.89 254	10	20	
41	9.79 589	16	9.90 346	25	10.09 654	9.89 244	11	19	
42	9.79 605	16	9.90 371	26	10.09 629	9.89 233	10	18	
43	9.79 621	15	9.90 397	26	10.09 603	9.89 223	10	17	
44	9.79 636	16	9.90 423	26	10.09 577	9.89 213	10	16	11 10 9
45	9.79 652	16	9.90 449	26	10.09 551	9.89 203	10	15	
46	9.79 668	16	9.90 475	26	10.09 525	9.89 193	10	14	1 | 1.1 1.0 0.9
47	9.79 684	15	9.90 501	26	10.09 499	9.89 183	10	13	2 | 2.2 2.0 1.8
48	9.79 699	16	9.90 527	26	10.09 473	9.89 173	11	12	3 | 3.3 3.0 2.7
49	9.79 715	16	9.90 553	25	10.09 447	9.89 162	10	11	4 | 4.4 4.0 3.6
50	9.79 731	15	9.90 578	26	10.09 422	9.89 152	10	10	5 | 5.5 5.0 4.5
51	9.79 746	16	9.90 604	26	10.09 396	9.89 142	10	9	6 | 6.6 6.0 5.4
52	9.79 762	16	9.90 630	26	10.09 370	9.89 132	10	8	7 | 7.7 7.0 6.3
53	9.79 778	15	9.90 656	26	10.09 344	9.89 122	10	7	8 | 8.8 8.0 7.2
54	9.79 793	16	9.90 682	26	10.09 318	9.89 112	11	6	9 | 9.9 9.0 8.1
55	9.79 809	16	9.90 708	26	10.09 292	9.89 101	10	5	
56	9.79 825	15	9.90 734	25	10.09 266	9.89 091	10	4	
57	9.79 840	16	9.90 759	26	10.09 241	9.89 081	10	3	
58	9.79 856	16	9.90 785	26	10.09 215	9.89 071	11	2	
59	9.79 872	15	9.90 811	26	10.09 189	9.89 060	10	1	
60	9.79 887		9.90 837		10.09 163	9.89 050		0	
	L. Cos.	d.	L. Cot.	c. d.	L. Tan.	L. Sin.	d.	′	P. P.

'	L. Sin.	d.	L. Tan.	c. d.	L. Cot.	L. Cos.	d.		P. P.
0	9.79 887	16	9.90 837	26	10.09 163	9.89 050	10	60	
1	9.79 903	15	9.90 863	26	10.09 137	9.89 040	10	59	
2	9.79 918	16	9.90 889	25	10.09 111	9.89 030	10	58	
3	9.79 934	16	9.90 914	26	10.09 086	9.89 020	11	57	
4	9.79 950	15	9.90 940	26	10.09 060	9.89 009	10	56	
5	9.79 965	16	9.90 966	26	10.09 034	8.88 999	10	55	
6	9.79 981	15	9.90 992	26	10.09 008	9.88 989	11	54	
7	9.79 996	16	9.91 018	25	10.08 982	9.88 978	10	53	
8	9.80 012	15	9.91 043	26	10.08·957	9.88 968	10	52	
9	9.80 027	16	9.91 069	26	10.08 931	9.88 958	10	51	26 25
10	9.80 043	15	9.91 095	26	10.08 905	9.88 948	11	50	1 2.6 2.5
11	9.80 058	16	9.91 121	26	10.08 879	9.88 937	10	49	2 5.2 5.0
12	9.80 074	15	9.91 147	25	10.08 853	9.88 927	10	48	3 7.8 7.5
13	9.80 089	16	9.91 172	26	10.08 828	9.88 917	11	47	4 10.4 10.0
14	9.80 105	15	9.91 198	26	10.08 802	9.88 906	10	46	5 13.0 12.5
15	9.80 120	16	9.91 224	26	10.08 776	9.88 896	10	45	6 15.6 15.0
16	9.80 136	15	9.91 250	26	10.08 750	9.88 886	11	44	7 18.2 17.5
17	9.80 151	15	9.91 276	25	10.08 724	9.88 875	10	43	8 20.8 20.0
18	9.80 166	16	9.91 301	26	10.08 699	9.88 865	10	42	9 23.4 22.5
19	9.80 182	15	9.91 327	26	10.08 673	9.88 855	11	41	
20	9.80 197	16	9.91 353	26	10.08 647	9.88 844	10	40	
21	9.80 213	15	9.91 379	25	10.08 621	9.88 834	10	39	
22	9.80 228	16	9.91 404	26	10.08 596	9.88 824	11	38	
23	9.80 244	15	9.91 430	26	10.08 570	9.88 813	10	37	
24	9.80 259	15	9.91 456	26	10.08 544	9.88 803	10	36	
25	9.80 274	16	9.91 482	25	10.08 518	9.88 793	11	35	
26	9.80 290	15	9.91 507	26	10.08 493	9.88 782	10	34	16 15
27	9.80 305	15	9.91 533	26	10.08 467	9.88 772	11	33	1 1.6 1.5
28	9.80 320	16	9.91 559	26	10.08 441	9.88 761	10	32	2 3.2 3.0
29	9.80 336	15	9.91 585	25	10.08 415	9.88 751	10	31	3 4.8 4.5
30	9.80 351	15	9.91 610	26	10.08 390	9.88 741	11	30	4 6.4 6.0
31	9.80 366	16	9.91 636	26	10.08 364	9.88 730	10	29	5 8.0 7.5
32	9.80 382	15	9.91 662	26	10.08 338	9.88 720	11	28	6 9.6 9.0
33	9.80 397	15	9.91 688	25	10.08 312	9.88 709	10	27	7 11.2 10.5
34	9.80 412	16	9.91 713	26	10.08 287	9.88 699	11	26	8 12.8 12.0
35	9.80 428	15	9.91 739	26	10.08 261	9.88 688	10	25	9 14.4 13.5
36	9.80 443	15	9.91 765	26	10.08 235	9.88 678	10	24	
37	9.80 458	15	9.91 791	25	10.08 209	9.88 668	11	23	
38	9.80 473	16	9.91 816	26	10.08 184	9.88 657	10	22	
39	9.80 489	15	9.91 842	26	10.08 158	8.88 647	11	21	
40	9.80 504	15	9.91 868	25	10.08 132	9.88 636	10	20	
41	9.80 519	15	9.91 893	26	10.08˙107	9.88 626	11	19	
42	9.80 534	16	9.91 919	26	10.08 081	9.88 615	10	18	
43	9.80 550	15	9.91 945	26	10.08 055	9.88 605	11	17	11 10
44	9.80 565	15	9.91 971	25	10.08 029	9.88 594	10	16	1 1.1 1.0
45	9.80 580	15	9.91 996	26	10.08 004	9.88 584	11	15	2 2.2 2.0
46	9.80 595	15	9.92 022	26	10.07 978	9.88 573	10	14	3 3.3 3.0
47	9.80 610	15	9.92 048	25	10.07 952	9.88 563	11	13	4 4.4 4.0
48	9.80 625	16	9.92 073	26	10.07 927	9.88 552	10	12	5 5.5 5.0
49	9.80 641	15	9.92 099	26	10.07 901	9.88 542	11	11	6 6.6 6.0
50	9.80 656	15	9.92 125	25	10.07 875	9.88 531	10	10	7 7.7 7.0
51	9.80 671	15	9.92 150	26	10.07 850	9.88 521	11	9	8 8.8 8.0
52	9.80 686	15	9.92 176	26	10.07 824	9.88 510	10	8	9 9.9 9.0
53	9.80 701	15	9.92 202	25	10.07 798	9.88 499	10	7	
54	9.80 716	15	9.92 227	26	10.07 773	9.88 489	11	6	
55	9.80 731	15	9.92 253	26	10.07 747	9.88 478	10	5	
56	9.80 746	16	9.92 279	25	10.07 721	9.88 468	11	4	
57	9.80 762	15	9.92 304	26	10.07 696	9.88 457	10	3	
58	9.80 777	15	9.92 330	26	10.07 670	9.88 447	11	2	
59	9.80 792	15	9.92 356	25	10.07 644	9.88 436	11	1	
60	9.80 807		9.92 381		10.07 619	9.88 425		0	
	L. Cos.	d.	L. Cot.	c. d.	L. Tan.	L. Sin.	d.	'	P. P.

40°

'	L. Sin.	d.	L. Tan.	c. d.	L. Cot.	L. Cos.	d.		P. P.
0	9.80 807	15	9.92 381	26	10.07 619	9.88 425	10	60	
1	9.80 822	15	9.92 407	26	10.07 593	9.88 415	11	59	
2	9.80 837	15	9.92 433	25	10.07 567	9.88 404	10	58	
3	9.80 852	15	9.92 458	26	10.07 542	9.88 394	11	57	
4	9.80 867	15	9.92 484	26	10.07 516	9.88 383	11	56	
5	9.80 882	15	9.92 510	25	10.07 490	9.88 372	10	55	
6	9.50 897	15	9.92 535	26	10.07 465	9.88 362	11	54	
7	9.80 912	15	9.92 561	26	10.07 439	9.88 351	11	53	
8	9.80 927	15	9.92 587	25	10.07 413	9.88 340	10	52	
9	9.80 942	15	9.92 612	26	10.07 388	9.88 330	11	51	26 25
10	9.80 957	15	9.92 638	25	10.07 362	9.88 319	11	50	1 | 2.6 2.5
11	9.80 972	15	9.92 663	26	10.07 337	9.88 308	10	49	2 | 5.2 5.0
12	9.80 987	15	9.92 689	26	10.07 311	9.88 298	11	48	3 | 7.8 7.5
13	9.81 002	15	9.92 715	25	10.07 285	9.88 287	11	47	4 | 10.4 10.0
14	9.81 017	15	9.92 740	26	10.07 260	9.88 276	10	46	5 | 13.0 12.5
15	9.81 032	15	9.92 766	26	10.07 234	9.88 266	11	45	6 | 15.6 15.0
16	9.81 047	14	9.92 792	25	10.07 208	9.88 255	11	44	7 | 18.2 17.5
17	9.81 061	15	9.92 817	26	10.07 183	9.88 244	10	43	8 | 20.8 20.0
18	9.81 076	15	9.92 843	25	10.07 157	9.88 234	11	42	9 | 23.4 22.5
19	9.81 091	15	9.92 868	26	10.07 132	9.88 223	11	41	
20	9.81 106	15	9.92 894	26	10.07 106	9.88 212	11	40	
21	9.81 121	15	9.92 920	25	10.07 080	9.88 201	10	39	
22	9.81 136	15	9.92 945	26	10.07 055	9.88 191	11	38	
23	9.81 151	15	9.92 971	25	10.07 029	9.88 180	11	37	
24	9.81 166	14	9.92 996	26	10.07 004	9.88 169	11	36	
25	9.81 180	15	9.93 022	26	10.06 978	9.88 158	10	35	
26	9.81 195	15	9.93 048	25	10.06 952	9.88 148	11	34	15 14
27	9.81 210	15	9.93 073	26	10.06 927	9.88 137	11	33	1 | 1.5 1.4
28	9.81 225	15	9.93 099	25	10.06 901	9.88 126	11	32	2 | 3.0 2.8
29	9.81 240	14	9.93 124	26	10.06 876	9.88 115	10	31	3 | 4.5 4.2
30	9.81 254	15	9.93 150	25	10.06 850	9.88 105	11	30	4 | 6.0 5.6
31	9.81 269	15	9.93 175	26	10.06 825	9.88 094	11	29	5 | 7.5 7.0
32	9.81 284	15	9.93 201	26	10.06 799	9.88 083	11	28	6 | 9.0 8.4
33	9.81 299	15	9.93 227	25	10.06 773	9.88 072	11	27	7 | 10.5 9.8
34	9.81 314	14	9.93 252	26	10.06 748	9.88 061	10	26	8 | 12.0 11.2
35	9.81 328	15	9.93 278	25	10.06 722	9.88 051	11	25	9 | 13.5 12.6
36	9.81 343	15	9.93 303	26	10.06 697	9.88 040	11	24	
37	9.81 358	14	9.93 329	25	10.06 671	9.88 029	11	23	
38	9.81 372	15	9.93 354	26	10.06 646	9.88 018	11	22	
39	9.81 387	15	9.93 380	26	10.06 620	9.88 007	11	21	
40	9.81 402	15	9.93 406	25	10.06 594	9.87 996	11	20	
41	9.81 417	14	9.93 431	26	10.06 569	9.87 985	10	19	
42	9.81 431	15	9.93 457	25	10.06 543	9.87 975	11	18	
43	9.81 446	15	9.93 482	26	10.06 518	9.87 964	11	17	
44	9.81 461	14	9.93 508	25	10.06 492	9.87 953	11	16	11 10
45	9.81 475	15	9.93 533	26	10.06 467	9.87 942	11	15	1 | 1.1 1.0
46	9.81 490	15	9.93 559	25	10.06 441	9.87 931	11	14	2 | 2.2 2.0
47	9.81 505	14	9.93 584	26	10.06 416	9.87 920	11	13	3 | 3.3 3.0
48	9.81 519	15	9.93 610	26	10.06 390	9.87 909	11	12	4 | 4.4 4.0
49	9.81 534	15	9.93 636	25	10.06 364	9.87 898	11	11	5 | 5.5 5.0
50	9.81 549	14	9.93 661	26	10.06 339	9.87 887	10	10	6 | 6.6 6.0
51	9.81 563	15	9.93 687	25	10.06 313	9.87 877	11	9	7 | 7.7 7.0
52	9.81 578	14	9.93 712	26	10.06 288	9.87 866	11	8	8 | 8.8 8.0
53	9.81 592	15	9.93 738	25	10.06 262	9.87 855	11	7	9 | 9.9 9.0
54	9.81 607	15	9.93 763	26	10.06 237	9.87 844	11	6	
55	9.81 622	14	9.93 789	25	10.06 211	9.87 833	11	5	
56	9.81 636	15	9.93 814	26	10.06 186	9.87 822	11	4	
57	9.81 651	14	9.93 840	25	10.06 160	9.87 811	11	3	
58	9.81 665	15	9.93 865	26	10.06 135	9.87 800	11	2	
59	9.81 680	14	9.93 891	25	10.06 109	9.87 789	11	1	
60	9.81 694		9.93 916		10.06 084	9.87 778		0	
	L. Cos.	d.	L. Cot.	c. d.	L. Tan.	L. Sin.	d.	'	P. P.

49°

288

′	L. Sin.	d.	L. Tan.	c. d.	L. Cot.	L. Cos.	d.		P. P.
0	9.81 694	15	9.93 916	26	10.06 084	9.87 778	11	60	
1	9.81 709	14	9.93 942	25	1J.06 058	9.87 767	11	59	
2	9.81 723	15	9.93 967	26	10.06 033	9.87 756	11	58	
3	9.81 738	14	9.93 993	25	10.06 007	9.87 745	11	57	
4	9.81 752	15	9.94 018	26	10.05 982	9.87 734	11	56	
5	9.81 767	14	9.94 044	25	10.05 956	9.87 723	11	55	
6	9.81 781	15	9.94 069	26	10.05 931	9.87 712	11	54	
7	9.81 796	14	9.94 095	25	10.05 905	9.87 701	11	53	
8	9.81 810	15	9.94 120	26	10.05 880	9.87 690	11	52	
9	9.81 825	14	9.94 146	25	10.05 854	9.87 679	11	51	26 / 25
10	9.81 839	15	9.94 171	26	10.05 829	9.87 668	11	50	1 | 2.6 2.5
11	9.81 854	14	9.94 197	25	10.05 803	9.87 657	11	49	2 | 5.2 5.0
12	9.81 868	14	9.94 222	26	10.05 778	9.87 646	11	48	3 | 7.8 7.5
13	9.81 882	15	9.94 248	25	10.05 752	9.87 635	11	47	4 | 10.4 10.0
14	9.81 897	14	9.94 273	26	10.05 727	9.87 624	11	46	5 | 13.0 12.5
15	9.81 911	15	9.04 299	25	10.05 701	9.87 613	12	45	6 | 15.6 15.0
16	9.81 926	14	9.94 324	26	10.05 676	9.87 601	11	44	7 | 18.2 17.5
17	9.81 940	15	9.94 350	25	10.05 650	9.87 590	11	43	8 | 20.8 20.0
18	9.81 955	14	9.94 375	26	10.05 625	9.87 579	11	42	9 | 23.4 22.5
19	9.81 969	14	9.94 401	25	10.05 599	9.87 568	11	41	
20	9.81 983	15	9.94 426	26	10.05 574	9.87 557	11	40	
21	9.81 998	14	9.94 452	25	10.05 548	9.87 546	11	39	
22	9.82 012	14	9.94 477	26	10.05 523	9.87 535	11	38	
23	9.82 026	15	9.94 503	25	10.05 497	9.87 524	11	37	
24	9.82 041	14	9.94 528	26	10.05 472	9.87 513	12	36	
25	9.82 055	14	9.94 554	25	10.05 446	9.87 501	11	35	
26	9.82 069	15	9.94 579	25	10.05 421	9.87 490	11	34	15 / 14
27	9.82 084	14	9.94 604	26	10.05 396	9.87 479	11	33	1 | 1.5 1.4
28	9.82 098	14	9.94 630	25	10.05 370	9.87 468	11	32	2 | 3.0 2.8
29	9.82 112	14	9.94 655	26	10.05 345	9.87 457	11	31	3 | 4.5 4.2
30	9.82 126	15	9.94 681	25	10.05 319	9.87 446	12	30	4 | 6.0 5.6
31	9.82 141	14	9.94 706	26	10.05 294	9.87 434	11	29	5 | 7.5 7.0
32	9.82 155	14	9.94 732	25	10.05 268	9.87 423	11	28	6 | 9.0 8.4
33	9.82 169	15	9.94 757	26	10.05 243	9.87 412	11	27	7 | 10.5 9.8
34	9.82 184	14	9.94 783	25	10.05 217	9.87 401	11	26	8 | 12.0 11.2
35	9.82 198	14	9.94 808	26	10.05 192	9.87 390	12	25	9 | 13.5 12.6
36	9.82 212	14	9.94 834	25	10.05 166	9.87 378	11	24	
37	9.82 226	14	9.94 859	25	10.05 141	9.87 367	11	23	
38	9.82 240	15	9.94 884	26	10.05 116	9.87 356	11	22	
39	9.82 255	14	9.94 910	25	10.05 090	9.87 345	11	21	
40	9.82 269	14	9.94 935	26	10.05 065	9.87 334	12	20	
41	9.82 283	14	9.94 961	25	10.05 039	9.87 322	11	19	
42	9.82 297	14	9.94 986	26	10.05 014	9.87 311	11	18	
43	9.82 311	15	9.95 012	25	10.04 988	9.87 300	12	17	12 / 11
44	9.82 326	14	9.95 037	25	10.04 963	9.87 288	11	16	
45	9.82 340	14	9.95 062	26	10.04 938	9.87 277	11	15	1 | 1.2 1.1
46	9.82 354	14	9.95 088	25	10.04 912	9.87 266	11	14	2 | 2.4 2.2
47	9.82 368	14	9.95 113	26	10.04 887	9.87 255	12	13	3 | 3.6 3.3
48	9.82 382	14	9.95 139	25	10.04 861	9.87 243	11	12	4 | 4.8 4.4
49	9.82 396	14	9.95 164	26	10.04 836	9.87 232	11	11	5 | 6.0 5.5
50	9.82 410	14	9.95 190	25	10.04 810	9.87 221	12	10	6 | 7.2 6.6
51	9.82 424	15	9.95 215	25	10.04 785	9.87 209	11	9	7 | 8.4 7.7
52	9.82 439	14	9.95 240	26	10.04 760	9.87 198	11	8	8 | 9.6 8.8
53	9.82 453	14	9.95 266	25	10.04 734	9.87 187	12	7	9 | 10.8 9.9
54	9.82 467	14	9.95 291	26	10.04 709	9.87 175	11	6	
55	9.82 481	14	9.95 317	25	10.04 683	9.87 164	11	5	
56	9.82 495	14	9.95 342	26	10.04 658	9.87 153	12	4	
57	9.82 509	14	9.95 368	25	10.04 632	9.87 141	11	3	
58	9.82 523	14	9.95 393	25	10.04 607	9.87 130	11	2	
59	9.82 537	14	9.95 418	26	10.04 582	9.87 119	12	1	
60	9.82 551		9.95 444		10.04 556	9.87 107		0	
	L. Cos.	d.	L. Cot.	c. d.	L. Tan.	L. Sin.	d.	′	P. P.

′	L. Sin.	d.	L. Tan.	c. d.	L. Cot.	L. Cos.	d.		P. P.
0	9.82 551	14	9.95 444	25	10.04 556	9.87 107	11	60	
1	9.82 565	14	9.95 469	26	10.04 531	9.87 096	11	59	
2	9.82 579	14	9.95 495	25	10.04 505	9.87 085	12	58	
3	9.82 593	14	9.95 520	25	10.04 480	9.87 073	11	57	
4	9.82 607	14	9.95 545	26	10.04 455	9.87 062	12	56	
5	9.82 621	14	9.95 571	25	10.04 429	9.87 050	11	55	
6	9.82 635	14	9.95 596	26	10.04 404	9.87 039	11	54	
7	9.82 649	14	9.95 622	25	10.04 378	9.87 028	12	53	
8	9.82 663	14	9.95 647	25	10.04 353	9.87 016	11	52	
9	9.82 677	14	9.95 672	26	10.04 328	9.87 005	12	51	26 25
10	9.82 691	14	9.95 698	25	10.04 302	9.86 993	11	50	1 | 2.6 2.5
11	9.82 705	14	9.95 723	25	10.04 277	9.86 982	12	49	2 | 5.2 5.0
12	9.82 719	14	9.95 748	26	10.04 252	9.86 970	11	48	3 | 7.8 7.5
13	9.82 733	14	9.95 774	25	10.04 226	9.86 959	12	47	4 | 10.4 10.0
14	9.82 747	14	9.95 799	26	10.04 201	9.86 947	11	46	5 | 13.0 12.5
15	9.82 761	14	9.95 825	25	10.04 175	9.86 936	12	45	6 | 15.6 15.0
16	9.82 775	13	9.95 850	25	10.04 150	9.86 924	11	44	7 | 18.2 17.5
17	9.82 788	14	9.95 875	26	10.04 125	9.86 913	11	43	8 | 20.8 20.0
18	9.82 802	14	9.95 901	25	10.04 099	9.86 902	12	42	9 | 23.4 22.5
19	9.82 816	14	9.95 926	26	10.04 074	9.86 890	11	41	
20	9.82 830	14	9.95 952	25	10.04 048	9.86 879	12	40	
21	9.82 844	14	9.95 977	25	10.04 023	9.86 867	12	39	
22	9.82 858	14	9.96 002	26	10.03 998	9.86 855	11	38	
23	9.82 872	13	9.96 028	25	10.03 972	9.86 844	12	37	
24	9.82 885	14	9.96 053	25	10.03 947	9.86 832	11	·36	
25	9.82 899	14	9.96 078	26	10.03 922	9.86 821	12	35	
26	9.82 913	14	9.96 104	25	10.03 896	9.86 809	11	34	14 13
27	9.82 927	14	9.96 129	26	10.03 871	9.86 798	12	33	1 | 1.4 1.3
28	9.82 941	14	9.96 155	25	10.03 845	9.86 786	11	32	2 | 2.8 2.6
29	9.82 955	13	9.96 180	25	10.03 820	9.86 775	12	31	3 | 4.2 3.9
30	9.82 968	14	9.96 205	26	10.03 795	9.86 763	11	30	4 | 5.6 5.2
31	9.82 982	14	9.96 231	25	10.03 769	9.86 752	12	29	5 | 7.0 6.5
32	9.82 996	14	9.96 256	25	10.03 744	9.86 740	12	28	6 | 8.4 7.8
33	9.83 010	13	9.96 281	26	10.03 719	9.86 728	11	27	7 | 9.8 9.1
34	9.83 023	14	9.96 307	25	10.03 693	9.86 717	12	26	8 | 11.2 10.4
35	9.83 037	14	9.96 332	25	10.03 668	9.86 705	11	25	9 | 12.6 11.7
36	9.83 051	14	9.96 357	26	10.03 643	9.86 694	12	24	
37	9.83 065	13	9.96 383	25	10.03 617	9.86 682	12	23	
38	9.83 078	14	9.96 408	25	10.03 592	9.86 670	11	22	
39	9.83 092	14	9.96 433	26	10.03 567	9.86 659	12	21	
40	9.83 106	14	9.96 459	25	10.03 541	9.86 647	12	20	
41	9.83 120	13	9.96 484	26	10.03 516	9.86 635	11	19	
42	9.83 133	14	9.96 510	25	10.03 490	9.86 624	12	18	
43	9.83 147	14	9.96 535	25	10.03 465	9.86 612	12	17	
44	9.83 161	13	9.96 560	26	10.03 440	9.86 600	11	16	12 11
45	9.83 174	14	9.96 586	25	10.03 414	9.86 589	12	15	1 | 1.2 1.1
46	9.83 188	14	9.96 611	25	10.03 389	9.86 577	12	14	2 | 2.4 2.2
47	9.83 202	13	9.96 636	26	10.03 364	9.86 565	11	13	3 | 3.6 3.3
48	9.83 215	14	9.96 662	25	10.03 338	9.86 554	12	12	4 | 4.8 4.4
49	9.83 229	13	9.96 687	25	10.03 313	9.86 542	12	11	5 | 6.0 5.5
50	9.83 242	14	9.96 712	26	10.03 288	9.86 530	12	10	6 | 7.2 6.6
51	9.83 256	14	9.96 738	25	10.03 262	9.86 518	11	9	7 | 8.4 7.7
52	9.83 270	13	9.96 763	25	10.03 237	9.86 507	12	8	8 | 9.6 8.8
53	9.83 283	14	9.96 788	26	10.03 212	9.86 495	12	7	9 | 10.8 9.9
54	9.83 297	13	9.96 814	25	10.03 186	9.86 483	11	6	
55	9.83 310	14	9.96 839	25	10.03 161	9.86 472	12	5	
56	9.83 324	14	9.96 864	26	10.03 136	9.86 460	12	4	
57	9.83 338	13	9.96 890	25	10.03 110	9.86 448	12	3	
58	9.83 351	14	9.96 915	25	10.03 085	9.86 436	11	2	
59	9.83 365	13	9.96 940	26	10.03 060	9.86 425	12	1	
60	9.83 378		9.96 966		10.03 034	9.86 413		0	
	L. Cos.	d.	L. Cot.	c. d.	L. Tan.	L. Sin.	d.	′	P. P.

′	L. Sin.	d.	L. Tan.	c. d.	L. Cot.	L. Cos.	d.		P. P.
0	9.83 378	14	9.96 966	25	10.03 034	9.86 413	12	60	
1	9.83 392	13	9.96 991	25	10.03 009	9.86 401	12	59	
2	9.83 405	14	9.97 016	26	10.02 984	9.86 389	12	58	
3	9.83 419	13	9.97 042	25	10.02 958	9.86 377	11	57	
4	9.83 432	14	9.97 067	25	10.02 933	9.86 366	12	56	
5	9.83 446	13	9.97 092	26	10.02 908	9.86 354	12	55	
6	9.83 459	14	9.97 118	25	10.02 882	9.86 342	12	54	
7	9.83 473	13	9.97 143	25	10.02 857	9.86 330	12	53	
8	9.83 486	14	9.97 168	25	10.02 832	9.86 318	12	52	
9	9.83 500	13	9.97 193	26	10.02 807	9.86 306	11	51	
									26 25
10	9.83 513	14	9.97 219	25	10.02 781	9.86 295	12	50	1 2.6 2.5
11	9.83 527	13	9.97 244	25	10.02 756	9.86 283	12	49	2 5.2 5.0
12	9.83 540	14	9.97 269	26	10.02 731	9.86 271	12	48	3 7.8 7.5
13	9.83 554	13	9.97 295	25	10.02 705	9.86 259	12	47	4 10.4 10.0
14	9.83 567	14	9.97 320	25	10.02 680	9.86 247	12	46	5 13.0 12.5
									6 15.6 15.0
15	9.83 581	13	9.97 345	26	10.02 655	9.86 235	12	45	7 18.2 17.5
16	9.83 594	14	9.97 371	25	10.02 629	9.86 223	12	44	8 20.8 20.0
17	9.83 608	13	9.97 396	25	10.02 604	9.86 211	11	43	9 23.4 22.5
18	9.83 621	13	9.97 421	26	10.02 579	9.86 200	12	42	
19	9.83 634	14	9.97 447	25	10.02 553	9.86 188	12	41	
20	9.83 648	13	9.97 472	25	10.02 528	9.86 176	12	40	
21	9.83 661	13	9.97 497	26	10.02 503	9.86 164	12	39	
22	9.83 674	14	9.97 523	25	10.02 477	9.86 152	12	38	
23	9.83 688	13	9.97 548	25	10.02 452	9.86 140	12	37	
24	9.83 701	14	9.97 573	25	10.02 427	9.86 128	12	36	
25	9.83 715	13	9.97 598	26	10.02 402	9.86 116	12	35	
26	9.83 728	13	9.97 624	25	10.02 376	9.86 104	12	34	
27	9.83 741	14	9.97 649	25	10.02 351	9.86 092	12	33	**14 13**
28	9.83 755	13	9.97 674	26	10.02 326	9.86 080	12	32	1 1.4 1.3
29	9.83 768	13	9.97 700	25	10.02 300	9.86 068	12	31	2 2.8 2.6
									3 4.2 3.9
30	9.83 781	14	9.97 725	25	10.02 275	9.86 056	12	30	4 5.6 5.2
31	9.83 795	13	9.97 750	26	10.02 250	9.86 044	12	29	5 7.0 6.5
32	9.83 808	13	9.97 776	25	10.02 224	9.86 032	12	28	6 8.4 7.8
33	9.83 821	13	9.97 801	25	10.02 199	9.86 020	12	27	7 9.8 9.1
34	9.83 834	14	9.97 826	25	10.02 174	9.86 008	12	26	8 11.2 10.4
									9 12.6 11.7
35	9.83 848	13	9.97 851	26	10.02 149	9.85 996	12	25	
36	9.83 861	13	9.97 877	25	10.02 123	9.85 984	12	24	
37	9.83 874	13	9.97 902	25	10.02 098	9.85 972	12	23	
38	9.83 887	14	9.97 927	26	10.02 073	9.85 960	12	22	
39	9.83 901	13	9.97 953	25	10.02 047	9.85 948	12	21	
40	9.83 914	13	9.97 978	25	10.02 022	9.85 936	12	20	
41	9.83 927	13	9.98 003	26	10.01 997	9.85 924	12	19	
42	9.83 940	14	9.98 029	25	10.01 971	9.85 912	12	18	
43	9.83 954	13	9.98 054	25	10.01 946	9.85 900	12	17	**12 11**
44	9.83 967	13	9.98 079	25	10.01 921	9.85 888	12	16	1 1.2 1.1
45	9.83 980	13	9.98 104	26	10.01 896	9.85 876	12	15	2 2.4 2.2
46	9.83 993	13	9.98 130	25	10.01 870	9.85 864	13	14	3 3.6 3.3
47	9.84 006	14	9.98 155	25	10.01 845	9.85 851	12	13	4 4.8 4.4
48	9.84 020	13	9.98 180	26	10.01 820	9.85 839	12	12	5 6.0 5.5
49	9.84 033	13	9.98 206	25	10.01 794	9.85 827	12	11	6 7.2 6.6
50	9.84 046	13	9.98 231	25	10.01 769	9.85 815	12	10	7 8.4 7.7
51	9.84 059	13	9.98 256	25	10.01 744	9.85 803	12	9	8 9.6 8.8
52	9.84 072	13	9.98 281	26	10.01 719	9.85 791	12	8	9 10.8 9.9
53	9.84 085	13	9.98 307	25	10.01 693	9.85 779	13	7	
54	9.84 098	14	9.98 332	25	10.01 668	9.85 766	12	6	
55	9.84 112	13	9.98 357	26	10.01 643	9.85 754	12	5	
56	9.84 125	13	9.98 383	25	10.01 617	9.85 742	12	4	
57	9.84 138	13	9.98 408	25	10.01 592	9.85 730	12	3	
58	9.84 151	13	9.98 433	25	10.01 567	9.85 718	12	2	
59	9.84 164	13	9.98 458	26	10.01 542	9.85 706	13	1	
60	9.84 177		9.98 484		10.01 516	9.85 693		0	
′	L. Cos.	d.	L. Cot.	c. d.	L. Tan.	L. Sin.	d.	′	P. P.

′	L. Sin.	d.	L. Tan.	c. d.	L. Cot.	L. Cos.	d.		P. P.
0	9.84 177	13	9.98 484	25	10.01 516	9.85 693	12	60	
1	9.84 190	13	9.98 509	25	10.01 491	9.85 681	12	59	
2	9.84 203	13	9.98 534	26	10.01 466	9.85 669	12	58	
3	9.84 216	13	9.98 560	25	10.01 440	9.85 657	12	57	
4	9.84 229	13	9.98 585	25	10.01 415	9.85 645	13	56	
5	9.84 242	13	9.98 610	25	10.01 390	9.85 632	12	55	
6	9.84 255	14	9.98 635	26	10.01 365	9.85 620	12	54	
7	9.84 269	13	9.98 661	25	10.01 339	9.85 608	12	53	
8	9.84 282	13	9.98 686	25	10.01 314	9.85 596	13	52	
9	9.84 295	13	9.98 711	26	10.01 289	9.85 583	12	51	
10	9.84 308	13	9.98 737	25	10.01 263	9.85 571	12	50	
11	9.84 321	13	9.98 762	25	10.01 238	9.85 559	12	49	
12	9.84 334	13	9.98 787	25	10.01 213	9.85 547	13	48	
13	9.84 347	13	9.98 812	26	10.01 188	9.85 534	12	47	
14	9.84 360	13	9.98 838	25	10.01 162	9.85 522	12	46	**26 25 14**
15	9.84 373	12	9.98 863	25	10.01 137	9.85 510	13	45	
16	9.84 385	13	9.98 888	25	10.01 112	9.85 497	12	44	1| 2.6 2.5 1.4
17	9.84 398	13	9.98 913	26	10.01 087	9.85 485	12	43	2| 5.2 5.0 2.8
18	9.84 411	13	9.98 939	25	10.01 061	9.85 473	13	42	3| 7.8 7.5 4.2
19	9.84 424	13	9.98 964	25	10.01 036	9.85 460	12	41	4| 10.4 10.0 5.6
20	9.84 437	13	9.98 989	26	10.01 011	9.85 448	12	40	5| 13.0 12.5 7.0
21	9.84 450	13	9.99 015	25	10.00 985	9.85 436	13	39	6| 15.6 15.0 8.4
22	9.84 463	13	9.99 040	25	10.00 960	9.85 423	12	38	7| 18.2 17.5 9.8
23	9.84 476	13	9.99 065	25	10.00 935	9.85 411	12	37	8| 20.8 20.0 11.2
24	9.84 489	13	9.99 090	26	10.00 910	9.85 399	13	36	9| 23.4 22.5 12.6
25	9.84 502	13	9.99 116	25	10.00 884	9.85 386	12	35	
26	9.84 515	13	9.99 141	25	10.00 859	9.85 374	13	34	
27	9.84 528	12	9.99 166	25	10.00 834	9.85 361	12	33	
28	9.84 540	13	9.99 191	26	10.00 809	9.85 349	12	32	
29	9.84 553	13	9.99 217	25	10.00 783	9.85 337	13	31	
30	9.84 566	13	9.99 242	25	10.00 758	9.85 324	12	30	
31	9.84 579	13	9.99 267	26	10.00 733	9.85 312	13	29	
32	9.84 592	13	9.99 293	25	10.00 707	9.85 299	12	28	
33	9.84 605	13	9.99 318	25	10.00 682	9.85 287	13	27	
34	9.84 618	12	9.99 343	25	10.00 657	9.85 274	12	26	
35	9.84 630	13	9.99 368	26	10.00 632	9.85 262	12	25	
36	9.84 643	13	9.99 394	25	10.00 606	9.85 250	13	24	
37	9.84 656	13	9.99 419	25	10.00 581	9.85 237	12	23	
38	9.84 669	13	9.99 444	25	10.00 556	9.85 225	13	22	
39	9.84 682	12	9.99 469	26	10.00 531	9.85 212	12	21	
40	9.84 694	13	9.99 495	25	10.00 505	9.85 200	13	20	**13 12**
41	9.84 707	13	9.99 520	25	10.00 480	9.85 187	12	19	1| 1.3 1.2
42	9.84 720	13	9.99 545	25	10.00 455	9.85 175	13	18	2| 2.6 2.4
43	9.84 733	12	9.99 570	26	10.00 430	9.85 162	12	17	3| 3.9 3.6
44	9.84 745	13	9.99 596	25	10.00 404	9.85 150	13	16	4| 5.2 4.8
45	9.84 758	13	9.99 621	25	10.00 379	9.85 137	12	15	5| 6.5 6.0
46	9.84 771	13	9.99 646	26	10.00 354	9.85 125	13	14	6| 7.8 7.2
47	9.84 784	12	9.99 672	25	10.00 328	9.85 112	12	13	7| 9.1 8.4
48	9.84 796	13	9.99 697	25	10.00 303	9.85 100	13	12	8| 10.4 9.6
49	9.84 809	13	9.99 722	25	10.00 278	9.85 087	13	11	9| 11.7 10.8
50	9.84 822	13	9.99 747	26	10.00 253	9.85 074	12	10	
51	9.84 835	12	9.99 773	25	10.00 227	9.85 062	13	9	
52	9.84 847	13	9.99 798	25	10.00 202	9.85 049	12	8	
53	9.84 860	13	9.99 823	25	10.00 177	9.85 037	13	7	
54	9.84 873	12	9.99 848	26	10.00 152	9.85 024	12	6	
55	9.84 885	13	9.99 874	25	10.00 126	9.85 012	13	5	
56	9.84 898	13	9.99 899	25	10.00 101	9.84 999	13	4	
57	9.84 911	12	9.99 924	25	10.00 076	9.84 986	12	3	
58	9.84 923	13	9.99 949	26	10.00 051	9.84 974	13	2	
59	9.84 936	13	9.99 975	25	10.00 025	9.84 961	12	1	
60	9.84 949		10.00 000		10.00 000	9.84 949		0	
	L. Cos.	d.	L. Cot.	c. d.	L. Tan.	L. Sin.	d.	′	P. P.

TABLE III. NATURAL TRIGONOMETRIC FUNCTIONS TO FOUR PLACES

	Sin	Cos	Tan	Cot	
0° 00′	.0000	1.0000	.0000		**90° 00′**
10	029	000	029	343.8	50
20	058	000	058	171.9	40
30	.0087	1.0000	.0087	114.6	30
40	116	.9999	116	85.94	20
50	145	999	145	68.75	10
1° 00′	.0175	.9998	.0175	57.29	**89° 00′**
10	204	998	204	49.10	50
20	233	997	233	42.96	40
30	.0262	.9997	.0262	38.19	30
40	291	996	291	34.37	20
50	320	995	320	31.24	10
2° 00′	.0349	.9994	.0349	28.64	**88° 00′**
10	378	993	378	26.43	50
20	407	992	407	24.54	40
30	.0436	.9990	.0437	22.90	30
40	465	989	466	21.47	20
50	494	988	495	20.21	10
3° 00′	.0523	.9986	.0524	19.08	**87° 00′**
10	552	985	553	18.07	50
20	581	983	582	17.17	40
30	.0610	.9981	.0612	16.35	30
40	640	980	641	15.60	20
50	669	978	670	14.92	10
4° 00′	.0698	.9976	.0699	14.30	**86° 00′**
10	727	974	729	13.73	50
20	756	971	758	13.20	40
30	.0785	.9969	.0787	12.71	30
40	814	967	816	12.25	20
50	843	964	846	11.83	10
5° 00′	.0872	.9962	.0875	11.43	**85° 00′**
10	901	959	904	11.06	50
20	929	957	934	10.71	40
30	.0958	.9954	.0963	10.39	30
40	987	951	992	10.08	20
50	.1016	948	.1022	9.788	10
6° 00′	.1045	.9945	.1051	9.514	**84° 00′**
10	074	942	080	9.255	50
20	103	939	110	9.010	40
30	.1132	.9936	.1139	8.777	30
40	161	932	169	8.556	20
50	190	929	198	8.345	10
7° 00′	.1219	.9925	.1228	8.144	**83° 00′**
10	248	922	257	7.953	50
20	276	918	287	7.770	40
30	.1305	.9914	.1317	7.596	30
40	334	911	346	7.429	20
50	363	907	376	7.269	10
8° 00′	.1392	.9903	.1405	7.115	**82° 00′**
10	421	899	435	6.968	50
20	449	894	465	6.827	40
30	.1478	.9890	.1495	6.691	30
40	507	886	524	6.561	20
50	536	881	554	6.435	10
9° 00′	.1564	.9877	.1584	6.314	**81° 00′**
	Cos	Sin	Cot	Tan	

	Sin	Cos	Tan	Cot	
9° 00′	.1564	.9877	.1584	6.314	**81° 00′**
10	593	872	614	197	50
20	622	868	644	084	40
30	.1650	.9863	.1673	5.976	30
40	679	858	703	871	20
50	708	853	733	769	10
10° 00′	.1736	.9848	.1763	5.671	**80° 00′**
10	765	843	793	576	50
20	794	838	823	485	40
30	.1822	.9833	.1853	5.396	30
40	851	827	883	309	20
50	880	822	914	226	10
11° 00′	.1908	.9816	.1944	5.145	**79° 00′**
10	937	811	974	066	50
20	965	805	.2004	4.989	40
30	.1994	.9799	.2035	4.915	30
40	.2022	793	065	843	20
50	051	787	095	773	10
12° 00′	.2079	.9781	.2126	4.705	**78° 00′**
10	108	775	156	638	50
20	136	769	186	574	40
30	.2164	.9763	.2217	4.511	30
40	193	757	247	449	20
50	221	750	278	390	10
13° 00′	.2250	.9744	.2309	4.331	**77° 00′**
10	278	737	339	275	50
20	306	730	370	219	40
30	.2334	.9724	.2401	4.165	30
40	363	717	432	113	20
50	391	710	462	061	10
14° 00′	.2419	.9703	.2493	4.011	**76° 00′**
10	447	696	524	3.962	50
20	476	689	555	914	40
30	.2504	.9681	.2586	3.867	30
40	532	674	617	821	20
50	560	667	648	776	10
15° 00′	.2588	.9659	.2679	3.732	**75° 00′**
10	616	652	711	689	50
20	644	644	742	647	40
30	.2672	.9636	.2773	3.606	30
40	700	628	805	566	20
50	728	621	836	526	10
16° 00′	.2756	.9613	.2867	3.487	**74° 00′**
10	784	605	899	450	50
20	812	596	931	412	40
30	.2840	.9588	.2962	3.376	30
40	868	580	994	340	20
50	896	572	.3026	305	10
17° 00′	.2924	.9563	.3057	3.271	**73° 00′**
10	952	555	089	237	50
20	979	546	121	204	40
30	.3007	.9537	.3153	3.172	30
40	035	528	185	140	20
50	062	520	217	108	10
18° 00′	.3090	.9511	.3249	3.078	**72° 00′**
	Cos	Sin	Cot	Tan	

	Sin	Cos	Tan	Cot	
18° 00′	.3090	.9511	.3249	3.078	**72° 00′**
10	118	502	281	047	50
20	145	492	314	018	40
30	.3173	.9483	.3346	2.989	30
40	201	474	378	960	20
50	228	465	411	932	10
19° 00′	.3256	.9455	.3443	2.904	**71° 00′**
10	283	446	476	877	50
20	311	436	508	850	40
30	.3338	.9426	.3541	2.824	30
40	365	417	574	798	20
50	393	407	607	773	10
20° 00′	.3420	.9397	.3640	2.747	**70° 00′**
10	448	387	673	723	50
20	475	377	706	699	40
30	.3502	.9367	.3739	2.675	30
40	529	356	772	651	20
50	557	346	805	628	10
21° 00′	.3584	.9336	.3839	2.605	**69° 00′**
10	611	325	872	583	50
20	638	315	906	560	40
30	.3665	.9304	.3939	2.539	30
40	692	293	973	517	20
50	719	283	.4006	496	10
22° 00′	.3746	.9272	.4040	2.475	**68° 00′**
10	773	261	074	455	50
20	800	250	108	434	40
30	.3827	.9239	.4142	2.414	30
40	854	228	176	394	20
50	881	216	210	375	10
23° 00′	.3907	.9205	.4245	2.356	**67° 00′**
10	934	194	279	337	50
20	961	182	314	318	40
30	.3987	.9171	.4348	2.300	30
40	.4014	159	383	282	20
50	041	147	417	264	10
24° 00′	.4067	.9135	.4452	2.246	**66° 00′**
10	094	124	487	229	65 50
20	120	112	522	211	40
30	.4147	.9100	.4557	2.194	30
40	173	088	592	177	20
50	200	075	628	161	10
25° 00′	.4226	.9063	.4663	2.145	**65° 00′**
10	253	051	699	128	64 50
20	279	038	734	112	40
30	.4305	.9026	.4770	2.097	30
40	331	013	806	081	20
50	358	001	841	066	10
26° 00′	.4384	.8988	.4877	2.050	**64° 00′**
10	410	975	913	035	50
20	436	962	950	020	40
30	.4462	.8949	.4986	2.006	30
40	488	936	.5022	1.991	20
50	514	923	059	977	10
27° 00′	.4540	.8910	.5095	1.963	**63° 00′**
	Cos	Sin	Cot	Tan	

	Sin	Cos	Tan	Cot	
27° 00′	.4540	.8910	.5095	1.963	**63° 00′**
10	566	897	132	949	50
20	592	884	169	935	40
30	.4617	.8870	.5206	1.921	30
40	643	857	243	907	20
50	669	843	280	894	10
28° 00′	.4695	.8829	.5317	1.881	**62° 00′**
10	720	816	354	868	50
20	746	802	392	855	40
30	.4772	.8788	.5430	1.842	30
40	797	774	467	829	20
50	823	760	505	816	10
29° 00′	.4848	.8746	.5543	1.804	**61° 00′**
10	874	732	581	792	50
20	899	718	619	780	40
30	.4924	.8704	.5658	1.767	30
40	950	689	696	756	20
50	975	675	735	744	10
30° 00′	.5000	.8660	.5774	1.732	**60° 00′**
10	025	646	812	720	50
20	050	631	851	709	40
30	.5075	.8616	.5890	1.698	30
40	100	601	930	686	20
50	125	587	969	675	10
31° 00′	.5150	.8572	.6009	1.664	**59° 00′**
10	175	557	048	653	50
20	200	542	088	643	40
30	.5225	.8526	.6128	1.632	30
40	250	511	168	621	20
50	275	496	208	611	10
32° 00′	.5299	.8480	.6249	1.600	**58° 00′**
10	324	465	289	590	50
20	348	450	330	580	40
30	.5373	.8434	.6371	1.570	30
40	398	418	412	560	20
50	422	403	453	550	10
33° 00′	.5446	.8387	.6494	1.540	**57° 00′**
10	471	371	536	530	50
20	495	355	577	520	40
30	.5519	.8339	.6619	1.511	30
40	544	323	661	501	20
50	568	307	703	1.492	10
34° 00′	.5592	.8290	.6745	1.483	**56° 00′**
10	616	274	787	473	50
20	640	258	830	464	40
30	.5664	.8241	.6873	1.455	30
40	688	225	916	446	20
50	712	208	959	437	10
35° 00′	.5736	.8192	.7002	1.428	**55° 00′**
10	760	175	046	419	50
20	783	158	089	411	40
30	.5807	.8141	.7133	1.402	30
40	831	124	177	393	20
50	854	107	221	385	10
36° 00′	.5878	.8090	.7265	1.376	**54° 00′**
	Cos	Sin	Cot	Tan	

	Sin	Cos	Tan	Cot	
36° 00′	.5878	.8090	.7265	1.376	**54° 00′**
10	901	073	310	368	50
20	925	056	355	360	40
30	.5948	.8039	.7400	1.351	30
40	972	021	445	343	20
50	995	004	490	335	10
37° 00′	.6018	.7986	.7536	1.327	**53° 00′**
10	041	969	581	319	50
20	065	951	627	311	40
30	.6088	.7934	.7673	1.303	30
40	111	916	720	295	20
50	134	898	766	288	10
38° 00′	.6157	.7880	.7813	1.280	**52° 00′**
10	180	862	860	272	50
20	202	844	907	265	40
30	.6225	.7826	.7954	1.257	30
40	248	808	.8002	250	20
50	271	790	050	242	10
39° 00′	.6293	.7771	.8098	1.235	**51° 00′**
10	316	753	146	228	50
20	338	735	195	220	40
30	.6361	.7716	.8243	1.213	30
40	383	698	292	206	20
50	406	679	342	199	10
40° 00′	.6428	.7660	.8391	1.192	**50° 00′**
10	450	642	441	185	50
20	472	623	491	178	40
30	.6494	.7604	8541	1.171	30
40	517	585	591	164	20
50	539	566	642	157	10
41° 00′	.6561	.7547	.8693	1.150	**49° 00′**
10	583	528	744	144	50
20	604	509	796	137	40
30	.6626	.7490	.8847	1.130	30
40	648	470	899	124	20
50	670	451	952	117	10
42° 00′	.6691	.7431	.9004	1.111	**48° 00′**
10	713	412	057	104	50
20	734	392	110	098	40
30	.6756	.7373	.9163	1.091	30
40	777	353	217	085	20
50	799	333	271	079	10
43° 00′	.6820	.7314	.9325	1.072	**47° 00′**
10	841	294	380	066	50
20	862	274	435	060	40
30	.6884	.7254	.9490	1.054	30
40	905	234	545	048	20
50	926	214	601	042	10
44° 00′	.6947	.7193	.9657	1.036	**46° 00′**
10	967	173	713	030	50
20	988	153	770	024	40
30	.7009	.7133	.9827	1.018	30
40	030	112	884	012	20
50	050	092	942	006	10
45° 00′	.7071	.7071	1.000	1.000	**45° 00′**
	Cos	Sin	Cot	Tan	

ANSWERS

Exercise 1.1, page 6

13. $(7, -5)$ **14.** $(3, -3)$ **15.** $(3, -8)$ **17.** $(6, -7)$ **18.** $(5, -7)$ **19.** $(4, 0)$
21. 5 **22.** 13 **23.** 17 **25.** $5, 5, 5\sqrt{2}$ **26.** $13, 5, 9\sqrt{2}$ **27.** $17, 13, 3\sqrt{2}$

Exercise 1.2, page 10

1. $\pi/10$ **2.** $\pi/3$ **3.** $\pi/30$ **5.** $5\pi/4$ **6.** $\pi/5$ **7.** $\pi/4$ **9.** $\pi/15$ **10.** $\pi/36$
11. $\pi/20$ **13.** $\pi/60$ **14.** $2\pi/15$ **15.** $\pi/9$ **17.** $5\pi/48$ **18.** $\pi/16$ **19.** $13\pi/54$
21. $\pi/27$ **22.** $5\pi/24$ **23.** $7\pi/50$ **25.** $7\pi/960$ **26.** $269\pi/1500$ **27.** $73\pi/6000$
29. $3\pi/2$ **30.** $5\pi/2$ **31.** 1.15π **33.** 10π **34.** 6π **35.** 7.4π **37.** $20°$ **38.** $30°$
39. $12°$ **41.** $108°$ **42.** $150°$ **43.** $105°$ **45.** $3°45'$ **46.** $78°45'$ **47.** $82°30'$
49. $16°52'30''$ **50.** $14°3'45''$ **51.** $13°20'$ **53.** $177°37'2''$ **54.** $246°22'20''$ **55.**
$154°41'55''$ **57.** $156°25'31''$ **58.** $105°25'27''$ **59.** $221°44'5''$

Exercise 1.3, page 13

1. .913 ft   **2.** 4.34 ft   **3.** 16.4 cm   **5.** 4.35 m   **6.** 9.49 cm   **7.** 9.92 in.   **9.** 35.97 in.   **10.** 17.39 in.   **11.** 15.90 m   **13.** 1.50   **14.** .659   **15.** .798   **17.** .6527   **18.** 1.532   **19.** .2126   **21.** $5\pi/6$ **22.** $7\pi/6$ **23.** $(7\pi/3)$ in.   **25.** 1.13 ft   **26.** 466.9   **27.** 70.4   **29.** 92.15 ft   **30.** 7π in.   **31.** 16.6 sec   **33.** 4.4 ft per sec, 17.6 radians per sec   **34.** 1047   **35.** 88/3   **37.** 571.2   **38.** 3.8 in.   **39.** 38°40′ N, 30°5′ W

Exercise 1.4, page 19

1. 5   **2.** 13   **3.** 17   **5.** ±24   **6.** ±15   **7.** ±8   **9.** ±15   **10.** ±24   **11.** ±7   **13.** 3   **14.** −8   **15.** −12   **17.** −8   **18.** $8\sqrt{15}$ **19.** −15   **21.** 8   **22.** −10   **23.** −16   **25.** 2.5   **26.** −24   **27.** 48

In problems 29 to 39, the function values are given in the order sin θ, cos θ, tan θ, cot θ, sec θ, and csc θ, and the given value, if any, is included.

29. 12/13, 5/13, 12/5, 5/12, 13/5, 13/12   **30.** −4/5, 3/5, −4/3, −3/4, 5/3, −5/4   **31.** 15/17, −8/17, −15/8, −8/15, −17/8, 17/15   **33.** 4/5, −3/5, −4/3, −3/4, −5/3, 5/4   **34.** −15/17, 8/17, −15/8, −8/15, 17/8, −17/15   **35.** 7/25, 24/25, 7/24, 24/7, 25/24, 25/7   **37.** −8/17, 15/17, −8/15, −15/8, 17/15, −17/8   **38.** −24/25, −7/25, 24/7, 7/24, −25/7, −25/24   **39.** 5/13, 12/13, 5/12, 12/5, 13/12, 13/5

Exercise 1.5, page 26

1. {(2, 7), (3, 9), (4, 11), (5, 13)}   **2.** {(−1, −7), (0, −4), (1, −1), (2, 2), (3, 5)}   **3.** {(−3, 8), (−2, 3), (−1, 0), (0, −1), (1, 0), (2, 3)}   **5.** {−4, −1, 2, 5, 8, 11, 14}   **6.** {−11, −9, −7, −5, −3, −1, 1, 3, 5}   **7.** {6, 12, 20, 30, 42, 56}   **9.** Yes, exactly one y for each x.   **10.** No, more than one y for some x.   **11.** No, more than one y for some x.   **13.** −12/13, 5/13, −12/5, −5/12, 13/5, −13/12   **14.** −8/17, −15/17, 8/15, 15/8, −17/15, −17/8   **15.** 24/25, −7/25, −24/7, −7/24, −25/7, 25/24   **17.** −7/25, −24/25, 7/24, 24/7, −25/24, −25/7   **18.** 15/17, −8/17, −15/8, −8/15, −17/8, 17/15   **19.** −3/5, 4/5, −3/4, −4/3, 5/4, −5/3   **21.** 4/5, ±3/5, ±4/3, ±3/4, ±5/3, 5/4   **22.** ±15/7, −8/17, ∓15/8, ∓8/15, −17/8, ±15/7   **23.** ±7/25, ∓24/25, −7/24, −24/7, ∓25/24, ±25/7   **25.** ±4/5, 3/5, ±4/3, ±3/4, 5/3, ±5/4   **26.** −24/25, ±7/25, ∓24/7, ∓7/24, ±25/7, −25/24   **27.** ∓4/5, ±3/5, −4/3, −3/4, ±5/3, ∓5/4

Exercise 1.6, page 31

The order of the function values in problems 1 to 11 is sine, cosine, tangent, cotangent, secant, and cosecant.

1. $1/\sqrt{2}$, $-1/\sqrt{2}$, −1, −1, $-\sqrt{2}$, $\sqrt{2}$ **2.** $\sqrt{3}/2$, −1/2, $-\sqrt{3}$, $-1/\sqrt{3}$, −2, $2/\sqrt{3}$ **3.** −1/2, $-\sqrt{3}/2$, $1/\sqrt{3}$, $\sqrt{3}$, $-2/\sqrt{3}$, −2   **5.** $-\sqrt{3}/2$, −1/2, $\sqrt{3}$, $1/\sqrt{3}$, −2, $-2/\sqrt{3}$ **6.** $-1/\sqrt{2}$, $-1/\sqrt{2}$, 1, 1, $-\sqrt{2}$, $-\sqrt{2}$ **7.** 1, 0, no value,

0, no value, 1   **9.** 0, -1, 0, no value, -1, no value   **10.** $1/2$, $-\sqrt{3}/2$, $-1/\sqrt{3}$, $-\sqrt{3}$, $-2/\sqrt{3}$, 2   **11.** $-1/\sqrt{2}$, $1/\sqrt{2}$, -1, -1, $\sqrt{2}$, $-\sqrt{2}$ **37.** False   **38.** False   **39.** True   **41.** True   **42.** False   **43.** True   **45.** False   **46.** False   **47.** True   **49.** True   **50.** True   **51.** False

Exercise 2.1, page 39

1. .8327, .833, .83, .8   **2.** 3.142, 3.14, 3.1, 3   **3.** .009176, .00918, .0092, .009   **5.** 79.84, 79.8, 80, 8(10)   **6.** 582.4, 582, 5.8(10²), 6(10²)   **7.** 7.298, 7.30, 7.3, 7   **9.** $8.365 \leq 8.37 < 8.375$ **10.** $78.225 \leq 78.23 < 78.235$ **11.** $6.35 \leq 6.4 < 6.45$ **13.** $369.5 \leq 370 < 370.5$ **14.** $4.25(10^2) \leq 4.3(10^2) < 4.35(10^2)$ **15.** $5.85(10^{-1}) \leq 5.9(10^{-1}) < 5.95(10^{-1})$ **17.** 8.6   **18.** 9.5   **19.** 54   **21.** 230   **22.** 481   **23.** $1.89(10^3)$ **25.** 1.8   **26.** .64   **27.** 8.6   **29.** 8.02   **30.** .215   **31.** .00337   **33.** 6.49   **34.** 864.52   **35.** 376.36   **37.** 6.111   **38.** 5.94   **39.** 7.00   **41.** 15.51   **42.** 252.42   **43.** 3.47

Exercise 2.2, page 45

1. .4067   **2.** .8572   **3.** .9657   **5.** .7969   **6.** .3314   **7.** 1.117   **9.** .1495   **10.** 1.104   **11.** .4014   **13.** .4522   **14.** .9997   **15.** .2363   **17.** .7916   **18.** .6450   **19.** 2.675   **21.** .0204   **22.** 3.204   **23.** .9601   **25.** 4°   **26.** 13°10′   **27.** 16°20′   **29.** 17°30′   **30.** 16°10′   **31.** 20°40′   **33.** 22°   **34.** 37°40′   **35.** 42°   **37.** 76°30′   **38.** 84°20′   **39.** 81°50′   **41.** 38°40′   **42.** 48°10′   **43.** 21°30′   **45.** 34°40′   **46.** 31°20′   **47.** 75°20′   **49.** .6025   **50.** 2.324   **51.** .3179   **53.** .8297   **54.** 1.845   **55.** .7367   **57.** 2.952   **58.** .0940   **59.** .7649   **61.** 1.466   **62.** .2220   **63.** .8865   **65.** .8983   **66.** .2019   **67.** .9536   **69.** 2.140   **70.** 1.115   **71.** .7912   **73.** 74°58′   **74.** 23°24′   **75.** 33°8′   **77.** 9°25′   **78.** 43°30′   **79.** 64°55′   **81.** 51°23′   **82.** 54°16′   **83.** 77°47′   **85.** 7°29′   **86.** 6°16′   **87.** 66°26′   **89.** 54°16′   **90.** 53°56′   **91.** 88°1′   **93.** 59°43′   **94.** 29°17′   **95.** 34°21′   **97.** 13°34′   **98.** 54°37′   **99.** 67°0′   **101.** 26°   **102.** 52°   **103.** 40°

Exercise 2.3, page 53

1. $B = 30°$, $a = 8\sqrt{3}$, $b = 8$ **2.** $A = 60°$, $a = 16$, $b = 16\sqrt{3}$ **3.** $B = 45°$, $b = 24$, $c = 24\sqrt{2}$ **5.** $B = 30°$, $a = 12\sqrt{3}$, $c = 24$ **6.** $A = 30°$, $b = 24\sqrt{3}$, $c = 48$ **7.** $A = 45°, B = 45°, b = 25$ **9.** $B = 53°, a = 43, b = 57$ **10.** $B = 16°$, $a = 56$, $b = 16$ **11.** $A = 50°$, $b = 21$, $c = 33$ **13.** $A = 27°$, $a = 28$, $c = 61$ **14.** $B = 44°$, $a = 88$, $c = 1.2(10^2)$ **15.** $B = 58°$, $b = 2.9$, $c = 7.9$ **17.** $A = 30°10′$, $B = 59°50′$, $a = 420$ **18.** $A = 38°0′$, $B = 52°0′$, $a = 70.8$ **19.** $B = 57°40′$, $a = 1.86$, $b = 2.93$ **21.** $A = 20°20′$, $a = 329$, $b = 889$ **22.** $B = 67°20′$, $a = 44.0, b = 18.4$ **23.** $B = 33°40′, b = 280, c = 506$ **25.** $A = 48°50′, a = 8.73$, $c = 11.6$ **26.** $A = 8°30′, a = 4.23, c = 28.6$ **27.** $B = 60°30′, a = 49.3, c = 100$ **29.** $A = 21°50′$, $b = 24.5$, $c = 26.4$ **30.** $B = 61°10′$, $b = 1.32$, $c = 2.73$ **31.** $A = 30°20′$, $B = 59°40′$, $b = .731$ **33.** $A = 51°10′$, $B = 38°50′$, $c = .733$ **34.** $A = 35°30′$, $B = 54°30′$, $c = 7.63$ **35.** $A = 43°10′$, $B = 46°50′$, $c = 712$ **37.** $B = 26°46′$, $a = 14.37$, $b = 7.247$ **38.** $B = 19°33′$, $a = .1527$, $b = .05424$ **39.**

$A = 25°28'$, $a = .3362$, $b = .7058$ **41.** $B = 26°13'$, $b = 4.560$, $c = 10.32$ **42.**
$B = 57°38'$, $b = 1.203$, $c = 1.424$ **43.** $A = 28°38'$, $a = 3.179$, $c = 6.634$ **45.**
$A = 31°14'$, $B = 58°46'$, $c = 1.915$ **46.** $A = 41°13'$, $B = 48°17'$, $c = 129.4$
47. $B = 38°18'$, $a = 1.033$, $c = 1.317$ **49.** $A = 57°47'$, $b = .2485$, $c = .4662$
50. $A = 11°23'$, $b = 10.80$, $c = 11.02$ **51.** $A = 47°2'$, $B = 42°58'$, $b = 42.54$
53. $A = 73°42'$, $B = 16°18'$, $a = .8150$ **54.** $A = 23°41'$, $B = 66°19'$, $a = 18.61$
55. $A = 63°14'$, $B = 26°46'$, $c = 48.27$ **57.** $A = 21°$, $B = 69°$, $b = 1.6(10^3)$
58. $A = 71°$, $B = 19°$, $a = 2.2(10^3)$ **59.** $A = 41°$, $B = 49°$, $c = 4.7(10^2)$ **61.**
$B = 63°$, $b = 5.3(10^2)$, $c = 5.9(10^2)$ **62.** $A = 18°$, $a = 2.3(10^2)$, $c = 7.5(10^2)$
63. $B = 48°50'$, $a = 6.01(10^3)$, $b = 6.87(10^2)$

Exercise 2.4, page 60

1. 1.472 ft   **2.** 11,900 ft   **3.** 163 lb, 50°10' with the horizontal   **5.** 741 ft   **6.**
24°   **7.** 74.9 yd   **9.** 33°5', 19.70 ft   **10.** 8.943 cm, 22.74 cm   **11.** 3101 ft   **13.**
19.9 ft   **14.** 18.74 cm, 61°46', 118°14'   **15.** 2.805 ft   **17.** N 72°26' E   **18.** 243
ft   **19.** 21.1 ft   **21.** 60°   **22.** 804.5 yd, 1508 yd   **23.** 6.7 mi, 9.9 mi   **25.** 127.1
lb   **26.** 188°50', 198 mph   **27.** 595 yd   **29.** 90.6 ft   **30.** 4.53(10³) sq ft   **31.**
627 mi, 299°30'   **33.** 732 mi, 341°40'   **34.** N 38° E   **35.** 21 mph, S 76° E   **37.**
255 mph, 7°10'   **38.** 123 ft   **39.** 2149 ft

Exercise 4.1, page 82

1. False   **2.** True   **3.** True   **5.** True   **6.** False   **7.** False   **9.** True   **10.** False
11. True   **13.** $\sqrt{2}(1 - \sqrt{3})/4$ **14.** $\frac{1}{2}\sqrt{2 + \sqrt{2}}$ **15.** $\sqrt{2}(1 + \sqrt{3})/4$ **17.**
$-24/25, 0$ **18.** $33/65, -63/65$ **19.** $84/85, -36/85$ **21.** $-61/189, 3\sqrt{34}/34$
22. $-7/25, -1/\sqrt{5}$ **23.** $-119/169, -3\sqrt{13}/13$

Exercise 4.2, page 85

1. False   **2.** False   **3.** True   **5.** True   **6.** False   **7.** False   **9.** False   **10.**
False   **11.** True   **13.** $\sqrt{2}(\sqrt{3} + 1)/4$ **14.** $\frac{1}{2}\sqrt{2 + \sqrt{2}}$ **15.** $\sqrt{2}(\sqrt{3} - 1)/4$
17. $16/65, 56/65$ **18.** $16/65, 56/65$ **19.** $24/25, 0$ **21.** $-24/25, 3/\sqrt{10}$ **22.**
$-120/169, 1/\sqrt{26}$ **23.** $-120/169, 5/\sqrt{26}$

Exercise 4.3, page 89

1. False   **2.** True   **3.** True   **5.** True   **6.** False   **7.** False   **9.** True   **10.** False
11. True   **13.** $2 + \sqrt{3}$ **14.** $2 - \sqrt{3}$ **15.** $1/(2 + \sqrt{3})$ **17.** $-16/63, -56/33$
18. $36/77, -84/13$ **19.** $44/117, -4/3$ **21.** $-24/7, 3$ **22.** $-20/119, -1/5$
23. $120/119, -5$ **45.** $\sin A = 3/5$, $\cos A = 4/5$ **46.** $\sin A = 5/\sqrt{34}$, $\cos A =$
$3/\sqrt{34}$ **47.** $\sin A = 1/\sqrt{26}$, $\cos A = 5/\sqrt{26}$

Exercise 4.4, page 94

1. $\sin 46°$ **2.** $-\cos 17°$ **3.** $-\tan 32°$ **5.** $-\cos 21°$ **6.** $\tan 37°$ **7.** $\cot 4°$
9. $-\tan 77°$ **10.** $-\cot 51°$ **11.** $-\sin 5°$ **13.** $\cot 84°$ **14.** $\sin 69°$ **15.** $-\cos 22°$

17. $-\sin 72°$ 18. $-\cos 47°$ 19. $-\tan 42°$ 21. $\cos 17°$ 22. $-\tan 83°$
23. $-\cot 63°$ 25. $\sin 14°$ 26. $-\cot 40°$ 27. $\tan 23°$ 29. $\cot 37°$ 30. $\tan 28°$
31. $-\cos 13°$ 33. 1.072 34. $-.3839$ 35. $.1219$ 37. $-.1421$ 38. $.2931$ 39.
1.091 41. $-.6745$ 42. $-.6691$ 43. $-.5878$ 45. $\sin 45° + \sin 7°$ 46. $\sin 70° - \sin 24°$ 47. $-\cos 82° + \cos 60°$ 49. $\sin 150° - \sin 120°$ 50. $-\cos 90° + \cos 45°$ 51. $\cos 60° + \cos 30°$ 53. $-\cos 3\theta + \cos \theta$ 54. $\cos 8\theta + \cos 4\theta$ 55.
$\sin 4\theta - \sin 2\theta$ 57. $2 \sin 15° \cos 2°$ 58. $-2 \cos 45° \sin 2°$ 59. $-2 \sin 52° \sin 30°$

Exercise 5.2, page 110

1. $2\pi/3, 2$ 2. $\pi/2, \infty$ 3. $\pi/5, \infty$ 5. $2\pi/5, \infty$ 6. $\pi/3, 5$ 7. $2\pi, \infty$ 9. $3\pi,$
∞ 10. $4\pi/3, \infty$ 11. $5\pi/9, 4$ 13. $2, 6$ 14. $1, 5$ 15. $2, \infty$ 17. 2 18. 3
19. 4 21. 2 22. 1 23. $2/3$ 25. $2\pi/3, 2, 2$ units to the left   26. $\pi/2, 3, 3$
units to the left   27. $\pi/2, \infty, 1.5$ units to the right   29. $\pi, \infty, \pi/2$ units to the
right   30. $2\pi/5, \infty, 3\pi/5$ units to the right   31. $3\pi, 5, 9\pi/2$ units to the left

Exercise 6.1, page 116

1. $\log_8 64 = 2$ 2. $\log_4 64 = 3$ 3. $\log_2 64 = 6$ 5. $\log_{25} 5 = 1/2$ 6. $\log_{16}$
$2 = 1/4$ 7. $\log_{81} 27 = 3/4$ 9. $6^2 = 36$ 10. $3^4 = 81$ 11. $4^6 = 4096$ 13.
$25^{3/2} = 125$ 14. $8^{2/3} = 4$ 15. $1296^{1/2} = 36$ 17. 2 18. 3 19. 5 21. -1
22. -3 23. -3 25. $\frac{5}{3}$ 26. $\frac{3}{2}$ 27. $\frac{3}{2}$ 29. 2 30. 2 31. 3 33. 8 34.
$\frac{1}{9}$ 35. $\frac{1}{125}$ 37. 4 38. 8 39. 9 41. 4 42. $\frac{1}{27}$ 43. $\frac{8}{125}$ 45. 6 46. 2
47. 3 49. $\frac{1}{6}$ 50. $\frac{1}{2}$ 51. $\frac{1}{3}$ 53. $\frac{2}{3}$ 54. $\frac{3}{5}$ 55. $\frac{4}{5}$ 57. 3 58. 4 59. 5 61. $\frac{3}{4}$
62. $\frac{2}{3}$ 63. $\frac{3}{2}$ 65. -2 66. -5 67. -3

Exercise 6.2, page 121

1. $\log a + \log b$ 2. $\log b + \log c + \log d$ 3. $\log 5 + \log p + \log q$ 5.
$\log a - \log b$ 6. $\log b + \log c - \log d$ 7. $\log 5 - \log p - \log q$ 9. $b \log a$
10. $d(\log b + \log c)$ 11. $pq \log 5$ 13. $.78$ 14. 1.33 15. 1.15 17. $.37$ 18.
$.67$ 19. $.07$ 21. 1.20 22. $.96$ 23. 1.86 25. 1 26. 2 27. 3 29. 1
30. 2 31. 0 33. -2 34. -1 35. -3 37. -3 38. -4 39. -2 41.
$.77815$ 42. $.84510$ 43. $9.95424 - 10$ 45. $8.86332 - 10$ 46. $9.90849 - 10$
47. $.53148$ 49. 1.41330 50. $.51455$ 51. $8.94399 - 10$ 53. 2.91073 54.
1.77217 55. $.78696$ 57. $.46746$ 58. 3.89020 59. 3.44731 61. $9.70174 - 10$
62. $8.09934 - 10$ 63. $7.39235 - 10$ 65. $8.71383 - 10$ 66. $7.84553 - 10$ 67.
$6.04179 - 10$

Exercise 6.3, page 123

1. 1.85951 2. $.88199$ 3. 4.11829 5. $.43835$ 6. 1.79914 7. 3.90487 9.
2.09153 10. $9.62681 - 10$ 11. $.53872$ 13. 1.632 14. 37.85 15. 8694 17.
$.02245$ 18. $.9999$ 19. 58.05 21. $7.390(10^6)$ 22. $7.947(10^5)$ 23. $3.316(10^6)$
25. $6.509(10^6)$ 26. $1.916(10^5)$ 27. $5.035(10^5)$ 29. $4.900(10^7)$ 30. $6.846(10^7)$
31. $8.235(10^7)$ 33. 29.66 34. 1933 35. 2.377 37. 42.06 38. 6364 39.
5999 41. $.05192$ 42. $.008302$ 43. $.007171$ 45. 300.77 46. 3881.7 47.

18.307   **49.** .052916   **50.** .83904   **51.** .0059946   **53.** .00030431   **54.** .000026481
55. .054487   **57.** 2.4125(10⁶)   **58.** 2.6205(10⁵)   **59.** 2.0487(10⁶)

Exercise 6.4, page 127

1. 17.32   **2.** 366.6   **3.** 223.0   **5.** .02271   **6.** 233.8   **7.** .5214   **9.** .5013   **10.**
22.19   **11.** .3068   **13.** 15.65   **14.** .001205   **15.** .3635   **17.** .00001494   **18.**
368.8   **19.** .9056   **21.** 2.913   **22.** 3.509   **23.** .3361   **25.** .1260   **26.** .9544
27. .8752   **29.** 1549   **30.** 167.5   **31.** .4158   **33.** .04453   **34.** 1.509   **35.** .7722
37. 6.8221   **38.** 129.21   **39.** 78.930   **41.** .26319   **42.** 1884.0   **43.** 700.97   **45.**
959.12   **46.** .0075723   **47.** 448.12   **49.** 103.7   **50.** 361.44   **51.** 2200   **53.**
13.63   **54.** .0002120   **55.** 3687   **57.** 1.8627   **58.** 23844   **59.** 45.183   **61.** 5.235
62. .72468   **63.** .0033718

Exercise 7.1, page 138

1. $C = 64°30'$, $a = 383$, $c = 363$, $K = 4.77(10^4)$ **2.** $C = 72°10'$, $b = 310$, $c = 318$,
$K = 3.14(10^4)$ **3.** $B = 62°10'$, $a = 5.99$, $c = 7.51$, $K = 19.9$ **5.** $C = 53°5'$,
$a = .6932$, $c = .5919$, $K = .1730$ **6.** $C = 96°57'$, $b = 149.8$, $c = 175.8$, $K = 5620$
7. $A = 42°41'$, $a = 6.439$, $c = 4.904$, $K = 15.16$ **9.** $B = 73°58'$, $a = 55.78$,
$b = 120.6$, $K = 3308$ **10.** $A = 137°47'$, $a = 115.9$, $b = 77.48$, $K = 1201$ **11.**
$C = 90°26'$, $b = 2.541(10^4)$, $c = 2.618(10^4)$, $K = 7.773(10^7)$ **13.** $A = 41°31'$,
$a = .3307$, $c = .1492$, $K = .02113$ **14.** $C = 74°21'$, $a = 3237$, $c = 5085$, $K = 7.622(10^6)$ **15.** $B = 14°55'$, $a = 891.1$, $b = 236.2$, $K = 1.052(10^5)$ **17.** $A = 69°49.2'$, $a = .024904$, $b = .011999$, $K = .00014839$ **18.** $B = 37°46.1'$, $a = .19307$,
$b = .23982$, $K = .021359$ **19.** $C = 42°58.0'$, $b = 826.34$, $c = 652.97$, $K = 2.6332(10^5)$ **21.** $B = 18°58.5'$, $b = 2.4896$, $c = 3.1856$, $K = 2.7327$ **22.** $A = 37°54.0'$, $a = .047240$, $c = .073717$, $K = .0016217$ **23.** $C = 87°39.8'$, $a = 11.582$,
$c = 24.437$, $K = 127.27$ **25.** First ranger has to go 1.000 mi   **26.** 48 ft   **27.**
2.38 mi   **29.** 4.3 mi   **30.** 14.9 ft   **31.** 178 ft

Exercise 7.2, page 145

1. $A = 43°40'$, $B = 59°40'$, $a = 700$, $K = 2.98(10^5)$ **2.** $B = 47°0'$, $C = 75°30'$,
$c = 14.5$, $K = 118$ **3.** No solution   **5.** $B = 41°35'$, $C = 82°12'$, $c = 36.38$, $K = 368.4$ **6.** $B = 122°37'$, $C = 21°5'$, $b = 1.225(10^4)$, $K = 1.898(10^7)$ **7.** No solution   **9.** $A = 20°44'$, $C = 116°4'$, $c = .8158$, $K = .08975$ **10.** $A = 77°18'$, $B = 24°26'$, $a = 215.5$, $K = 9640$ **11.** $A = 140°55'$, $C = 17°51'$, $a = 10.24$, $K = 9.225$
13. $B = 112°20.1'$, $C = 13°27.1'$, $b = .24723$, $K = .0062344$ **14.** $A = 106°54.2'$,
$C = 51°49.1'$, $a = 26005$, $K = 1.0080(10^8)$, $A' = 30°32.4'$, $C' = 128°10.9'$, $a' = 13811$, $K' = 5.3535(10^7)$ **15.** $A = 145°50.1'$, $B = 7°48.1'$, $a = 4954.0$, $K = 1.3173(10^6)$ **17.** $A = 8°12.1'$, $C = 145°26.2'$, $c = 12.983$, $K = 9.4104$ **18.** No
solution   **19.** $A = 106°54.2'$, $C = 51°49.1'$, $a = 52009$, $K = 4.0322(10^8)$, $A' = 30°32.4'$, $C' = 128°10.9'$, $a' = 27621$, $K' = 2.1414(10^8)$ **21.** 690 mi, 198°   **22.**
50.64 mi, N 84°46' W   **23.** 231 ft

Exercise 7.3, page 150

1. 44   **2.** 436   **3.** $1.0(10^2)$ **5.** 42   **6.** 5.8   **7.** 20   **9.** 35°   **10.** 39°   **11.** 14°
13. 38°   **14.** 35°   **15.** 81°   **17.** 18, one solution   **18.** 8, one solution   **19.** No
solution   **21.** $A = 54°, B = 60°, C = 66°$ **22.** $A = 41°, B = 53°, C = 86°$ **23.**
$A = 61°0', B = 53°20', C = 65°40'$ **25.** 1615 ft   **26.** 333 lb, 20°40'   **27.** 301
mph, 206°

Exercise 7.4, page 154

1. $a = 6.64, B = 15°20', C = 118°40', K = 7.11$ **2.** $c = 1.14(10^3), A = 47°0',$
$B = 33°40', K = 2.65(10^5)$ **3.** $a = 235, B = 40°10', C = 91°30', K = 2.39(10^4)$
5. $a = .02026, B = 41°25', C = 100°17', K = .0002156$ **6.** $b = 479.5, A = 67°5',$
$C = 40°31', K = 7.218(10^4)$ **7.** $c = 25.45, A = 42°43', B = 93°55', K = 319.1$
9. $a = 4.401, B = 55°27', C = 96°5', K = 16.64$ **10.** $a = .1485, B = 40°58',$
$C = 53°18', K = .005812$ **11.** $b = .5423, A = 35°45', C = 24°29', K = .04100$
13. $a = 25.954, B = 46°14.7', C = 97°24.1', K = 407.01$ **14.** $b = 4.1720, A =$
$61°5.7', C = 57°41.6', K = 7.3470$ **15.** $C = 231.24, A = 89°28.1', B = 29°9.1',$
$K = 14836$ **17.** $a = 3.6880, B = 53°37.5', C = 94°48.9', K = 10.424$ **18.** $C =$
$20263, A = 107°14.6', B = 45°15.6', K = 3.0165(10^8)$ **19.** $a = 90.346, B =$
$52°31.4', C = 38°44.2', K = 2027.2$ **21.** 741 lb, 48°50'   **22.** $A = 32°58', B =$
$40°34'$ **23.** $1.1(10^2)$ nautical mi   **25.** 61°50', 259 mph   **26.** 87.5 sq ft   **27.** 25 ft

Exercise 7.5, page 159

1. $A = 54°0', B = 28°20', C = 97°40', K = .0921$ **2.** $A = 30°40', B = 60°20',$
$C = 89°0', K = 8.30(10^4)$ **3.** $A = 40°20', B = 66°20', C = 73°20', K = 108$ **5.**
$A = 31°42', B = 50°12', C = 98°6', K = 3.309(10^4)$ **6.** $A = 14°52', B = 36°4',$
$C = 129°2', K = .08996$ **7.** $A = 75°58', B = 59°38', C = 44°24', K = 16.71$ **9.**
$A = 69°42', B = 59°20', C = 50°58', K = 3.314(10^7)$ **10.** $A = 66°22', B = 73°20',$
$C = 40°18', K = .1508$ **11.** $A = 77°46', B = 60°52', C = 41°22', K = .002453$
13. $A = 40°35.0', B = 83°19.8', C = 56°5.2', K = .048369$ **14.** $A = 58°43.2',$
$B = 19°17.6', C = 101°59.4', K = 7.6773(10^6)$ **15.** $A = 25°34.2', B = 53°2.4',$
$C = 101°23.6', K = .0084176$ **17.** $A = 38°21.2', B = 70°23.6', C = 71°15.2',$
$K = 496.60$ **18.** $A = 85°54.8', B = 50°18.2', C = 43°47.0', K = 1.7017(10^5)$ **19.**
$A = 22°35.8', B = 90°54.2', C = 66°30.0', K = 1.4905$ **21.** S 6°0' W, S 48°20' W,
22. 280°20'   **23.** 101°20'   **25.** 3.28 ft   **26.** 2.13 ft   **27.** 44 cu ft

Exercise 7.6, page 161

1. $C = 84°8', b = 538.8, c = 776.7, K = 1.654(10^5)$ **2.** $B = 68°10', a = 4.056,$
$c = 2.687, K = 5.058$ **3.** $B = 89°4', b = 1826, c = 848.5, K = 6.919(10^5)$ **5.**
No solution   **6.** $A = 56°10', B = 84°54', a = 453.4, K = 1.353(10^4)$ **7.** $A =$
$83°55', B = 29°53', a = 2.252, K = 1.162$ **9.** $A = 42°4', C = 58°25', b = 234.5,$
$K = 1.596(10^4)$ **10.** $B = 42°1', C = 89°45', a = .4192, K = .07886$ **11.** $B =$
$63°27', C = 104°19', a = 135.0, K = 3.726(10^4)$ **13.** $A = 70°16', B = 50°46',$

$C = 58°58'$, $K = 1211$ **14.** $A = 108°34'$, $B = 29°14'$, $C = 42°14'$, $K = .0002277$
15. $A = 34°26'$, $B = 50°46'$, $C = 94°48'$, $K = .02937$ **17.** $B = 68°19.9'$, $a = $
54313, $c = 66326$, $K = 1.6739(10^9)$ **18.** $B = 79°15.6'$, $a = 3529.6$, $c = 4214.2$,
$K = 7.3067(10^6)$ **19.** $C = 32°16.8'$, $a = 7.6428$, $b = 7.3480$, $K = 14.996$ **21.**
$A = 49°50.7'$, $C = 73°55.5'$, $a = 1.0837$, $K = .61364$; $A' = 17°41.7'$, $C' = 106°4.5'$,
$a' = .43095$, $K' = .24403$ **22.** No solution   **23.** $B = 40°1.3'$, $C = 73°37.5'$, $c = $
7.2452, $K = 16.115$ **25.** $A = 70°23.7'$, $B = 53°1.5'$, $c = 57.263$, $K = 1478.4$
26. $A = 100°47.3'$, $C = 51°1.1'$, $b = 174.30$, $K = 24{,}552$ **27.** $B = 61°45.6'$, $C = $
$42°0.6'$, $a = 9.6098$, $K = 28.030$ **29.** $A = 107°54.6'$, $B = 24°44.8'$, $C = 47°20.6'$,
$K = .0002407$ **30.** $A = 47°17.4'$, $B = 36°46.2'$, $C = 95°56.6'$, $K = 418.66$ **31.**
$A = 83°31.8'$, $B = 52°30.4'$, $C = 43°57.8'$, $K = 1.6360(10^7)$ **33.** $1.93(10^4)$ sq ft
34. 296 ft, $2.95(10^4)$ sq ft   **35.** 99 paces   **37.** \$ 5080   **38.** $83°20'$ **39.** 25.3 tons,
$81°10'$ **41.** 217 lb, S $86°10'$ E   **42.** 65.6 ft   **43.** 4.6 ft   **45.** 126 ft   **46.** $B = $
$32°8'$, $C = 111°37'$, $a = 1034$ ft, $b = 930.3$ ft   **47.** 330 ft   **49.** 229.0 lb   **50.** 7290
yd, S $15°48'$ W   **51.** 1.76   **53.** $1.13(10^4)$ sq ft   **54.** 956 cu ft   **55.** 140 sq ft
57. 56.8 ft, 56.8 ft, 34.4 ft, $72°20'$ **58.** 445 lb, N $8°50'$ E   **59.** No, 1 ft

Exercise 8.1, page 169

1. $\pi/3, 5\pi/3$ **2.** $\pi/6, 5\pi/6$ **3.** $2\pi/3, 5\pi/3$ **5.** $\pi/3, 2\pi/3, 4\pi/3, 5\pi/3$ **6.** $\pi/6$,
$5\pi/6, 7\pi/6, 11\pi/6$ **7.** $\pi/3, 2\pi/3, 4\pi/3, 5\pi/3$ **9.** $0, 3\pi/4, \pi, 7\pi/4$ **10.** $\pi/6$,
$\pi/2, 3\pi/2, 11\pi/6$ **11.** $0, 5\pi/6, \pi, 11\pi/6$ **13.** $\pi/4, \pi/2, 3\pi/4, 5\pi/4, 3\pi/2, 7\pi/4$
14. $\pi/3, 2\pi/3, 4\pi/3, 5\pi/3$ **15.** $\pi/6, 5\pi/6, 7\pi/6, 11\pi/6$ **17.** $\pi/2, 7\pi/6, 11\pi/6$
18. $7\pi/6, 11\pi/6$ **19.** $\pi/3, 5\pi/3$ **21.** $0, 2\pi/3, 4\pi/3$ **22.** $\pi/3, 5\pi/6, 4\pi/3, 11\pi/6$
23. $\pi/6, 5\pi/6$ **25.** $\pi/4, \pi/3, 5\pi/4, 4\pi/3$ **26.** $\pi/3, 3\pi/4, 4\pi/3, 7\pi/4$ **27.** $\pi/6$,
$\pi/3, 5\pi/3, 11\pi/6$ **29.** $0, 2\pi/3, \pi, 4\pi/3$ **30.** $0, \pi/6, 5\pi/6, \pi$ **31.** $\pi/2, 7\pi/6$,
$3\pi/2, 11\pi/6$ **33.** $\pi/3, 3\pi/4, 5\pi/3, 7\pi/4$ **34.** $\pi/6, \pi/3, 5\pi/6, 5\pi/3$ **35.** $0, \pi/2$
37. $\pi/6, \pi/3, 5\pi/6, 4\pi/3$ **38.** $\pi/3, \pi, 4\pi/3$ **39.** $\pi/6, \pi/3, 7\pi/6, 4\pi/3$ **41.**
$\pi/6, \pi/2, 5\pi/6, 3\pi/2$ **42.** $3\pi/4, 7\pi/4$ **43.** $\pi/9, 4\pi/9, 7\pi/9, 10\pi/9, 7\pi/6, 13\pi/9$,
$16\pi/9, 11\pi/6$

Exercise 8.2, page 173

1. $0, \pi/3, 5\pi/3$ **2.** $\pi/3, 2\pi/3, 4\pi/3, 5\pi/3$ **3.** $0, 2\pi/3, 4\pi/3$ **5.** $\pi/3, \pi, 5\pi/3$
6. $\pi/2, 4\pi/3, 5\pi/3$ **7.** $\pi/6, 3\pi/4, 7\pi/4, 7\pi/6$ **9.** $\pi/6, 5\pi/6, 3\pi/2$ **10.** $\pi/4$,
$3\pi/4, 5\pi/4, 7\pi/4$ **11.** $\pi/6, 5\pi/6, 7\pi/6, 11\pi/6$ **13.** $\pi/8, 5\pi/8, 9\pi/8, 13\pi/8$
14. $\pi/12, 5\pi/12, 7\pi/12, 11\pi/12, 13\pi/12, 17\pi/12, 19\pi/12, 23\pi/12$ **15.** $\pi/4, \pi/2$,
$3\pi/4, 5\pi/4, 3\pi/2, 7\pi/4$ **17.** 0 **18.** $\pi/2, 2\pi/3, 4\pi/3, 3\pi/2$ **19.** $2\pi/3, 4\pi/3$
21. $0, \pi$ **22.** $0, \pi/2, \pi, 3\pi/2$ **23.** $\pi/18, 5\pi/18, 13\pi/18, 17\pi/18, 25\pi/18, 29\pi/18$
25. $\pi/4, 3\pi/4, 5\pi/4, 7\pi/4, \pi/12, 5\pi/12, 13\pi/12, 17\pi/12$ **26.** $\pi/6, \pi/2, 5\pi/6$,
$7\pi/6, 3\pi/2, 11\pi/6$ **27.** $\pi/6, 2\pi/3, 7\pi/6, 5\pi/3$ **29.** $0, \pi/2, \pi, 3\pi/2$ **30.** $\pi/5$,
$3\pi/5, \pi, 7\pi/5, 9\pi/5$ **31.** $\pi/2, 3\pi/2, \pi/3, 2\pi/3, 4\pi/3, 5\pi/3$ **33.** $\pi/6, 3\pi/2$ **34.**
No solution   **35.** $\pi/2, 7\pi/6$ **37.** $7\pi/12, 23\pi/12$ **38.** $\pi/12, 5\pi/12$ **39.** $\pi/6$

Exercise 9.1, page 184

1. $\pi/6$ **2.** $\pi/6$ **3.** $\pi/3$ **5.** $2\pi/3$ **6.** $-\pi/3$ **7.** $-\pi/6$ **9.** $2\pi/3$ **10.** $-\pi/3$
11. $-\pi/4$ **13.** $.3$ **14.** $.7$ **15.** -2 **17.** $1/3$ **18.** $.2$ **19.** $1/4$ **21.** $-\sqrt{3}$
22. $4/5$ **23.** $13/12$ **25.** u **26.** $1/u$ **27.** u **29.** $u/\sqrt{u^2+1}$ **30.** $\sqrt{u^2-1}/u$
31. $u/\sqrt{u^2+1}$ **33.** $\sqrt{1-u^2}$ **34.** $1/\sqrt{u^2+1}$ **35.** $-\sqrt{1-u^2}/u$ **37.**
$(2-u^2)/u^2$ **38.** $2u/(u^2-1)$ **39.** $-2u\sqrt{1-u^2}$ **41.** $-u$ **42.** $-1/\sqrt{1+u^2}$
43. $\sqrt{1-u^2}/u$ **45.** $\sqrt{(1+u)/2}$ **46.** $\sqrt{(1+u)/2}$ **47.** $(\sqrt{1+u^2}-1)/u$
49. $uv-\sqrt{1-u^2}\sqrt{1-v^2}$ **50.** $uv+\sqrt{1-u^2}\sqrt{1-v^2}$ **51.** $(uv+1)/(v-u)$
53. θ **54.** $\pi/2-\theta$ **55.** $\pi/2-\theta$ **57.** $\pi/2+\theta$ **58.** $-\theta$ **59.** θ **61.** Arcsin
y/r **62.** $\pi-$ Arcsin y/r **63.** Arcsin y/r **65.** Arcsin 1 **66.** Arcsin $(-16/65)$
67. Arcsin $(-16/65)$

Exercise 10.1, page 194

25. 5, arctan $(4/3)$ **26.** $\sqrt{29}$, arctan $(-5/2)$ **27.** $\sqrt{17}$, arctan $(1/(-4))$ **29.**
$\sqrt{29}$, arctan $(-2/5)$ **30.** $\sqrt{41}$, arctan $(5/4)$ **31.** $\sqrt{58}$, arctan $(-7/(-3))$ **33.**
$\sqrt{2}$, $7\pi/4$ **34.** 2, $2\pi/3$ **35.** $2\sqrt{2}$, $5\pi/4$ **37.** 2, $\pi/6$ **38.** $\sqrt{2}$, $3\pi/4$ **39.** 2,
$4\pi/3$ **41.** $x=2, y=3$ **42.** $x=2, y=-6$ **43.** $x=2, y=-4$ **45.** $x=1$,
$y=2$ **46.** $x=3, y=-2$ **47.** $x=3, y=2$ **49.** $x=1, y=2$ **50.** $x=1$,
$y=-1$ **51.** $x=-2, y=1$

Exercise 10.2, page 196

1. $6+8i$ **2.** $8+2i$ **3.** $2-8i$ **5.** $(8, 3)$ **6.** $(2, -2)$ **7.** $(11, -1)$ **9.**
$1-2i$ **10.** $3-4i$ **11.** $2+6i$ **13.** $(1, 2)$ **14.** $(3, -4)$ **15.** $(1, -9)$ **17.**
$9+19i$ **18.** $26-2i$ **19.** $-6+33i$ **21.** $(7, 17)$ **22.** $(0, 6)$ **23.** $(0, 25)$
25. $(18-i)/13$ **26.** $(16+11i)/29$ **27.** $(10-11i)/13$ **29.** $(\frac{18}{25}, \frac{1}{25})$ **30.**
$(\frac{16}{13}, -\frac{11}{13})$ **31.** $(\frac{10}{17}, \frac{11}{17})$ **33.** $x=-6, y=4$ **34.** $x=3, y=5$ **35.** $x=15$,
$y=5$ **37.** $x=1, y=7$ **38.** $x=2, y=5$ **39.** $x=24, y=4$

Exercise 10.3, page 202

1. $\sqrt{2}$ cis $45°$ **2.** 2 cis $120°$ **3.** 2 cis $330°$ **5.** 2 cis $150°$ **6.** 2 cis $300°$ **7.**
$\sqrt{2}$ cis $225°$ **9.** $\sqrt{5}$ cis $333°30'$ **10.** $\sqrt{13}$ cis $213°40'$ **11.** $\sqrt{29}$ cis $68°10'$ **13.**
$\sqrt{10}$ cis $251°30'$ **14.** $\sqrt{13}$ cis $56°20'$ **15.** $\sqrt{34}$ cis $121°0'$ **17.** $5\sqrt{2}(1+i)$ **18.**
$3(\sqrt{3}+i)$ **19.** $6(1+\sqrt{3}i)$ **21.** $15(-1+\sqrt{3}i)$ **22.** $2\sqrt{2}(-1+i)$ **23.**
$-12(\sqrt{3}+i)$ **25.** $\sqrt{2}(1+i)$ **26.** $2.5(1+\sqrt{3}i)$ **27.** $2(\sqrt{3}+i)$ **29.**
$-2\sqrt{2}(1+i)$ **30.** $-1+\sqrt{3}i$ **31.** $-1.25(\sqrt{3}+i)$ **33.** $(\sqrt{3}+i)/2$ **34.** $2i$
35. $(1+i)/\sqrt{2}$ **37.** $\sqrt{3}+i$ **38.** $1+\sqrt{3}i$ **39.** $(\sqrt{3}-i)/2$ **41.** $2\sqrt{2}$ cis $15°$
42. 4 cis $150°$ **43.** $2\sqrt{2}$ cis $15°$ **45.** $(1/\sqrt{2})$ cis $195°$ **46.** $\sqrt{2}$ cis $285°$ **47.**
cis $150°$

Exercise 10.4, page 206

1. $2\sqrt{2}$ *cis* $135°$ **2.** 4 *cis* $180°$ **3.** $4\sqrt{2}$ *cis* $45°$ **5.** 8 *cis* $180°$ **6.** 4 *cis* $240°$
7. 32 *cis* $300°$ **9.** 8 *cis* $90°$ **10.** 16 *cis* $120°$ **11.** 32 *cis* $30°$ **13.** 125 *cis* $249°30'$
14. 169 *cis* $134°40'$ **15.** 17^3 *cis* $264°$ **17.** $2^{1/3}$ *cis* $(110° + n\,120°)$, $n = 0, 1, 2$
18. $2^{1/6}$ *cis* $(45° + n\,120°)$, $n = 0, 1, 2$ **19.** $2^{1/3}$ *cis* $(80° + n\,120°)$, $n = 0, 1, 2$
21. $2^{1/4}$ *cis* $(15° + n\,90°)$, $n = 0, 1, 2, 3$ **22.** $2^{1/4}$ *cis* $(75° + n\,90°)$, $n = 0, 1, 2, 3$
23. 2 *cis* $(0° + n\,90°)$, $n = 0, 1, 2, 3$ **25.** $2^{1/10}$ *cis* $(9° + n\,72°)$, $n = 0, 1, 2, 3, 4$
26. $2^{1/5}$ *cis* $(66° + n\,72°)$, $n = 0, 1, 2, 3, 4$ **27.** $2^{1/6}$ *cis* $(35° + n\,60°)$, $n = 0, 1, 2,$
$3, 4, 5$ **29.** $2^{1/8}$ *cis* $(15° + n\,45°)$, $n = 0, 1, 2, 3, \ldots, 7$ **30.** $2^{1/8}$ *cis* $(30° + n\,45°)$,
$n = 0, 1, 2, 3, \ldots, 7$ **31.** $2^{1/18}$ *cis* $(25° + n\,40°)$, $n = 0, 1, 2, \ldots, 8$ **33.** 3
cis $(0° + n\,180°)$, $n = 0, 1$ **34.** 3 *cis* $(0° + n\,120°)$, $n = 0, 1, 2$ **35.** 2 *cis*
$(45° + n\,90°)$, $n = 0, 1, 2, 3$ **37.** 2 *cis* $(15° + n\,60°)$, $n = 0, 1, 2, 3, 4, 5$ **38.**
4 *cis* $(30° + n\,120°)$, $n = 0, 1, 2$ **39.** 1 *cis* $(22°30' + n\,90°)$, $n = 0, 1, 2, 3$

Exercise 11.1, page 215

1. $(3, -315°)$, $(-3, 225°)$, $(-3, -135°)$; $(3, 300°)$, $(-3, 120°)$, $(-3, -240°)$;
$(-2, -285°)$, $(2, 255°)$, $(2, -105°)$; $(-4, 315°)$, $(4, 135°)$, $(4, -225°)$ **2.** $(5, -250°)$,
$(-5, 290°)$, $(-5, -70°)$; $(-5, -290°)$, $(5, 250°)$, $(5, -110°)$; $(6, 240°)$, $(-6, 60°)$,
$(-6, -300°)$; $(-5, 330°)$, $(5, 150°)$, $(5, -210°)$ **3.** $(-2, 310°)$, $(2, 130°)$,
$(2, -230°)$; $(-7, 160°)$, $(7, 340°)$, $(7, -20°)$; $(5, 220°)$, $(-5, 40°)$, $(-5, -320°)$;
$(2, 430°)$, $(-2, 290°)$, $(-2, -110°)$ **5.** $(-7, -45°)$, $(7, -225°)$, $(7, 135°)$;
$(7, 210°)$, $(-7, 30°)$, $(7, -150°)$; $(-4, 240°)$, $(4, 60°)$, $(4, -300°)$; $(6, -90°)$,
$(-6, -270°)$, $(-6, 90°)$ **6.** $(9, 260°)$, $(-9, 80°)$, $(-9, -280°)$; $(-2, 340°)$,
$(2, 160°)$, $(2, -200°)$; $(5, -50°)$, $(-5, -230°)$, $(-5, 130°)$; $(-3, -225°)$, $(3, 315°)$,
$(3, -45°)$ **7.** $(-3, 135°)$, $(3, -45°)$, $(3, 315°)$; $(4, -320°)$, $(-4, 220°)$,
$(-4, -140°)$; $(-2, -130°)$, $(2, -310°)$, $(2, 50°)$; $(5, 190°)$, $(-5, 10°)$, $(-5, -350°)$

Exercise 11.2, page 218

Problems 25 to 48 are the polar forms of the rectangular equations given in problems
1 to 24, respectively.

INDEX